Aggregation Operators

Studies in Fuzziness and Soft Computing

Editor-in-chief
Prof. Janusz Kacprzyk
Systems Research Institute
Polish Academy of Sciences
ul. Newelska 6
01-447 Warsaw, Poland
E-mail: kacprzyk@ibspan.waw.pl
http://www.springer.de/cgi-bin/search_book.pl?series=2941

Further volumes of this series can be found at our homepage.

Vol. 78. U. Seiffert and L.C. Jain (Eds.)
Self-Organizing Neural Networks, 2002
ISBN 3-7908-1417-2

Vol. 79. A. Osyczka
Evolutionary Algorithms for Single and Multicriteria Design Optimization, 2002
ISBN 3-7908-1418-0

Vol. 80. P. Wong, F. Aminzadeh and M. Nikravesh (Eds.)
Soft Computing for Reservoir Characterization and Modeling, 2002
ISBN 3-7908-1421-0

Vol. 81. V. Dimitrov and V. Korotkich (Eds.)
Fuzzy Logic, 2002
ISBN 3-7908-1425-3

Vol. 82. Ch. Carlsson and R. Fullér
Fuzzy Reasoning in Decision Making and Optimization, 2002
ISBN 3-7908-1428-8

Vol. 83. S. Barro and R. Marín (Eds.)
Fuzzy Logic in Medicine, 2002
ISBN 3-7908-1429-6

Vol. 84. L.C. Jain and J. Kacprzyk (Eds.)
New Learning Paradigms in Soft Computing, 2002
ISBN 3-7908-1436-9

Vol. 85. D. Rutkowska
Neuro-Fuzzy Architectures and Hybrid Learning, 2002
ISBN 3-7908-1438-5

Vol. 86. M.B. Gorzałczany
Computational Intelligence Systems and Applications, 2002
ISBN 3-7908-1439-3

Vol. 87. C. Bertoluzza, M.Á. Gil and D.A. Ralescu (Eds.)
Statistical Modeling, Analysis and Management of Fuzzy Data, 2002
ISBN 3-7908-1440-7

Vol. 88. R.P. Srivastava and T.J. Mock (Eds.)
Belief Functions in Business Decisions, 2002
ISBN 3-7908-1451-2

Vol. 89. B. Bouchon-Meunier, J. Gutiérrez-Ríos, L. Magdalena and R.R. Yager (Eds.)
Technologies for Constructing Intelligent Systems 1, 2002
ISBN 3-7908-1454-7

Vol. 90. B. Bouchon-Meunier, J. Gutiérrez-Ríos, L. Magdalena and R.R. Yager (Eds.)
Technologies for Constructing Intelligent Systems 2, 2002
ISBN 3-7908-1455-5

Vol. 91. J.J. Buckley, E. Eslami and T. Feuring
Fuzzy Mathematics in Economics and Engineering, 2002
ISBN 3-7908-1456-3

Vol. 92. P.P. Angelov
Evolving Rule-Based Models, 2002
ISBN 3-7908-1457-1

Vol. 93. V.V. Cross and T.A. Sudkamp
Similarity and Compatibility in Fuzzy Set Theory, 2002
ISBN 3-7908-1458-X

Vol. 93. M. MacCrimmon and P. Tillers (Eds.)
The Dynamics of Judical Proof, 2002
ISBN 3-7908-1459-8

Vol. 94. T. Young Lin, Y.Y. Yao and L.A. Zadeh (Eds)
Data Mining, Rough Sets and Granular Computing, 2002
ISBN 3-7908-1461-X

Vol. 95. M. Schmitt, H.-N. Teodorescu, A. Jain, A. Jain, S. Jain and L.C. Jain (Eds.)
Computational Intelligence Processing in Medical Diagnosis, 2002
ISBN 3-7908-1463-6

Tomasa Calvo
Gaspar Mayor
Radko Mesiar
Editors

Aggregation Operators

New Trends and Applications

With 18 Figures
and 9 Tables

Physica-Verlag
A Springer-Verlag Company

Professor Tomasa Calvo
University of the Balearic Islands
Department of Mathematics
and Computer Science
07071 Palma de Mallorca
Spain
tomasa.calvo@uib.es
or
University of Alcalá
Department of Computer Science
28871 Alcalá de Henares
Spain
tomasa.calvo@uah.es

Professor Gaspar Mayor
University of the Balearic Islands
Department of Mathematics
and Computer Science
07071 Palma de Mallorca
Spain
gmayor@uib.es

Professor Radko Mesiar
Slovak Technical University
Faculty of Civil Engineering
Radlinského 11
81368 Bratislava
Slovakia
mesiar@vox.svf.stuba.sk

ISSN 1434-9922
ISBN 978-3-662-00319-0 ISBN 978-3-7908-1787-4 (eBook)
DOI 10.1007/978-3-7908-1787-4

Cataloging-in-Publication Data applied for
Die Deutsche Bibliothek – CIP-Einheitsaufnahme
Aggregation operators: new trends and applications; with 9 tables / Tomasa Calvo ... (ed.). – Heidelberg; New York: Physica-Verl., 2002
(Studies in fuzziness and soft computing; Vol. 97)
ISBN 978-3-662-00319-0

Physica-Verlag Heidelberg New York
a member of BertelsmannSpringer Science+Business Media GmbH

Softcover reprint of the hardcover 1st edition 2002

Hardcover Design: Erich Kirchner, Heidelberg

SPIN 10863361 88/2202-5 4 3 2 1 0 – Printed on acid-free paper

Foreword

1. The increasing number of research papers appeared in the last years that either make use of aggregation functions or contribute to its theoretical study asses its growing importance in the field of Fuzzy Logic and in others where uncertainty and imprecision play a relevant role. Since these papers are published in many journals, few books and several proceedings of conferences, books on aggregation are particularly welcome. To my knowledge, "Agregation Operators. New Trends and Applications" is the first book aiming at generality , and I take it as a honour to write this Foreword in response to the gentle demand of its editors, Radko Mesiar, Tomasa Calvo and Gaspar Mayor. My pleasure also derives from the fact that twenty years ago I was one of the first Spaniards interested in the study of aggregation functions, and this book includes work by several Spanish authors.

The book contains nice and relevant original papers, authored by some of the most outstanding researchers in the field, and since it can serve, as the editors point out in the **Preface**, as a small handbook on aggregation, the book is very useful for those entering the subject for the first time. The book also contains a part dealing with potential areas of application, so it can be helpful in gaining insight on the future developments. The book is also a good presentation of the current knowledge and terminology on aggregation operators, and it gives a good survey of the kind of applications to be foreseeing in the future.

2. Although closed by the geometric mean, the set $\{0, 1\}$ is not closed, in particular, for the arithmetic mean. The set $\{0, 1/2, 1\}$, obtained adding the arithmetical mean of numbers 0 and 1, is neither closed for the arithmetic nor for the geometric mean, and the same happens with the set $\{0, 1/2, \sqrt{2}/2, 3/4, 1\}$ obtained after adding the new values $\sqrt{2}/2$ and $3/4$, similarly obtained. The unit interval $[0, 1]$ is, however closed for any quasi-linear mean. Accordingly, $[0, 1]^X$ can be viewed as a necessary enlargement of $\{0, 1\}^X$ to aggregate several complete crisp informations.

Thus, fuzzy sets appear as entities necessary to synthesize complete or incomplete linguistically given information and, of course, linguistically given imprecise information. This way aggregation could be seen as "genetic material" for the introduction of fuzzy sets, and the growing amount of the research on aggregation within the field of fuzzy logic can not be seen as a rear phenomenon.

The usefulness of fuzzy sets in order to represent incomplete or imprecise information, led to the early study of imprecise relations and non-necessarily additive measures in the field of fuzzy logic. This fact, jointly with the suitability of synthesizing relations or measures coming from several sources of information, contrived to different types of aggregation that later became themselves subjects of research giving way, for example, to mathematical characterizations of the n-dimensional aggregation operators assigning to n inputs of a given type an output of the same type.

Although non-rarity not always yields original or interesting results, most of the studies on aggregation published till now are indeed original and interesting. Here we have a nice example of the influence of real problems into theoretical research and, in fact, only the experience in coming years can show if the new concepts and results will be useful in applications, i.e. fertile in the way of Karl Menger's thought: "The only justification for scientific definitions-for the way words are used in science-lies in their *fertility*; that is, in the possibility of using them in a great number of important propositions, preferably practical proposition" **Morality, Decision and Social Organization**; Reidel, Dordrecht, 1974).

3. In the applications of fuzzy logic it is usual to consider variables taking their values in the real line, endowed with its natural total ordering. Nevertheless, once variables are qualified by a linguistic label (P), the resulting atomic statements "x is P" are not always ordered in the same way as numbers in the real line, i.e., it can be $x \leq y$ but "y is P" less or non-comparable with "x is P". The order $\leq_P$ that a linguistic label P defines in the universe of discourse X is not always the same than that of the universe (if there is any) and, in general, it is a partial ordering.

When there are several linguistic labels $P, Q, \ldots$, several partial orderings $\leq_P, \leq_Q, \ldots$, are to be considered in X, and problem arises of knowing which is the ordering which results from a given logical operation between the linguistic labels. For example, it seems clear that the logical operation $\neg$ gives the orderings $\leq_{\neg P} = \leq_P^{-1}$, but still it is not clear enough what happens with the logical operations $\wedge, \vee$, and with the linguistic operation "antonym": Is it always $\leq_{antP} \subset \leq_P^{-1}$, and $\leq_{P \wedge Q} = \leq_P \cap \leq_Q$? What happens when P and Q are not in the same universe of discourse?

The problem is important in itself because membership functions $\mu_P, \mu_P, \ldots$ should reflect the "meaning" of $P, Q, \ldots$ in their respective universes of discourse and, following Ludwig Wittgenstein (**Philosophical Investigations**, Blackwell, Oxford, 1958), "the meaning of a word is its use in language". Hence, membership functions should reflect the current use of their corresponding linguistic labels in the universe of discourse, a use that is twofold: the comparative use (x is less P than y, that is $x \leq_P y$), and the extensional use (that is the fuzzy set **P**, characterized by the membership function μ_P). And, of course, the only way in which a function $\mu_P : X \longrightarrow [0, 1]$ is always non-decreasing is relatively to the pair of orderings $(\leq_P, \leq)$, which are, respectively, the one generated by P on X and the natural order on the unit interval. Only then membership functions can be considered as evaluations (or generalized measures) of the extend to which each x in X is P.

So it could be of interest to study aggregation operators behaving monotonically for a partial order in the real line and not only for its natural total ordering, together with the logical properties which aggregation operators introduce between fuzzy sets. A type of properties whose character is not clear yet, as they cannot be generated within classical logic frameworks, but that certainly arise when "considering an enlarged scheme in which different quantities variously interact. We should pass from a static situation in which we study more and more general and sophisticated one-notion theories to a situation in which we pay attention to the *Information-dynamics* in which the measurements of different aspects of a changing information are quantitatively related and connected" (Settimo Termini, **On Some Vagaries of Vagueness**, forthcoming in **Annals of Mathematics and Artificial Intelligence** (Special Issue, Guest Editor R. Scozzafava), 2001).

As fuzzy logic deals with the representation of either incomplete or imprecise information, a future new theory of information (different from the well known theory of Shannon, a theory of the transmission of information), will consider the problem of aggregating information as a basic one. In fact an abstract of a piece of news is an aggregation of information, and I would think that right now we are witnessing the birth of a new scientific subject: a Theory of Information Dynamics. The content of this book could be one of the founding stones of a part of such a theory.

With the aggregation operators, fuzzy logic carries on with the task (iniciated with t-norms and t-conorms) of introducing mathematical analysis into logic. A task that John von Neumann seemed to foresee as early as 1948, when he wrote: "Everybody who has worked in formal logic will confirm that it is one of the technically most refractory parts of mathematics. The reason for this is that it deals with rigid, all-or-none concepts, and has very little contact with the continuous concept of the real or of the complex number, that is with mathematical analysis" (**The General and Logical Theory of Automata, in Cerebral Mechanisms in Behaviour - The Hixon Symposium**, pp. 288-328, Wiley (New York), 1951). And, essentially, fuzzy logic deals with flexible concepts.

4. In this short Foreword I have tried to distillate some of the reasons for the interest I attribute to the study of aggregation operators and the challenge the theory faces. I hope this book will cause a deep interest in the reader thanks to the efforts of the editors and the quality of the contributions in

its pages. I also express my wishes for the best success for "Aggregation Operators. New Trends and Applications"!.

Boadilla del Monte, Madrid.

Enric Trillas
September, 2001.

Preface

The theory of aggregation operators is a rapidly developing domain, supported by the use of aggregation concepts when modelling different engineering, economical, social, etc., problems. However, there is no unified notation in this field. Most of the results are disseminated in different papers or specialized monographs, where usually a specific class of aggregation operators is discussed. This is, for example, the case of R.R. Yager's and J. Kacprzyk's edited volume dedicated to OWA operators in 1997 and of the monograph of E.P. Klement, R. Mesiar and E. Pap (2000) devoted to triangular norms. Some other edited volumes are devoted to the latest trends and developments in aggregation techniques, such as the monograph of B. Bouchon-Meunier (1998). Aggregation operators are traditionally presented at international conferences IFSA, IEEE, IPMU, EUSFLAT, EUROFUSE, FSTA, etc.

This edited volume aims at describing the current status of major parts of the theory of aggregation operators and also at showing potential areas for its applications. In overview parts, we have gathered self-contained papers, presenting essential material on selected topics in a comprehensive form, avoiding proofs and unnecessary details. This is not the case of several contributions bringing new results and ideas, which aim to show the latest streaming of aggregation theory and potential areas for its applications.

The book is divided in four parts. The first part brings the state-of-art overview of aggregation operators, including several construction methods and discussion about distinguished classes of aggregation operators. The second part is devoted to the theoretical aspects of aggregation. In the following part, several applications and potential fields of applications of aggregation operators are treated. The fourth part offers an alternative look on aggregation of verbal variables as fuzzy quantities, different from currently used Zadeh's extension principle based fuzzy arithmetics or interval arithmetics based approach.

The first part is by Calvo, Kolesárová, Komorníková and Mesiar. It has an introductory and overview character and it can serve also as a small hand–book on aggregation. In six chapters, first aggregation operators are introduced and then several properties of these operators are discussed and illustrated by numerous examples. The second chapter is devoted to various construction methods for obtaining new types of aggregation operators from given ones. Chapters 3 – –6 deal with several distinguished classes of aggregation operators. First, operators related to the arithmetic mean are discussed, such as weighted means, OWA operators, quasi-arithmetic means, etc. Chapter 4 shows how several types of integrals are applied in aggregation theory, and it shows the Lebesgue, Choquet, Sugeno and other integral-based aggregation operators. In chapter 5, triangular norms and related aggregation operators are presented, such as triangular conorms, uninorms, nullnorms, gamma operators, etc. Finally, generated aggregation operators are treated.

The second theoretical part consists of four chapters. It opens with an overview paper by Grabisch dealing with aggregation based on integrals, stressing the recent results related to the use of Choquet and Sugeno integrals in aggregation of real inputs (including both positive and negative values), as well as in aggregation of ordinal information.

The second paper has also an overview character and is written by Sander. It presents the first complete characterization of special associative aggregation operators, including the full proofs. This chapter generalizes several well-known characterizations and representations of continuous associative aggregation operators.

The next paper by Torra and Godo introduces continuous WOWA operators, generalizing both OWA operators and weighted mean operators. An application of these operators to defuzzification is shown.

In the last paper of this part, written by Yager, weighted uninorms are treated. A linear method of assigning weights to defend against strategic manipulation is proposed. Next, an advanced OWA based method to determine strategic manipulation is discussed.

The third part of the book deals with several potential areas of application of aggregation operators. It consists of five chapters.

In the first one, Pap discusses the role of aggregation operators in the engineering design. It starts with an axiomatization of engineering design based on von Neumann and Morgenstern approach. Then a general definition of decision making problems is introduced, allowing to apply also fuzzy measures and fuzzy systems. Some optimization and identification methods are also included.

The second paper by Marichal deals with aggregation of interacting criteria in decision making. The role of the discrete Choquet integral in this context is stressed. Special attention is paid to the aggregation problems based on 2-order Choquet integrals allowing to model the intersection of two but not more criteria.

The third paper, by Herrera, Herrera-Viedma and Martínez, analyzes the aggregation of linguistic information. It brings a comparative analysis of some aggregation approaches according to the results obtained in decision-making applications.

The fourth contribution by Domingo-Ferrer and Torra, describes microaggregation, i.e., a technique for statistical confidentiality based on aggregation operators. The goals of statistical confidentiability data are given, including some practical problems of reviewing the application of microaggregation to statistical disclosure problem.

In the last contribution in this part, Blanco, Delgado, Martínez-Bautista, Sánchez and Vila study a new method to evaluate quantified sentences. This quantifier-based method is shown to extend the quantified aggregation related to the Choquet integral.

The fourth part of the book was written by Mareš an Mesiar. In six chapters they present and discuss quantitative verbal variables and their aggregation.

In the first chapter, fuzzy quantities and their Zadeh's extension principle based aggregation is recalled. the second chapter describes the structure of verbal variables and brings several examples of shapes and scales. The third chapter is devoted to the elementary processing of verbal variables. The proposed methods are illustrated in the fourth chapter.

The fifth chapter is related to a short overview of triangular norms and the aggregation of fuzzy quantities based on T-extension principle. Finally, the sixth chapter brings the triangular norm-based processing of verbal quantities.

Note also that the increasing interest in the theory and application of aggregation operators has resulted into the creation of the AGOP working group (AGgregation OPerators) in the framework of the international EUSFLAT association, and this edited volume is one of the outcomes of the AGOP work.

Contributions contained in this edited volume show the state-of-art and new streaming of aggregation operators theory and its applications. The editors wish this volume will contribute to promote various applications of aggregation operators toward the new millenium.

Alcalá and Bratislava
September, 2001

T. Calvo
G. Mayor
R. Mesiar

Contents

Part 3
Applications of Aggregation Operators

Part 4
Fuzzy Quantities and Their Aggregation

Part 1

Aggregation Operators: Basic Concepts, Issues and Properties

Aggregation Operators: Properties, Classes and Construction Methods

Tomasa Calvo[1], Anna Kolesárová[2], Magda Komorníková[3], and Radko Mesiar[3]

[1]University of the Balearic Islands,
E-07071 Palma de Mallorca, Spain
[1]University of Alcalá,
E-28871 Alcalá de Henares, (Madrid, Spain)
e-mail: tomasa.calvo@{uib,uah}.es
[2]ChTF STU, Radlinského 9
81237 Bratislava, Slovak Republic
e-mail: kolesaro@cvt.stuba.sk
[3]SvF STU, Radlinského 11
81368 Bratislava, Slovak Republic
e-mail: {magda, mesiar}@vox.svf.stuba.sk
[3]ÚTIA AVČR
18208 Prague 8, Czech Republic

1 Introduction

Aggregation (fusion) of several input values into a single output value is an indispensable tool not only of mathematics or physics, but of majority of engineering, economical, social and other sciences. The problems of aggregation are very broad and heterogeneous, in general. Therefore we restrict ourselves in this contribution to the specific topic of the aggregation of finite number of real inputs only. Closely related topics of aggregating infinitely many real inputs [23,109,64,52,43,42,44,99], of aggregating inputs from some ordinal scales [41,50], of aggregating complex inputs (such as probability distributions [107,114], fuzzy sets [143]), etc., are treated, among others, in the quoted papers, and we will not deal with them. In this spirit, if the number of input values is fixed, say n, an aggregation operator is a real function of n variables. This is still a too general topic. Therefore we restrict our considerations regarding inputs as well as outputs to some fixed interval (scale) $I = [a, b] \subseteq [-\infty, \infty]$. It is a matter of rescaling to fix $I = [0, 1]$. Therefore, if not explicitly stated, we will assume throughout that both inputs and outputs are from the unit interval $I = [0, 1]$, and hence each n–ary aggregation operator $\mathbf{A}_{(n)}$ is always a mapping

$$\mathbf{A}_{(n)} : [0,1]^n \to [0,1] . \tag{1}$$

In general, the number of the input values to be aggregated is not known. Therefore an aggregation operator $\mathbf{A}$ should be a mapping

$$\mathbf{A} : \bigcup_{n \in \mathbb{N}} [0,1]^n \to [0,1] . \tag{2}$$

Note that $\mathbf{A}|_{[0,1]^n} = \mathbf{A}_{(n)}$ for all $n \in \mathbb{N}$, and hence a global aggregation operator from (2) can be introduced as a family $\mathbf{A} = (\mathbf{A}_{(n)})_{n\in\mathbb{N}}$ of n–ary operators, where, in general, operators $\mathbf{A}_{(n)}$ and $\mathbf{A}_{(m)}$ for different n and m need not be related. This possible defect of general aggregation operators will be discussed on appropriate places in the next chapters. Sometimes only partial operators $\mathbf{A}_{(n)}$ will be discussed, depending on the topic. For the illustration and the next use we now give some examples of operators in the spirit of (1) or (2).

Example 1 (i) Consider the arithmetic mean given by

$$\mathbf{M}(x_1, \ldots, x_n) = \frac{1}{n} \sum_{i=1}^{n} x_i . \tag{3}$$

Note that the arithmetic mean can be defined by the formula (3) on any interval $I = [a, b] \subseteq [-\infty, \infty]$. In case $I = [-\infty, \infty]$ we have to adopt some convention with respect to $+\infty + (-\infty)$. Throughout this contribution we will assume $+\infty + (-\infty) = -\infty$.

(ii) Consider the operators

$$\mathbf{max}(x_1, \ldots, x_n) = \max(x_1, \ldots, x_n), \tag{4}$$

and

$$\mathbf{min}(x_1, \ldots, x_n) = \min(x_1, \ldots, x_n) . \tag{5}$$

As before, the formulae (4) and (5) can be applied to define operators acting on an arbitrary interval $I = [a, b]$.

(iii) Let

$$\mathbf{\Pi}(x_1, \ldots, x_n) = x_1\, x_2 \ldots x_n = \prod_{i=1}^{n} x_i \tag{6}$$

be the product operator. Note that the formula (6) if applied to inputs from some (general) interval $[a, b]$ need not result in an output from $[a, b]$. Further, if necessary, the convention $0.\infty = 0.(-\infty) = 0$ will be adopted.

(iv) Consider the Gödel implication $\mathbf{I}_G : [0,1]^2 \to [0,1]$,

$$\mathbf{I}_G(x, y) = \begin{cases} 1 & \text{if } x \leq y , \\ y & \text{else,} \end{cases} \tag{7}$$

which is defined as a binary operator only. For more details we recommend [54,45]. Note that the binary operator $\mathbf{I}_G$ can be extended to a global operator, see (2), in several ways. For example (using the same notation for all involved operators), if $n > 2$,

$$\begin{aligned}\mathbf{I}_G(x_1, \ldots, x_n) &= \mathbf{I}_G\left(x_1, \mathbf{I}_G(x_2, \ldots, x_n)\right) \\ &= \mathbf{I}_G\left(x_1, \mathbf{I}_G(\ldots, \mathbf{I}_G(x_{n-1}, x_n)\ldots)\right),\end{aligned}$$

that is,

$$\mathbf{I}_G(x_1, \ldots, x_n) = \begin{cases} x_n & \text{if } \min(x_1, \ldots, x_{n-1}) > x_n \\ 1 & \text{else.} \end{cases} \tag{8}$$

For $n = 1$, we can put $\mathbf{I}_G(x) = x$.

The contribution is organized as follows. In the next Chapter, we first introduce boundary and monotonicity properties of operators acting on $\bigcup_{n \in \mathbb{N}} [0,1]^n$, and subsequently, we define aggregation operators which will be investigated and discussed. Then several other properties of aggregation operators discussed and illustrated. Chapter 3 is devoted to several construction methods for aggregation operators. Chapters 4–7 characterize and discuss several distinguished classes of aggregation operators. First, aggregation operators related to the arithmetic mean are treated, such as weighted arithmetic means, quasi–arithmetic means, OWA operators, etc. Next, integrals related to aggregation operators are investigated, stressing the role of the Lebesgue, Choquet, Sugeno, etc., integrals in aggregation. Also an important class of triangular norms and related operators is included in a special Chapter. Finally, the class of generated aggregation operators is introduced. The contribution ends with some concluding remarks.

Note that the aim of this contribution is to give an overview of major aggregation operators and techniques. Though it cannot be exhaustive, we believe that it will give a good look at the aggregation domain, hopefully useful for potential users in the spirit of a small handbook of aggregation. Note that nice overviews of Dubois and Prade [27] or of Mizumoto [100,101] contain a lot of important examples and other information related to aggregation operators. Similarly, there are special chapters of distinguished monographs related to aggregation operators, such as [64,34,7,61,140]. The mentioned publications are very good sources of information in some specific aggregation domain (e.g. [61] is devoted to the triangular norms, [140] deals with OWA operators), but they lack a general point of view, or they do not already contain the recent results or classes of the aggregation operators, such as OWA operators, for example.

2 Basic properties of aggregation operators

2.1 Boundary conditions

Following with [64,96,69] as an indispensable property of any n–ary aggregation operator $\mathbf{A}_{(n)} : [0,1]^n \to [0,1]$ we assume preservation of the bounds of the domain and of the range, i.e.,

$$\mathbf{A}_{(n)}(0,\ldots,0) = 0 \quad \text{and} \quad \mathbf{A}_{(n)}(1,\ldots,1) = 1\,.$$

In other words, if we have only minimal (maximal) possible inputs then we should obtain the minimal (maximal) possible output. For $\mathbf{A} : \bigcup_{n\in\mathbb{N}} [0,1]^n \to [0,1]$ this means

$$\mathbf{A}(0,\ldots,0) = 0 \quad \text{and} \quad \mathbf{A}(1,\ldots,1) = 1 \tag{9}$$

independently of the number of inputs. Note that all operators introduced in Chapter 1 fulfill the boundary condition (9) up to the Gödel implication $\mathbf{I}_G$ given in (7) and (8). Another example of an operator violating (9) is the collapsed output operator $\mathbf{C}_c : \bigcup_{n\in\mathbb{N}} [0,1]^n \to [0,1]$, given by $\mathbf{C}_c(x_1,\ldots,x_n) = c$, where $c \in [0,1]$ is a fixed constant.

A specific case is the aggregation of a singleton, i.e., the unary operator $\mathbf{A}_{(1)} : [0,1] \to [0,1]$. The relation (9) means that $\mathbf{A}_{(1)}(0) = 0$ and $\mathbf{A}_{(1)}(1) = 1$. For many scientists the aggregation (fusion) of a singleton is not an (true) aggregation, and they propose as a convention:

$$\mathbf{A}_{(1)}(x) = x, \quad x \in [0,1]\,. \tag{10}$$

Throughout this contribution, we will only deal with operators complying with this convention. Unless not explicitly mentioned, we adopt this convention, for t–norms and t–conorms in particular. Note that all examples in Chater 1 fit the boundary condition (10). Only the operator $\mathbf{C}_c$ introduced above is an example of an operator violating (10).

2.2 Monotonicity

The standard monotonicity of an n–ary operator $\mathbf{A}_{(n)} : [0,1]^n \to [0,1]$ is the monotonicity of a function of n variables. Because of the boundary conditions (9) the monotonicity of an operator

$$\mathbf{A} : \bigcup_{n\in\mathbb{N}} [0,1]^n \to [0,1]$$

means the non–decreasingness of all partial n–ary operators $\mathbf{A}_{(n)}$, i.e.,

$$\forall n \in \mathbb{N} \;:\; x_1 \le y_1,\ldots,x_n \le y_n \Longrightarrow \mathbf{A}(x_1,\ldots,x_n) \le \mathbf{A}(y_1,\ldots,y_n)\,. \tag{11}$$

Evidently, the monotonicity (11) is related to the Cartesian partial order, when two input systems $(x_1, \ldots, x_n)$ and $(y_1, \ldots, y_m)$ are comparable only if $n = m$ and $x_i \le y_i$ for all $i = 1, \ldots, n$, $n \in \mathbb{N}$ (or $x_i \ge y_i$). As in the case of boundary conditions (9), also the monotonicity (11) is the property of relevant n–ary operators $\mathbf{A}_{(n)}$ and does not relate to two input–tuples with different cardinalities.

There are alternative approaches to the monotonicity of aggregation operators, see, e.g., α and β orders introduced in [86]. A deeper discussion of these types of monotonicity can be found in [11]. We recall here the definition and later also some results related to these orders.

Let $\mathbf{x} = (x_1, \ldots, x_n)$ and $\mathbf{y} = (y_1, \ldots, y_m)$ be elements of $\bigcup_{n \in \mathbb{N}} [0, 1]^n$. Then the following partial orderings can be considered:

- $\mathbf{x} \le_\alpha \mathbf{y}$ if and only if $n \le m$ and $x_1 \le y_1, \ldots, x_n \le y_n$, $\max_{i=1}^{n}(x_i) \le \min_{n+1}^{m}(y_i)$, with the convention $\min(\emptyset) = 1$.
- $\mathbf{x} \le_\beta \mathbf{y}$ if and only if $n \ge m$ and $x_1 \le y_1, \ldots, x_n \le y_n$, $\max_{m+1}^{n}(x_i) \le \min_{i=1}^{m}(y_i)$, with the convention $\max(\emptyset) = 0$.

To get an impression about the Cartesian partial order $\le$, the α–order $\le_\alpha$ and the β–order $\le_\beta$, we give the next example:

$$\begin{aligned}&(0.2, 0.1, 0.7, 0.1) \le_\beta (0.2, 0.1, 0.7) \le_\alpha \\ &(0.2, 0.2, 0.7, 0.7) \le_\beta (0.8, 0.7) \le (0.8, 0.8).\end{aligned}$$

Obviously, an operator non–decreasing with respect to the α–order (β–order) is necessarily also non–decreasing with respect to the Cartesian partial order, i.e., (11) is satisfied. As an example of an operator fitting (11) which is not α–monotone, it is enough to take the product $\mathbf{\Pi}$.

Note that among the operators we have already discussed, the Gödel implication $\mathbf{I}_G$ introduced in (7), see also (8), is not monotone in the sense of (11). In fact, $\mathbf{I}_G$ is (as a binary operator) non–decreasing in the second coordinate, but non–increasing in the first coordinate.

Note also that we cannot modify (11) in the following sense:

$$\forall n \in \mathbb{N}: \quad x_1 < y_1, \ldots, x_n < y_n \Longrightarrow \mathbf{A}(x_1, \ldots, x_n) \le \mathbf{A}(y_1, \ldots, y_n). \tag{12}$$

For $n \ge 2$, define, e.g.,

$$\mathbf{A}(x_1, \ldots, x_n) = \begin{cases} (1 - x_n)x_n & \text{if } x_1 = 0, \\ \frac{1+x_n}{2} & \text{else.} \end{cases} \tag{13}$$

Then $\mathbf{A}$ fits (12) but not (11). We introduce two additional types of monotonicity sometimes required in applications.

A monotone operator $\mathbf{A} : \bigcup_{n\in\mathbb{N}} [0,1]^n \to [0,1]$ fulfilling (11) is called

(i) *strictly monotone* if

$$\forall n \in \mathbb{N}, i \in \{1,\ldots,n\} : (x_i \leq y_i \ \wedge (x_1,\ldots,x_n) \neq (y_1,\ldots,y_n)) \\ \Downarrow \\ \mathbf{A}(x_1,\ldots,x_n) < \mathbf{A}(y_1,\ldots,y_n)\,, \tag{14}$$

(ii) *jointly strictly monotone* if

$$\forall n \in \mathbb{N} : x_1 < y_1,\ldots,x_n < y_n \Longrightarrow \mathbf{A}(x_1,\ldots,x_n) < \mathbf{A}(y_1,\ldots,y_n)\,. \tag{15}$$

Recall that strict monotonicity ensures the joint strict monotonicity but not vice–versa. Further note that the strict monotonicity is equivalent to *cancelativity*, which means that if $\mathbf{A}(x_1,\ldots,x_n) = \mathbf{A}(y_1,\ldots,y_n)$ and $x_i = y_i$ for all $i \in \{1,\ldots,n\} \setminus \{i_0\}$ for some $i_0 \in \{1,\ldots,n\}$, then $x_{i_0} = y_{i_0}$. For example, the arithmetic mean $\mathbf{M}$ is strictly monotone, i.e., cancelative, and hence also jointly strictly monotone. Operators **max** and **min** are jointly strictly monotone but not strictly monotone. Similarly, the product $\mathbf{\Pi}$ is jointly strictly monotone, however if 0 occurs among inputs, cancelativity (which is equivalent to strict monotonicity) is violated.

2.3 Aggregation operators

Following [64,96,69], throughout this contribution we will deal with operators fulfilling boundary conditions (9), (10) and the monotonicity condition (11).

Definition 1. An operator $\mathbf{A} : \bigcup_{n\in\mathbb{N}} [0,1]^n \to [0,1]$ is called an aggregation operator if it fulfills the boundary conditions (9),(10) and the monotonicity condition (11).

Note that if an operator $\mathbf{A}$ acting on $I = [a,b] \subseteq [-\infty,\infty]$ fulfills relevantly modified conditions (9), (10), (11), it will be called an aggregation operator on I.

Remember that among till now introduced operators, the arithmetic mean $\mathbf{M}$, minimum **min**, maximum **max**, product $\mathbf{\Pi}$ are aggregation operators in the sense of Definition 1 2.1, while the Gödel implication $\mathbf{I}_G$, collapsed output operator $\mathbf{C}_c$ for arbitrary $c \in [0,1]$, and the operator $\mathbf{A}$ introduced in (13) are examples which are not aggregation operators in the sense of Definition 1. Our framework for aggregation operators is enough general to include most of the relevant operators used for the fusion of input data.

The standard comparison of functions of n variables allows us to compare aggregation operators.

Definition 2. Let $\mathbf{A}, \mathbf{B} : \bigcup_{n\in\mathbb{N}} [0,1]^n \to [0,1]$ be two aggregation operators. We say that $\mathbf{A}$ is weaker than $\mathbf{B}$, with notation $\mathbf{A} \le \mathbf{B}$, if

$$\forall n \in \mathbb{N}, \forall (x_1,\dots,x_n) \in [0,1]^n : \quad \mathbf{A}(x_1,\dots,x_n) \le \mathbf{B}(x_1,\dots,x_n). \quad (16)$$

If $\mathbf{A} \le \mathbf{B}$, we also say that $\mathbf{B}$ is stronger than $\mathbf{A}$.

It is easy to find the weakest aggregation operator $\mathbf{A}_w$ and the strongest aggregation operator $\mathbf{A}_s$:

$$\forall n \ge 2, (x_1,\dots,x_n) \ne (1,\dots,1) \quad : \quad \mathbf{A}_w(x_1,\dots,x_n) = 0, \quad (17)$$

and

$$\forall n \ge 2, (x_1,\dots,x_n) \ne (0,\dots,0) \quad : \quad \mathbf{A}_s(x_1,\dots,x_n) = 1. \quad (18)$$

For any aggregation operator $\mathbf{A}$ we have

$$\mathbf{A}_w \le \mathbf{A} \le \mathbf{A}_s .$$

For already introduced aggregation operators we have

$$\mathbf{A}_w \le \mathbf{\Pi} \le \mathbf{min} \le \mathbf{M} \le \mathbf{max} \le \mathbf{A}_s .$$

Important examples of aggregation operators are projections $\mathbf{P}_F$ (the projection to the first coordinate) and $\mathbf{P}_L$ (the projection to the last coordinate):

$$\mathbf{P}_F(x_1,\dots,x_n) = x_1, \qquad \mathbf{P}_L(x_1,\dots,x_n) = x_n . \quad (19)$$

Observe that $\mathbf{P}_F$ and $\mathbf{P}_L$ are incomparable, as well as the pairs $\mathbf{P}_L$ and $\mathbf{M}$, $\mathbf{P}_F$ and $\mathbf{M}$. Moreover,

$$\mathbf{min} \le \mathbf{P}_F \le \mathbf{max}, \quad \mathbf{min} \le \mathbf{P}_L \le \mathbf{max}.$$

As already mentioned, some authors when thinking of aggregation (fusion) require at least two input values. Often only a binary form $\mathbf{A}_{(2)}$ of an aggregation operator is known. The ternary form $\mathbf{A}_{(3)}$ (or n–ary for $n > 2$) of that operator needs not to have any relationship with $\mathbf{A}_{(2)}$, in general. However, if only $\mathbf{A}_{(2)}$ is known, we have several ways for extending $\mathbf{A}_{(2)}$ to a complete aggregation operator $\mathbf{A}$. One possibility (compare also Example 1 (iv)) is the backward inductive extension of the binary operator, i.e.,

$$\mathbf{A}^*(x_1,\dots,x_n) = \mathbf{A}_{(2)}\left(x_1, \mathbf{A}_{(2)}\left(\dots, \mathbf{A}_{(2)}(x_{n-1}, x_n)\dots\right)\right), \ n > 2 \quad (20)$$

and of course, $\mathbf{A}^*(x) = x$ because of (10). An alternative approach is the forward inductive extension of the binary operator, i.e.,

$$\mathbf{A}_*(x_1,\dots,x_n) = \mathbf{A}_{(2)}\left(\dots(\mathbf{A}_{(2)}(\mathbf{A}_{(2)}(x_1,x_2),x_3))\dots,x_n\right), \ n > 2 \quad (21)$$

and $\mathbf{A}_*(x) = x$.

For example, let $\mathbf{M}_{(2)}(x,y) = \frac{x+y}{2}$. Then

$$\mathbf{M}^*(x_1,\ldots,x_n) = \frac{x_1}{2} + \frac{x_2}{2^2} + \ldots + \frac{x_{n-2}}{2^{n-2}} + \frac{x_{n-1}}{2^{n-1}} + \frac{x_n}{2^{n-1}}, \tag{22}$$

while

$$\mathbf{M}_*(x_1,\ldots,x_n) = \frac{x_1}{2^{n-1}} + \frac{x_2}{2^{n-1}} + \frac{x_3}{2^{n-2}} + \ldots + \frac{x_{n-1}}{2^2} + \frac{x_n}{2}. \tag{23}$$

Observe that $\mathbf{A}_* = \mathbf{A}^*$ if and only if $\mathbf{A}_{(2)}$ is associative, see also Section 2.7.

2.4 Idempotency

In Algebra, idempotency is an algebraic property related to a binary operation $*$, i.e., x is an idempotent element with respect to an operation $*$ if $x * x = x$. Extending this notion to n–ary operators and rewriting it for aggregation operators, we can introduce idempotent aggregation operators as follows.

Definition 3. Let $\mathbf{A} : \bigcup_{n\in\mathbb{N}} [0,1]^n \to [0,1]$ be an aggregation operator. An element $x \in [0,1]$ is called $(\mathbf{A})$–idempotent element whenever $\mathbf{A}_{(n)}(x,\ldots,x) = x$ for all $n \in \mathbb{N}$. $\mathbf{A}$ is called an idempotent aggregation operator if each $x \in [0,1]$ is an idempotent element of $\mathbf{A}$.

Note that the boundary condition (9) means that 0 and 1 are $\mathbf{A}$–idempotent elements for each aggregation operator $\mathbf{A}$. Therefore 0 and 1 are called trivial idempotent elements. Observe that the idempotency of an aggregation operator $\mathbf{A}$ is also called unanimity, and in multi–criteria decision making reads as follows: if all criteria are satisfied in the same degree x, then also the global score should be x. This property is in some areas supposed to be a genuine property of aggregation operators, e.g., in already mentioned multi–criteria decision making [34]. Idempotent aggregation operators are also called averaging operators.

Notice that for monotone operators (in the sense of (11)) the idempotency of an operator $\mathbf{A}$ is equivalent to so called compensation property:

$$\mathbf{min} \leq \mathbf{A} \leq \mathbf{max}. \tag{24}$$

Now it is evident that **min**, **M**, **max** are idempotent aggregation operators, while $\mathbf{A}_w$, $\mathbf{\Pi}$, $\mathbf{A}_s$ are not. Observe also that the operators $\mathbf{M}^*$, $\mathbf{M}_*$ introduced in (22) and (23), respectively, are idempotent aggregation operators, as well as all aggregation operators constructed in the spirit given in (20) and (21), based on an idempotent binary operator $\mathbf{A}_{(2)}$. The compensation property (24) ensures another important feature of idempotent operators: for any interval $[c,d] \subset [0,1]$, any idempotent aggregation operator A and any n–tuple $(x_1,\ldots,x_n) \in [c,d]^n$, also the value $\mathbf{A}(x_1,\ldots,x_n) \in [c,d]$. Consequently, $\mathbf{A}|_{\bigcup\limits_{n\in\mathbb{N}}[c,d]^n}$ is an idempotent aggregation operator acting on $[c,d]$. Obviously, for a general aggregation operator $\mathbf{A}$ and for fixed c,d, the last claim (without idempotency) is true if and only if c and d are idempotent elements of $\mathbf{A}$.

In Algebra, idempotency is an algebraic property related to a binary operation Note that non–idempotent aggregation operators $\mathbf{A}_w$, $\mathbf{\Pi}$, $\mathbf{A}_s$ have only trivial idempotents. As an example of an operator which is not idempotent but has also a non–trivial idempotent element, take an arbitrarily chosen element $c \in]0,1[$ and define the aggregation operator $\mathbf{A}_{\{c\}} : \bigcup\limits_{n\in\mathbb{N}}[0,1]^n \to [0,1]$ as follows:

$$\mathbf{A}_{\{c\}}(x_1,\ldots,x_n) = \max\left(0, \min\left(1, c + \sum_{i=1}^{n}(x_i - c)\right)\right). \tag{25}$$

By means of a straightforward computation it is easy to see that the only idempotent elements of $\mathbf{A}_{\{c\}}$ are 0, 1 and c.

2.5 Continuity

The continuity of an aggregation operator $\mathbf{A}$ is simply the continuity of all n–ary operators $\mathbf{A}_{(n)}$ in the standard sense of the continuity of real functions of n variables.

Definition 4. An aggregation operator $\mathbf{A} : \bigcup\limits_{n\in\mathbb{N}}[0,1]^n \to [0,1]$ is called a continuous aggregation operator if for all $n \in \mathbb{N}$ the operators $\mathbf{A}_{(n)} : [0,1]^n \to [0,1]$ are continuous, that is, if

$\forall x_1,\ldots,x_n \in [0,1], \forall (x_{1j})_{j\in\mathbb{N}},\ldots,(x_{nj})_{j\in\mathbb{N}} \in [0,1]^{\mathbb{N}} : \lim\limits_{j\to\infty} x_{ij} = x_i$, for $i = 1,\ldots,n$ then $\lim\limits_{j\to\infty} \mathbf{A}_{(n)}(x_{1j},\ldots,x_{nj}) = \mathbf{A}_{(n)}(x_1,\ldots,x_n)$.

In engineering applications continuous aggregation operators are usually applied, reflecting the property that a "small" error in inputs cannot cause a "big" error in the output. From the mathematical point of view, because of the compactness of domains $[0,1]^n$, $n \in \mathbb{N}$, the continuity of an aggregation operator $\mathbf{A}$ is equivalent to its uniform continuity expressed by:

$$\forall \epsilon > 0, \forall n \in \mathbb{N}, \exists \delta > 0 : |x_i - y_i| < \delta, i = 1,\ldots,n$$

$$\Downarrow$$

$$|\mathbf{A}(x_1,\ldots,x_n) - \mathbf{A}(y_1,\ldots,y_n)| < \epsilon. \tag{26}$$

Because of the monotonicity condition (11) the continuity of an aggregation operator $\mathbf{A}$ is also equivalent to the intermediate value property.

Definition 5. Let $(x_1, \ldots, x_n)$, $(y_1, \ldots, y_n) \in [0,1]^n$, $n \in \mathbb{N}$ be any n–tuples such that $x_i \le y_i$, $i = 1, \ldots, n$. An aggregation operator $\mathbf{A}$ has the intermediate value property if

$$\forall z \in [\mathbf{A}(x_1, \ldots, x_n), \mathbf{A}(y_1, \ldots, y_n)]$$
$$\exists\, z_i \in [x_i, y_i],\ i = 1, \ldots, n : \ \mathbf{A}(z_1, \ldots, z_n) = z\,. \tag{27}$$

Note that the intermediate value property (27) allows to introduce the equivalent of continuity in the case of aggregation operators acting on ordinal (discrete) scales. This property is sometimes called smoothness [40,87], though it has no relationship with differentiability.

An important analytical property of functions of n variables allowing to estimate the error when dealing with imprecise input data is the Lipschitz property. Recall that an aggregation operator $\mathbf{A}$ fulfills the *Lipschitz property* with constant $c \in]0, \infty[$ ($\mathbf{A}$ is c–Lipschitz for short) if

$$\forall n \in \mathbb{N},\ \forall (x_1, \ldots, x_n),\ (y_1, \ldots, y_n) \in [0,1]^n$$

$$|\ \mathbf{A}(x_1, \ldots, x_n) - \mathbf{A}(y_1, \ldots, y_n)\ | \le c \sum_{i=1}^{n} |x_i - y_i|\,. \tag{28}$$

Clearly, the Lipschitz property (with arbitrary c) ensures continuity but not vice–versa.

Within the already introduced aggregation operators, the operators $\mathbf{A}_w$ and $\mathbf{A}_s$ are examples of non–continuous operators. Operators $\mathbf{M}$, $\mathbf{\Pi}$, $\mathbf{min}$, $\mathbf{max}$, $\mathbf{M}^*$ and $\mathbf{M}_*$ from (22) and (23), respectively, $\mathbf{P}_F$, $\mathbf{P}_L$ and $\mathbf{A}_{\{c\}}$ from (25) are continuous aggregation operators which all also fulfill the Lipschitz property. Note that all considered till now continuous operators are 1–Lipschitz. As an example of a continuous aggregation operator which is not Lipschitz for any $c \in]0, \infty[$ we recall the geometric mean $\mathbf{G}$,

$$\mathbf{G}(x_1, \ldots, x_n) = \left(\prod_{i=1}^{n} x_i\right)^{1/n}\,. \tag{29}$$

From the differentiability point of view, a continuous aggregation operator $\mathbf{A}$ is c–Lipschitz if and only if all partial derivatives of all operators $\mathbf{A}_{(n)}$, $n \in \mathbb{N}$ are bounded from above by c in all points where they exist. Observe also that for various $n \in \mathbb{N}$, the Lipschitz constant c can vary, and hence it should be (if needed) denoted by c_n. For example, the arithmetic mean $\mathbf{M}$ is 1–Lipschitz and because of the boundary condition (10) no global aggregation operator $\mathbf{A}$ can be c–Lipschitz with $c < 1$. Because of the boundary condition (9), the smallest Lipschitz constant c_n related to $\mathbf{A}_{(n)}$ is $c_n = 1/n$. Indeed, we have always

$$1 = |\ \mathbf{A}_{(n)}(1, \ldots, 1) - \mathbf{A}_{(n)}(0, \ldots, 0)\ | = \frac{1}{n} \sum_{i=1}^{n} |1 - 0|\,.$$

It can be easily shown that the unique aggregation operator $\mathbf{A}$, such that $\mathbf{A}_{(n)}$ is $\frac{1}{n}$– Lipschitz for all $n \in \mathbb{N}$, is just the arithmetic mean $\mathbf{M}$. Hence remarkably, $\mathbf{M}$ is the most stable aggregation operator under possible input errors. Further note that for aggregation operators $\mathbf{A}^*$ and $\mathbf{A}_*$ constructed from some given $\mathbf{A}_{(2)}$ as indicated in (20) and (21), respectively, $\mathbf{A}^*_{(n)}$ and $\mathbf{A}_{*(n)}$ for $n \geq 2$ are c–Lipschitz if and only if $\mathbf{A}_{(2)}$ is c–Lipschitz. Therefore $\mathbf{M}^*$ given in (22) and $\mathbf{M}_*$ given in (23) are 0.5–Lipschitz for any number of inputs exceeding 1.

1–Lipschitz aggregation operators are intensively studied, e.g., in [70,17]. Observe that for $n = 2$, $\mathbf{A}_{(2)}$ is 1–Lipschitz if and only if $\tilde{\mathbf{A}} : [0,1]^2 \longrightarrow [0,1]$, $\tilde{\mathbf{A}}(x,y) = x+y-\mathbf{A}(x,y)$, is an aggregationoperator [70]. Then the operator $\sim$ acts on 1–Lipschitz binary aggregation operators as a kind of duality, compare also [36] or (194).

Finally, observe that all n–ary operators $\mathbf{A}_{(n)}$ of some aggregation operator $\mathbf{A}$ can be Lipschitz while the global operator $\mathbf{A}$ need not be (i.e., then $\sup_{n\in\mathbb{N}} c_n = +\infty$), see the following example.

Example 2 Define $\mathbf{Q} : \bigcup_{n\in\mathbb{N}} [0,1]^n \to [0,1]$ by

$$\mathbf{Q}(x_1,\ldots,x_n) = \prod_{i=1}^{n} x_i^i . \tag{30}$$

Then, e.g., $\mathbf{Q}_{(2)}(x,y) = xy^2$, i.e.,

$$\frac{\partial \mathbf{Q}_{(2)}(x,y)}{\partial x} = y^2, \qquad \frac{\partial \mathbf{Q}_{(2)}(x,y)}{\partial y} = 2xy\,,$$

and hence $\mathbf{Q}_{(2)}$ is 2–Lipschitz. Similarly, we can show that $\mathbf{Q}_{(n)}$ is n–Lipschitz but not c_n–Lipschitz for any $c_n < n$, $n \in \mathbb{N}$. Evidently, $\mathbf{Q}$ cannot fulfill the Lipschitz property for any $c \in]0,\infty[$.

Even stronger requirement than the Lipschitz property is related to the stability of aggregation operators, in which we assume that the output error of resulting aggregation cannot exceed any of possible input errors. Such aggregation operators are called kernel aggregation operators [70,103] or ∞–stable aggregation operators [17]. They fulfill, for all $n \in \mathbb{N}$, $(x_1,\ldots,x_n)$, $(y_1,\ldots,y_n) \in [0,1]^n$:

$$|\mathbf{A}(x_1,\ldots,x_n) - \mathbf{A}(y_1,\ldots,y_n)| \leq \max_i(|x_i - y_i|). \tag{31}$$

Obviously, each kernel aggregation operator is also 1–Lipschitz (but not vice–versa). Just for theoretical purposes we also recall other types of continuity. An aggregation operator $\mathbf{A} : \bigcup_{n\in\mathbb{N}} [0,1]^n \to [0,1]$ is called *lower (upper)*

semi–continuous if for all $n \in \mathbb{N}$, the operator $\mathbf{A}_{(n)} : [0,1]^n \to [0,1]$ is lower (upper) semi–continuous. Recall that
$\mathbf{A}_{(n)}$ is lower semi–continuous if

$$\forall\, (x_{1j})_{j\in\mathbb{N}}, \ldots, (x_{nj})_{j\in\mathbb{N}} \in [0,1]^{\mathbb{N}}:$$

$$\sup_j \mathbf{A}_{(n)}(x_{1j}, \ldots, x_{nj}) = \mathbf{A}_{(n)}(\sup_j x_{1j}, \ldots, \sup_j x_{nj}),$$

and upper semi–continuous if

$$\forall\, (x_{1j})_{j\in\mathbb{N}}, \ldots, (x_{nj})_{j\in\mathbb{N}} \in [0,1]^{\mathbb{N}}:$$

$$\inf_j \mathbf{A}_{(n)}(x_{1j}, \ldots, x_{nj}) = \mathbf{A}_{(n)}(\inf_j x_{1j}, \ldots, \inf_j x_{nj}).$$

Note that the weakest aggregation operator $\mathbf{A}_w$ is upper semi–continuous but not lower semi–continuous (the only discontinuity point of $\mathbf{A}_{w(n)}$, $n \geq 2$, is the maximal input $(1, \ldots, 1)$). Similarly, the strongest aggregation operator $\mathbf{A}_s$ is lower semi–continuous but not upper semi–continuous (with discontinuity point $(0, \ldots, 0)$ of $\mathbf{A}_{s(n)}$, $n \geq 2$).

As usually, the continuity of an aggregation operator $\mathbf{A}$ is equivalent to its simultaneous lower and upper semi–continuity. As a peculiar non–continuous aggregation operator recall the operator $\mathbf{T}_\diamond : \bigcup_{n\in\mathbb{N}} [0,1]^n \to [0,1]$ introduced by Smutná [118], see below.

First observe that each $x \in]0,1]$ can be uniquely represented in the form $x = \sum_{k=1}^{\infty} \frac{1}{2^{\xi_k}}$, where $(\xi_k) \subset \mathbb{N}^{\mathbb{N}}$ is a strictly increasing sequence of integers. We have (for any $n \in \mathbb{N}$):

$$\mathbf{T}_\diamond(x_1, \ldots, x_n) = \begin{cases} 0 & \text{if } \min(x_1, \ldots, x_n) = 0 \\ \sum_{k=1}^{\infty} \frac{1}{2^{u_k}} & \text{else,} \end{cases} \tag{32}$$

where $u_k = k + \sum_{i=1}^{n} (\xi_{i_k} - k)$. The aggregation operator $\mathbf{T}_\diamond$ is lower semi–continuous. However, for any $n \geq 2$, the set of discontinuity points is dense in $[0,1]^n$. For more details see [118] or [61]. Also note that there is no weakest nor strongest continuous aggregation operator. However, for each $n \in \mathbb{N}$ and $c_n \in [\frac{1}{n}, \infty[$ there can be found the weakest (strongest) n–ary aggregation operator $\mathbf{A}_{(n)}$ which is c_n–Lipschitz. For example, for $n = 2$ and $c_2 = 1$, the weakest 1–Lipschitz binary aggregation operator is the operator $\mathbf{T}_{L(2)}$, see (161), which is called the Łukasiewicz t–norm.

2.6 Symmetry

The standard commutativity of binary operators $x * y = y * x$ can be easily generalized for n–ary operators with $n > 2$. However, then we already speak about the symmetry (reflecting the geometric properties of relevant graphs). Similarly to the previous properties, the symmetry of an aggregation operator $\mathbf{A}$ means the symmetry of all n–ary operators $\mathbf{A}_{(n)}$, $n \in \mathbb{N}$.

Definition 6. An aggregation operator $\mathbf{A} : \bigcup_{n\in\mathbb{N}} [0,1]^n \to [0,1]$ is called a symmetric aggregation operator if

$$\forall n \in \mathbb{N}, \forall x_1,\ldots,x_n \in [0,1] : \mathbf{A}(x_1,\ldots,x_n) = \mathbf{A}(x_{\alpha(1)},\ldots,x_{\alpha(n)}) \quad (33)$$

for all permutations $\alpha = (\alpha(1),\ldots,\alpha(n))$ of $(1,\ldots,n)$.

All aggregation operators introduced till now up to the operator $\mathbf{Q}$, see (30), projections $\mathbf{P_F}$ and $\mathbf{P_L}$, $\mathbf{M}^*$, see (22), and $\mathbf{M}_*$, see (23), are symmetric. Prominent examples of non–symmetric aggregation operators are weighted arithmetic means which will be discussed in Section 4.1.

The symmetry of an aggregation operator reflects the same importance of single criteria in multi–criteria decision making, i.e., the knowledge of the order of the input score is irrelevant. Hence this property is sometimes called anonymity. There are several attempts how to introduce the weights into the symmetric aggregation [13,14,41,135,134,137], see also Section 3.3.

Vice–versa, any non–symmetric aggregation operator $\mathbf{A}$ can be symmetrized [111] in the following way: $\forall n \in \mathbb{N}$, $\forall x_1,\ldots,x_n \in [0,1]$ let $(x'_1,\ldots,x'_n)$ be a non–decreasing (non–increasing) permutation of the input vector $(x_1,\ldots,x_n)$, and define $\tilde{\mathbf{A}}(x_1,\ldots,x_n) = \mathbf{A}(x'_1,\ldots,x'_n)$. Obviously, $\tilde{\mathbf{A}}$ is a symmetric aggregation operator and $\mathbf{A} = \tilde{\mathbf{A}}$ if and only if $\mathbf{A}$ is symmetric. Note that symmetric sums introduced and investigated by Silvert [117] needs not be symmetric aggregation operators. Symmetric sums are related to the duality of aggregation operators and will be discussed in Section 3.2.

2.7 Associativity

The associativity of a binary operation $*$ defined on a domain I means that the couple $(I,*)$ is a semigroup. More important is the fact that then we have a genuine extension of $*$ as a binary operation to an n–ary operation on I for all $n > 2$. Recall that the associativity of a binary operator $\mathbf{A}_{(2)} : [0,1]^2 \to [0,1]$ means:

$$\forall x, y, z \in [0,1] : \quad \mathbf{A}_{(2)}\left(\mathbf{A}_{(2)}(x,y),z\right) = \mathbf{A}_{(2)}\left(x,\mathbf{A}_{(2)}(y,z)\right) .$$

Consequently, if we apply either (20) or (21) to an associative operator $\mathbf{A}_{(2)}$ we obtain the same aggregation operator $\mathbf{A}^* = \mathbf{A}_* = \mathbf{A}$. Formally, the associativity of an aggregation operator $\mathbf{A}$ is defined as follows.

Definition 7. An aggregation operator $\mathbf{A} : \bigcup_{n\in\mathbb{N}} [0,1]^n \to [0,1]$ is associative if

$$\forall n,m \in \mathbb{N}, \forall x_1,\dots,x_n,y_1,\dots,y_m \in [0,1] :$$

$$\mathbf{A}(x_1,\dots,x_n,y_1,\dots,y_m) = \mathbf{A}\left(\mathbf{A}(x_1,\dots,x_n),\mathbf{A}(y_1,\dots,y_m)\right) . \tag{34}$$

The associativity of an aggregation operator $\mathbf{A}$ allows to aggregate first some subsystems of all inputs, and then the partial outputs. For practical purposes we can start with aggregation procedure before knowing all inputs to be aggregated. New (additional) input data are then simply aggregated with the actual aggregated output. From the structural point of view, an associative aggregation operator $\mathbf{A}$ is uniquely determined by the corresponding binary operator $\mathbf{A}_{(2)} : [0,1]^2 \to [0,1]$, and then, as already mentioned, $\mathbf{A} = \mathbf{A}^* = \mathbf{A}_*$, where we can apply either (20) or (21). Therefore usually the same notation for $\mathbf{A}$ and $\mathbf{A}_{(2)}$ is commonly used.

As examples of associative aggregation operators recall $\mathbf{A}_w$, $\mathbf{A}_s$, **min**, **max**, $\mathbf{\Pi}$, $\mathbf{T}_\diamond$, $\mathbf{P}_F$, $\mathbf{P}_L$.

Non–associative aggregation operators are $\mathbf{M}$, $\mathbf{M}^*$ given in (22), $\mathbf{M}_*$ given in (23), operators $\mathbf{A}_{\{c\}}$ from (25), the geometric mean $\mathbf{G}$, and the operator $\mathbf{Q}$ given in (30).

As already mentioned, associativity leaves no freedom to n–ary operators $\mathbf{A}_{(n)}$, $n > 2$ when $\mathbf{A}_{(2)}$ is known. Associativity reduces efficiently the computational complexity of $\mathbf{A}$ when aggregating many (more than 2) inputs.

On the other hand, associativity is a very strong and rather restrictive property, especially together with continuity, see, for example, [85] or the chapter of W. Sander in this book [113]. Therefore sometimes some modifications of associativity preserving its advantages (from the computational point of view) and extending the freedom in the choice of $\mathbf{A}_{(n)}$, $n > 2$, are introduced. So, e.g., if two (or another number) associative operators $\mathbf{B}$, $\mathbf{C}$ and functions f, g, h are given such that $\mathbf{A} = f(\mathbf{B}\circ g, \mathbf{C}\circ h)$, i.e.,

$$\mathbf{A}(x_1,\dots,x_n) = f\left(\mathbf{B}(g(x_1),\dots,g(x_n)),\mathbf{C}(h(x_1),\dots,h(x_n))\right) ,$$

then $\mathbf{A}$ is called a *quasi–associative operator.* Note that first steps in this direction in the field of aggregation can be found in [139]. For example, the arithmetic mean $\mathbf{M}$ is quasi–associative operator with $\mathbf{B}(x_1,\dots,x_n) = \mathbf{C}(x_1,\dots,x_n) = \sum_{i=1}^{n} x_i$. The standard sum of real numbers is obviously associative. Further, $f(x,y) = \frac{x}{y}$, $g(x) = x$ and $h(x) = 1$. Consequently, $\mathbf{B}\circ g = \mathbf{B}$, $\mathbf{C}\circ h(x_1,\dots,x_n) = n$, thus $f(\mathbf{B}\circ g, \mathbf{C}\circ h) = \mathbf{M}$. Briefly, $\mathbf{M} = \frac{\text{SUM}}{\text{CARD}}$. Similarly, $\mathbf{G} = (\mathbf{\Pi})^{1/\text{CARD}}$.

On the other hand, the non–associative operator $\mathbf{A}_{\{c\}}$ given in (25) is also a quasi–associative aggregation operator. We have $\mathbf{A}_{\{c\}} = f(\mathbf{B})$, where $\mathbf{B}(x_1,\dots,x_n) = c + \sum_{i-1}^{n}(x_i - c)$ is an associative operator and $f(x) =$

$\max(0, \min(1, x)) = \text{med}(0, x, 1)$, see (44). Note that for $c \in \{0, 1\}$ this operator is associative.

2.8 Bisymmetry

An important property of aggregation operators, especially in multi–criteria decision making, is bisymmetry. This property generalizes the simultaneous associativity and symmetry and can be illustrated on the global evaluation of an alternative which is evaluated by m jurymen with respect to n criteria. Let x_{ij}, $i \in \{1, \dots, m\}$, $j \in \{1, \dots, n\}$, corresponds to a score given to our alternative by juryman number i according to the criterion j. To compute the global score $\mathbf{A}_{(mn)}(x_{11}, \dots, x_{mn})$, we can first evaluate the score given by the ith juryman, $x_{i.} = \mathbf{A}_{(n)}(x_{i1}, \dots, x_{in})$, and then to look for the global evaluation in the form of $\mathbf{A}_{(m)}(x_{1.}, \dots, x_{m.})$. However, also another way of computation can be chosen. We may first evaluate the score related to the jth criterion, $x_{.j} = \mathbf{A}_{(m)}(x_{1j}, \dots, x_{mj})$ and then as the global score take the value $\mathbf{A}_{(n)}(x_{.1}, \dots, x_{.n})$. The bisymmetry of $\mathbf{A}$ then simply means that all three types of computing the global evaluation (the first one is just $\mathbf{A}_{(mn)}(x_{11}, \dots, x_{mn})$) lead to the same result. The formal definition follows.

Definition 8. An aggregation operator $\mathbf{A} : \bigcup_{n \in \mathbb{N}} [0,1]^n \to [0,1]$ is bisymmetric if

$$\forall n, m \in \mathbb{N}, \forall x_{11}, \dots, x_{mn} \in [0,1] \ :$$

$$\begin{aligned}\mathbf{A}_{(mn)}(x_{11}, \dots, x_{mn}) &= \mathbf{A}_{(m)}\left(\mathbf{A}_{(n)}(x_{11}, \dots, x_{1n}), \dots, \mathbf{A}_{(n)}(x_{m1}, \dots, x_{mn})\right) \\ &= \mathbf{A}_{(n)}\left(\mathbf{A}_{(m)}(x_{11}, \dots, x_{m1}), \dots, \mathbf{A}_{(m)}(x_{1n}, \dots, x_{mn})\right). \qquad (35)\end{aligned}$$

If the input data are written in the form of the matrix (x_{ij}), then (35) means that the aggregation first over the columns and the following aggregation of these partial outputs give the same result as the aggregation first over the rows and the following aggregation of these partial results.

Observe that a symmetric associative operator is necessarily bisymmetric, for example, $\mathbf{A}_w$, $\mathbf{A}_s$, $\mathbf{\Pi}$, **min**, **max**, $\mathbf{T}_\diamond$. The arithmetic mean $\mathbf{M}$ and the geometric mean $\mathbf{G}$ are examples of symmetric and bisymmetric operators which are not associative. Similarly, $\mathbf{P}_F$, $\mathbf{P}_L$ are bisymmetric associative operators which are not symmetric. Weighted means generated by a quantifier (different from $\mathbf{M}$) discussed in Section 4.1. are examples of bisymmetric aggregation operators which are neither symmetric nor associative, e.g.,

$$\mathbf{A}(x_1, \dots, x_n) = \sum_{i=1}^{n} \frac{2i-1}{n^2} x_i \,.$$

Further, define the aggregation operator $\mathbf{A} = \mathbf{M}^2$. Evidently, $\mathbf{A}$ is continuous, symmetric but neither associative nor bisymmetric.

On the other hand, the operator

$$\mathbf{H}(x,y) = \begin{cases} 0 & \text{if } (x,y) \in [0,0.5] \times [0,0.3] \\ \min(x,y) & \text{else,} \end{cases} \tag{36}$$

is an associative operator which is neither symmetric nor bisymmetric. Note that because of its associativity it is enough to define the binary form of this operator.

The most applied form of aggregation operators is their binary form. Often the bisymmetry is only related to the simplest non–trivial case $n = m = 2$ and (35) is modified into the form

$$\forall\, x,\ y,\ u,\ v \in [0,1] \ : \mathbf{A}\left(\mathbf{A}(x,y), \mathbf{A}(u,v)\right) = \mathbf{A}\left(\mathbf{A}(x,u), \mathbf{A}(y,v)\right). \tag{37}$$

The property described by (37) is called 2–bisymmetry. Notice that the aggregation operator $\mathbf{Q}$ introduced in (30) is 2–bisymmetric, because of

$$\mathbf{Q}\left(\mathbf{Q}(x,y), \mathbf{Q}(u,v)\right) = \mathbf{Q}\left(\mathbf{Q}(x,u), \mathbf{Q}(y,v)\right) = xy^2u^2v^4.$$

Since $\mathbf{Q}(x,y,u,v) = xy^2u^3v^4$, $\mathbf{Q}$ is not bisymmetric.

Note that the operator $\mathbf{A} = \mathbf{M}^2$ introduced above is not 2–bisymmetric (and hence not bisymmetric).

Finally, recall once more the strength of associativity with continuity. Following the representation of continuous associative aggregation operators given in the chapter of this book written by W. Sander [113], the continuity and associativity of an aggregation operator $\mathbf{A}$ ensure also the bisymmetry of $\mathbf{A}$, but not necessarily the symmetry of $\mathbf{A}$.

2.9 Neutral element

The neutral element is again a well–known notion coming from the area of binary operations. Recall that for a binary operation $*$ defined on a domain X, an element $e \in X$ is called a *neutral element* (of the operation $*$) if

$$\forall\, x \in X : \quad x * e = e * x = x\,.$$

Clearly, any binary operation $*$ can have at most one neutral element. From the previous equalities we can see that the action of the neutral element of a binary operation has the same effect as its omitting. This idea is the background of the general definition.

Definition 9. Let $\mathbf{A} : \bigcup_{n \in \mathbb{N}} [0,1]^n \to [0,1]$ be an aggregation operator. An element $e \in [0,1]$ is called a neutral element of $\mathbf{A}$ if

$$\forall\, n \in \mathbb{N},\ \forall\, x_1, \ldots, x_n \in [0,1] \text{ if } x_i = e \text{ for some } i \in \{1, \ldots, n\} \text{ then}$$

$$\mathbf{A}(x_1, \ldots, x_n) = \mathbf{A}(x_1, \ldots, x_{i-1}, x_{i+1}, \ldots, x_n)\,. \tag{38}$$

So the neutral element can be omitted from aggregation inputs without influencing the final output. In multi-criteria decision making, assigning a score equal to the neutral element (if it exists) to some criterion means that only the other criteria fulfillments are decisive for the global evaluation.

A typical example is the product $\mathbf{\Pi}$ with neutral element $e = 1$. Similarly, **min**, $\mathbf{T}_\diamond$ given in (32), $\mathbf{H}$ given in (36) have neutral element 1, **max** has neutral element $e = 0$. The operator $\mathbf{A}_{\{c\}}$, $c \in [0,1]$ from (25) has neutral element $e = c$. From these examples we can see that the existence of the neutral element is not related to the previous properties as continuity, associativity, symmetry or bisymmetry. Aggregation operators $\mathbf{M}$, $\mathbf{G}$, $\mathbf{A}_w$, $\mathbf{A}_s$, $\mathbf{P}_F$, $\mathbf{P}_L$, etc., are examples of operators without neutral element.

Observe that if e is a neutral element of an aggregation $\mathbf{A}$ then it is necessarily also an $\mathbf{A}$–idempotent element, that is, e is a trivial idempotent of $\mathbf{A}$. Therefore for any aggregation operator $\mathbf{A}$ with neutral element $e \in]0,1[$, we can introduce two operators acting on $[0,e]$ and $[e,1]$, respectively. By increasing linear transformations of the scales $[0,e]$ and $[e,1]$ to $[0,1]$ we finally get two aggregation operators $\mathbf{A}_{[0]}$ and $\mathbf{A}_{[1]}$ with neutral elements $e_0 = 1$ and $e_1 = 0$, respectively. Take, for example, the aggregation operator $\mathbf{E} : \bigcup_{n\in\mathbb{N}} [0,1]^n \to [0,1]$ defined by

$$\mathbf{E}(x_1,\ldots,x_n) = \frac{\prod_{i=1}^{n} x_i}{\prod_{i=1}^{n} x_i + \prod_{i=1}^{n}(1-x_i)} \tag{39}$$

with convention $\frac{0}{0} = 0$. Note that this operator was introduced and discussed by several authors [59,139,35] and is sometimes called $3-\Pi$–operator. It is symmetric, associative and $e = 0.5$ is its neutral element. Note that this operator is a typical example of a generated uninorm [59,35,72], see Section 6.2.

Put $\varphi : [0,1] \to [0,0.5]$, $\varphi(x) = \frac{x}{2}$, where φ is the above mentioned linear transformation, and define $\mathbf{E}_{[0]} = \varphi^{-1} \circ \mathbf{E} \circ \varphi$, i.e.,

$$\mathbf{E}_{[0]}(x_1,\ldots,x_n) = \varphi^{-1}\left(\mathbf{E}(\varphi(x_1),\ldots,\varphi(x_n))\right) .$$

Because of the associativity of $\mathbf{E}$, also $\mathbf{E}_{[0]}$ is associative and hence we can discuss its binary form only. It holds:

$$\begin{aligned}\mathbf{E}_{[0]}(x,y) &= 2\frac{\frac{x}{2}\frac{y}{2}}{\frac{x}{2}\frac{y}{2} + (1-\frac{x}{2})(1-\frac{y}{2})} \\ &= \frac{2xy}{xy + (2-x)(2-y)} = \frac{xy}{2-x-y+xy}.\end{aligned} \tag{40}$$

$\mathbf{E}_{[0]}$ is so called Hamacher t–norm with parameter $\lambda = 2$, i.e., $\mathbf{E}_{[0]} = \mathbf{T}_2^H$, see Section 6.1.

Similarly, applying the transformation $\psi : [0,1] \to [0.5,1]$, $\psi(x) = \frac{x+1}{2}$, we get $\mathbf{E}_{[1]} = \psi^{-1} \circ \mathbf{E} \circ \psi$, i.e.,

$$\mathbf{E}_{[1]}(x,y) = \frac{x+y}{1+xy}, \tag{41}$$

which is the Hamacher t–conorm $\mathbf{S}_2^H$, see Section 6.1. $\mathbf{E}_{[1]}$ is sometimes called the Einstein sum (because it has the form of the formula for summing the speeds related to the absolute light speed).

As it was shown in (40) and (41), the decomposition of an aggregation $\mathbf{A}$ with neutral element $e \in]0,1[$ described above can be applied for constructing aggregation operators with neutral elements 0 and 1, respectively. Oppositely, if two aggregation operators with neutral elements 0 and 1, respectively, are given, then for any fixed $e \in]0,1[$ we can construct an aggregation operator $\mathbf{A}$ with neutral element e. More details on such constructions are given in Section 3.4 devoted to the ordinal sums of aggregation operators.

2.10 Annihilator

Analyzing properties of the standard product $\mathbf{\Pi}$, we can see that if 0 occurs among inputs to be aggregated, the final output is surely 0, regardless the other inputs. In multi–criteria decision making this means that 0 is the "veto" element.

Going back to Algebra, the situation described above corresponds to the existence of an annihilator (absorbing element). Recall that for a binary operation $*$ defined on a domain X, an element $a \in X$ is an annihilator (of the operation $*$) if

$$\forall x \in X \ : \quad a * x = x * a = a\,.$$

We extend this notion to the aggregation operators as follows.

Definition 10. Let $\mathbf{A} : \bigcup\limits_{n\in\mathbb{N}} [0,1]^n \to [0,1]$ be an aggregation operator. An element $a \in [0,1]$ is called an annihilator of $\mathbf{A}$ if

$$\forall n \in \mathbb{N}, \forall x_1,\ldots,x_n \in [0,1] : a \in \{x_1,\ldots,x_n\}$$

Downarrow

$$\mathbf{A}(x_1,\ldots,x_n) = a\,. \tag{42}$$

If the annihilator a occurs among the input values, the resulting output is necessarily equal to a. According to the famous product–zero property the annihilator a is also called a zero element. Besides the operator $\mathbf{\Pi}$ with annihilator $a = 0$, also operators **min**, $\mathbf{A}_w$, $\mathbf{E}$ given by (39), $\mathbf{E}_{[0]}$ from (40), $\mathbf{T}_\diamond$ given by (32), $\mathbf{H}$ defined by (36), $\mathbf{Q}$ defined by (30) and the geometric mean **G** have annihilator 0. The operators **max**, $\mathbf{A}_s$, the operator $\mathbf{E}_{[1]}$ given by (41) have annihilator 1. The operators **M**, $\mathbf{A}_{\{c\}}$ for $c \in]0,1[$, $\mathbf{P}_F$, $\mathbf{P}_L$ are examples of aggregation operators without annihilator.

Similarly to the case of the neutral element, the annihilator, if it exists, is determined uniquely, and can take any value in $[0,1]$. Indeed, fix an element $a \in [0,1]$ and define the operator $\mathbf{med}_a : \bigcup_{n \in \mathbb{N}} [0,1]^n \to [0,1]$ by

$$\mathbf{med}_a(x,y) = \mathbf{med}(x,y,a)\,. \tag{43}$$

The operator $\mathbf{med}_a$ is called an a–median [14,15,31,39] and it is also a nullnorm, see Section 6.3. It is a idempotent, symmetric, associative, continuous, bisymmetric operator, which has no neutral element. However, the element a is its annihilator. Because of its associativity the binary form of (43) is sufficient to describe $\mathbf{med}_a$. Observe that $\mathbf{med}_0 = \mathbf{min}$, $\mathbf{med}_1 = \mathbf{max}$. Note that the median operator **med** is the standard aggregation operator in statistics acting on real values and it is given by

$$\mathbf{med}(x_1,\ldots,x_n) = \begin{cases} \frac{x'_k + x'_{k+1}}{2} & \text{if } n = 2k, \\ x'_{k+1} & \text{if } n = 2k+1, \end{cases} \tag{44}$$

where $(x'_1,\ldots,x'_n)$ is a non–decreasing permutation of the inputs $(x_1,\ldots,x_n)$. The operator **med** can also be defined recursively by means of the boundary condition (10) $\mathbf{med}(x) = x$ and the binary formula $\mathbf{med}(x,y) = \mathbf{M}(x,y) = \frac{x+y}{2}$, as follows:

$$\text{for } n > 2 \;: \quad \mathbf{med}(x_1,\ldots,x_n) = \mathbf{med}(y_1,\ldots,y_{n-2}),$$

where $(y_1,\ldots,y_{n-2})$ is the $(n-2)$–tuple obtained from the n–tuple $(x_1,\ldots,x_n)$ erasing one maximal and one minimal element.

For an aggregation operator **A** with annihilator a, a is necessarily an **A**–idempotent element, that is, a is a trivial idempotent element. Therefore operators with annihilator have a property alike to the property of operators with neutral element regarding the structure of these operators. Suppose that **A** is an aggregation operator with annihilator $a \in]0,1[$. Then we can introduce two operators acting on inputs from $[0,a]$ and on inputs from $[a,1]$, respectively, and by appropriate linear transformations of the scales $[0,a]$ and $[a,1]$ into $[0,1]$ to construct two aggregation operators $\mathbf{A}_{[0]}$ and $\mathbf{A}_{[1]}$ with annihilators $a_0 = 1$ and $a_1 = 0$, respectively. Applying this construction method to a–medians $\mathbf{med}_a$ introduced in (43) we obtain

$$(\mathbf{med}_a)_{[0]} = \mathbf{max}\,, \qquad (\mathbf{med}_a)_{[1]} = \mathbf{min},$$

independently of $a \in]0,1[$.

The following observation applies specifically to aggregation operators with annihilator $a \in]0,1[$. Because of the monotonicity (11), for any aggregation operator $\mathbf{A} : \bigcup_{n\in\mathbb{N}} [0,1]^n \to [0,1]$ with annihilator $a \in]0,1[$ it holds:

$$\min(x_1,\ldots,x_n) \le a \le \max(x_1,\ldots,x_n) \Longrightarrow \mathbf{A}(x_1,\ldots,x_n) = a\,.$$

Note also that an aggregation operator $\mathbf{A}$ with annihilator $a \in]0,1[$ cannot have neutral element e. However, $\mathbf{A}$ may have a neutral element e if its annihilator $a \in \{0,1\}$, and $e \neq a$.

Aggregation operators $\mathbf{A}$ with annihilator $a \in \{0,1\}$ can have another special element $b \in \{0,1\} \setminus \{a\}$ with almost annihilating property:

$$\forall n \in \mathbb{N}, \forall x_1,\ldots,x_n \in [0,1](b \in \{x_1,\ldots,x_n\} \ \wedge \ a \notin \{x_1,\ldots,x_n\})$$

$$Downarrow$$

$$\mathbf{A}(x_1,\ldots,x_n) = b\,. \tag{45}$$

Such element b is called a *weak annihilator* [69]. As an example we can take the $3-\Pi$–operator $\mathbf{E}$ introduced in (39) with annihilator $a = 0$ and weak annihilator $b = 1$.

2.11 Some other properties

Some other specific properties of aggregation operators, not mentioned in previous sections, have been investigated in the area of aggregation operators. We briefly recall some of them. An interesting property of aggregation operators generalizing the existence of an annihilator is the temporary breakdown property [69].

Definition 11. Let $\mathbf{A} : \bigcup_{n\in\mathbb{N}} [0,1]^n \to [0,1]$ be an aggregation operator. $\mathbf{A}$ is said to have the temporary breakdown property if there is an element $t \in [0,1]$ such that

$$\forall n,m \in \mathbb{N}\cup\{0\}, \forall x_1,\ldots,x_n,y_1,\ldots,y_m \in [0,1] \ :$$

$$\mathbf{A}(x_1,\ldots,x_n,t,y_1,\ldots,y_m) = \mathbf{A}(t,y_1,\ldots,y_m)\,. \tag{46}$$

Note that if $m = 0$ we have

$$\mathbf{A}(x_1,\ldots,x_n,t) = \mathbf{A}(t) = t,$$

i.e., t acts as the annihilator. Moreover, if an aggregation operator $\mathbf{A}$ has annihilator a then $\mathbf{A}$ has the temporary breakdown property with $t = a$. The temporary breakdown property reflects the renewal of cumulated output after a total collapse. As an example of an aggregation operator without annihilator but having the temporary breakdown property we can choose any aggregation operator $\mathbf{A}_*$ defined by means of (21), starting from a binary operator $\mathbf{A}_{(2)} : [0,1]^2 \to [0,1]$ such that $\mathbf{A}_{(2)}(x,t) = t$ for some $t \in [0,1]$ and all $x \in [0,1]$. The simplest example of this type is the projection operator $\mathbf{P}_L$ (for which t can be chosen arbitrarily). Another example is the operator $\mathbf{A}$ given by

$$\mathbf{A}(x_1,\ldots,x_n) = \sum_{i=1}^{n}\prod_{j=i}^{n}\frac{x_j}{2} + \prod_{j=1}^{n}\frac{x_j}{2}. \tag{47}$$

Clearly, $\mathbf{A}$ given by the last formula has the temporary breakdown property with $t = 0$. Indeed,

$$\mathbf{A}(x_1,\ldots,x_n,0,y_1,\ldots,y_m) = \sum_{i=1}^{m}\prod_{j=i}^{m}\frac{x_j}{2} = \mathbf{A}(0,y_1,\ldots,y_m).$$

For example,

$$\begin{aligned}\mathbf{A}_{(n+m+1)}(1,\ldots,1,0,1,\ldots,1) &= \mathbf{A}_{(m+1)}(0,1,\ldots,1)\\ &= \frac{1}{2^m} + \ldots + \frac{1}{2} = 1 - \frac{1}{2^m}.\end{aligned}$$

Among operators not having the temporary breakdown property recall, e.g., $\mathbf{M}$, $\mathbf{A}_{\{c\}}$ for $c \notin \{0,1\}$ (see (25)), $\mathbf{P}_F$, etc.

Another property of aggregation operators linking different partial operators arising from an aggregation operator $\mathbf{A}$ is the self–identity property introduced by Yager [142].

Definition 12. An aggregation operator $\mathbf{A} : \bigcup_{n\in\mathbb{N}} [0,1]^n \to [0,1]$ is said to have the self–identity property if

$\forall n \in \mathbb{N}, \forall x_1,\ldots,x_n \in [0,1]$:

$$\mathbf{A}(x_1,\ldots,x_n) = \mathbf{A}\left(x_1,\ldots,x_n,\mathbf{A}(x_1,\ldots,x_n)\right). \tag{48}$$

This property reflects the stability of the aggregation operator $\mathbf{A}$ generalizing the next well known property of the arithmetic mean: if a sample of inputs $x_1,\ldots,x_n$ is given and $\bar{x}$ is the corresponding arithmetic mean, then adding new additional inputs all equal to $\bar{x}$ will not influence the final arithmetic mean.

Because of the boundary condition (10), we can inductively show that each aggregation operator possessing the self–identity property must be idempotent. Interestingly, weighted means possessing the self–identity property correspond to the collapse–invariant weighted means introduced in [66]. All such weighted means are constructed by means of (21), see, e.g., the operator $\mathbf{M}_*$ given in (23) or by a slight modification of (21), see Section 4.1.

As a prominent example recall the arithmetic mean **M**. Other examples of operators possessing the self–identity property are, e.g., projections $\mathbf{P}_F$, $\mathbf{P}_L$, **max**, **min**, **G**. On the other hand, the operators $\mathbf{\Pi}$, **E** given in (39), $\mathbf{A}_{\{c\}}$ from (25), etc., do not have the self–identity property.

Another property related to the above one, that we will call *strong self–identity*, was introduced in [12]. An aggregation operator **A** is said to have the strong self–identity property if and only if $\forall n \in \mathbb{N}$, $\forall x_1, \ldots, x_n \in [0,1]$:

$$\mathbf{A}(x_1, \ldots, x_n, b) \leq \mathbf{A}(x_1, \ldots, x_n) \text{ if } b \leq \mathbf{A}(x_1, \ldots, x_n),$$

and

$$\mathbf{A}(x_1, \ldots, x_n, b) \geq \mathbf{A}(x_1, \ldots, x_n) \text{ if } b \geq \mathbf{A}(x_1, \ldots, x_n).$$

Evidently, in the framework of aggregation operators, the self–identity and the strong self–identity are equivalent.

The next five properties are closely related to the linear structure of some aggregation operators.

Definition 13. An aggregation operator $\mathbf{A} : \bigcup_{n \in \mathbb{N}} [0,1]^n \to [0,1]$ is said to be

(i) shift–invariant if

$\forall n \in \mathbb{N}$, $\forall b \in]0,1[$, $\forall x_1, \ldots, x_n \in [0, 1-b]$:

$$\mathbf{A}(x_1 + b, \ldots, x_n + b) = \mathbf{A}(x_1, \ldots, x_n) + b\,, \tag{49}$$

(ii) homogeneous if

$\forall n \in \mathbb{N}$, $\forall b \in]0,1[$, $\forall x_1, \ldots, x_n \in [0,1]$:

$$\mathbf{A}(bx_1, \ldots, bx_n) = b\mathbf{A}(x_1, \ldots, x_n)\,, \tag{50}$$

(iii) linear if it is homogeneous and shift–invariant,

(iv) additive if

$\forall n \in \mathbb{N}, \forall x_1, \ldots, x_n, y_1, \ldots, y_n \in [0,1]$ such that $x_1 + y_1, \ldots, x_n + y_n \in [0,1]$:

$$\mathbf{A}(x_1 + y_1, \ldots, x_n + y_n) = \mathbf{A}(x_1, \ldots, x_n) + \mathbf{A}(y_1, \ldots, y_n)\,, \tag{51}$$

(v) comonotone additive if

$\forall n \in \mathbb{N}$, $\forall x_1, \ldots, x_n, y_1, \ldots, y_n \in [0,1]$ such that $(x_i - x_j)(y_i - y_j) \geq 0$ for all $i, j \in \{1, \ldots, n\}$ and $x_1 + y_1, \ldots, x_n + y_n \in [0,1]$:

$$\mathbf{A}(x_1 + y_1, \ldots, x_n + y_n) = \mathbf{A}(x_1, \ldots, x_n) + \mathbf{A}(y_1, \ldots, y_n)\,. \tag{52}$$

Note that additivity ensures comonotone additivity and that comonotone additivity ensures linearity. Deeper discussions related to these properties can be found, e.g., in [3,23,84]. These properties are important especially for characterizing some integral–based aggregation operators, see Chapter 5. So, e.g., additive aggregation operators are just operators related to the Lebesgue integral, that is, to the weighted means, while comonotone additivity is a genuine property of the Choquet integral–based aggregation operators.

Because of the boundary condition (9), any aggregation operator fulfilling at least one of the properties introduced in Definition 13 is idempotent. Each shift–invariant aggregation operator can be constructed (and characterized) as follows. Take any aggregation operator $\mathbf{B} : \bigcup_{n\in\mathbb{N}} [0,1]^n \to [0,1]$ and define the operator $\mathbf{S}^{\mathbf{B}} : \bigcup_{n\in\mathbb{N}} [0,1]^n \to [0,1]$ by

$$\mathbf{S}^{\mathbf{B}}(x_1,\ldots,x_n) = a + \mathbf{B}(x_1 - a,\ldots,x_n - a)\,, \tag{53}$$

where $a = \min(x_1,\ldots,x_n)$. Evidently, $\mathbf{S}^{\mathbf{B}}$ is a shift–invariant operator. Further, an aggregation operator $\mathbf{A} : \bigcup_{n\in\mathbb{N}} [0,1]^n \to [0,1]$ is shift–invariant if and only if $\mathbf{A} = \mathbf{S}^{\mathbf{A}}$. Take, for example, the operator $\mathbf{S}^{\Pi}$ given by

$$\mathbf{S}^{\Pi}(x_1,\ldots,x_n) = a + \prod_{i=1}^{n}(x_i - a) = a, \quad a = \min(x_1,\ldots,x_n)\,. \tag{54}$$

Hence $\mathbf{S}^{\Pi} = \mathbf{min}$.

However, the construction (53) may violate the monotonicity (11). Indeed, for $e \in]0,1[$ define the operator $\mathbf{B}_e : \bigcup_{n\in\mathbb{N}} [0,1]^n \to [0,1]$ by

$$\mathbf{B}_e(x_1,\ldots,x_n) = \begin{cases} \min(x_1,\ldots,x_n) & \text{if } \max(x_1,\ldots,x_n) \le e, \\ \max(x_1,\ldots,x_n) & \text{else.} \end{cases}$$

Note that $\mathbf{B}_e$ is a disjunctive uninorm, see Section 6.2, i.e., a symmetric associative (and idempotent) aggregation operator with neutral element e. The related shift–invariant operator $\mathbf{S}^{\mathbf{B}_e}$ is given by

$$\mathbf{S}^{\mathbf{B}_e} = \begin{cases} \min(x_1,\ldots,x_n) & \text{if } \max(x_1,\ldots,x_n) - \min(x_1,\ldots,x_n) \le e, \\ \max(x_1,\ldots,x_n) & \text{else.} \end{cases}$$

Observe that $\mathbf{S}^{\mathbf{B}_e}$ is neither homogeneous nor linear, and because of

$$\mathbf{S}^{\mathbf{B}_e}(0,1) = 1 > \mathbf{S}^{\mathbf{B}_e}(1-e,1) = 1 - e\,,$$

it is not a monotone operator. Recently, the structure of shift–invariant aggregation operators was characterized in [74]. In fact, each shift–invariant aggregation operators can be obtained by means of the construction (53) as $\mathbf{A} = \mathbf{S}^{\mathbf{B}}$ from some kernel aggregation operator, see (31) or [70,103] (observe that these operators are called ∞–stable in [17]). Further, $\mathbf{S}^{\mathbf{B}}$ is an aggregation operator if and only if $\mathbf{B}$ fulfills for any $n \in \mathbb{N}$, $n \geq 2$, $(x_1, \ldots, x_n)$, $(y_1, \ldots, y_n) \in [0,1]^n$ and for some $i \in \{1, \ldots, n\}$, $x_i = y_i = 0$:

$$| \mathbf{B}(x_1, \ldots, x_n) - \mathbf{B}(y_1, \ldots, y_n) | \leq \max_i (| x_i - y_i |) \quad . \tag{55}$$

Note that the operator $\mathbf{B}_e$ given about does not fit (55). Indeed, we have for all $\varepsilon \in]0, 1-e]$, $|\mathbf{B}_e(0, e+\varepsilon) - \mathbf{B}_e(0, e)| = e + \varepsilon > \varepsilon$.

Similarly, starting from an arbitrary aggregation operator $\mathbf{B} : \bigcup_{n \in \mathbb{N}} [0,1]^n \to [0,1]$, a homogeneous operator $\mathbf{H}^{\mathbf{B}}$ can be constructed by

$$\mathbf{H}^{\mathbf{B}}(x_1, \ldots, x_n) = b\mathbf{B}(\frac{x_1}{b}, \ldots, \frac{x_n}{b}), \tag{56}$$

where $b = \max(x_1, \ldots, x_n) > 0$. However, again the monotonicity (11) of $\mathbf{H}^{\mathbf{B}}$ can be violated. So, e.g., the homogeneous operator $\mathbf{H}^{\Pi} : \bigcup_{n \in \mathbb{N}} [0,1]^n \to [0,1]$ is given by

$$\mathbf{H}^{\Pi}(x_1, \ldots, x_n) = b \prod_{i=1}^{n} \frac{x_i}{b} = \frac{\prod_{i=1}^{n} x_i}{b^{n-1}} \tag{57}$$

with convention $\frac{0}{0} = 0$, i.e.,

$$\mathbf{H}^{\Pi}_{(n)} = \frac{\mathbf{\Pi}_{(n)}}{(\mathbf{max}_{(n)})^{n-1}}.$$

Then $\mathbf{H}^{\Pi}_{(3)}(0.5, 0.5, 0.5) = 0.5 > \mathbf{H}^{\Pi}_{(3)}(0.5, 0.5, 1) = 0.25$, that is, the monotonicity property (11) does not hold.

Anyway, homogeneous aggregation operators $\mathbf{A}$ are characterized by the equality $\mathbf{A} = \mathbf{H}^{\mathbf{A}}$.

Recall that an aggregation operator $\mathbf{A}$ is linear if and only if it is shift–invariant and homogeneous. Summarizing (53) and (56) for a given operator $\mathbf{B}$ we can construct a linear operator $\mathbf{L}^{\mathbf{B}} : \bigcup_{n \in \mathbb{N}} [0,1]^n \to [0,1]$ as follows:

$$\mathbf{L}^{\mathbf{B}}(x_1, \ldots, x_n) = \begin{cases} a & \text{if } a = b, \\ a + (b-a)\mathbf{B}\left(\frac{x_1 - a}{b-a}, \ldots, \frac{x_n - a}{b-a}\right) & \text{else,} \end{cases} \tag{58}$$

where $a = \min(x_1, \ldots, x_n)$, $b = \max(x_1, \ldots, x_n)$.

Obviously, $\mathbf{A}$ is a linear aggregation operator if and only if $\mathbf{A} = \mathbf{L}^{\mathbf{A}}$. Note that the characterization of all aggregation operators $\mathbf{A}$, for which $\mathbf{L}^{\mathbf{A}}$ is a monotone operator, is still an open problem. However, we expect that it will be solved in the near future in a similar spirit as in the case of shift–invariant aggregation operators [74].

Recall that $\mathbf{L}^{\Pi} = \mathbf{min}$. Moreover, $\mathbf{L}^{\mathbf{B}} = \mathbf{min}$ if and only if $\mathbf{B}$ is an aggregation operator with annihilator 0.

Operators $\mathbf{M}$, $\mathbf{P}_F$ and $\mathbf{P}_L$ fulfill all properties introduced in Definition 13. The operators **min, med** and **max** are linear and hence shift–invariant and homogeneous, comonotone additive but not additive. The geometric mean $\mathbf{G}$ is only homogeneous, while the product $\mathbf{\Pi}$ does not fulfill any of these properties, which evidently follows from the non–idempotency of $\mathbf{\Pi}$. An example of an idempotent aggregation operator violating all properties from Definition 13 is the a–median $\mathbf{med}_a$, $a \in]0,1[$, introduced in (43).

An alternative property of aggregation operators limiting the computational complexity in spirit similar to associativity is the decomposability property introduced by Kolmogorov [71] and Nagumo [106].

Definition 14. An aggregation operator $\mathbf{A} : \bigcup_{n\in\mathbb{N}} [0,1]^n \to [0,1]$ is called decomposable if $\forall\, k, n \in \mathbb{N}$, $\forall\, x_1, \ldots, x_k, y_1, \ldots, y_n \in [0,1]$:

$$\mathbf{A}_{(n+k)}(x_1,\ldots,x_k,y_1,\ldots,y_n) =$$
$$\mathbf{A}_{(n+k)}\left(\mathbf{A}_{(k)}(x_1,\ldots,x_k),\ldots,\mathbf{A}_{(k)}(x_1,\ldots,x_k),y_1,\ldots,y_n\right) . \qquad (59)$$

A deep discussion related to decomposable operators, together with several historical notes, can be found in [33,83,84]. Observe that decomposability of a continuous aggregation operator $\mathbf{A}$ forces its idempotency, and therefore it can also be reformulated for idempotent aggregation operators as follows:

$\forall n, m \in \mathbb{N} \cup \{0\}$ and $\forall x_1, \ldots, x_n, y_1, \ldots, y_m \in [0,1]$:

$$\mathbf{A}(x_1,\ldots,x_n,y_1,\ldots,y_m) - \mathbf{A}(z_1,\ldots,z_n,y_1,\ldots,y_m)$$

whenever

$$\mathbf{A}(x_1,\ldots,x_n) = \mathbf{A}(z_1,\ldots,z_n) .$$

Typical examples of decomposable aggregation operators are the arithmetic mean $\mathbf{M}$, the geometric mean $\mathbf{G}$, and also the operators **max** and **min**. Observe that a decomposable aggregation operator $\mathbf{A}$ is uniquely determined by its n–ary operators $\mathbf{A}_{(n)}$ restricted to input n–tuples of the form $(x,\ldots,x,y)$, with $x, y \in [0,1]$, $n > 2$, i.e., we need to know the operators $\mathbf{A}_{(n)}$ with the computational complexity corresponding to the binary operators only. As an example, suppose that a sequence $\mathcal{A} = (a_n)_{n\in\mathbb{N}} \in [0,1]^{\mathbb{N}}$ with $a_1 = 1$ is given and let $\mathbf{A}_{(n)}(x,\ldots,x,y) = (1-a_n)x + a_n y$, for each $n \in \mathbb{N}$. Then the relevant decomposable aggregation operator $\mathbf{A}_{\mathcal{A}} : \bigcup_{n\in\mathbb{N}} [0,1]^n \to [0,1]$ is given by

$$\mathbf{A}_{\mathcal{A}}(x_1,\ldots,x_n) = \sum_{i=1}^{n}\left(a_i x_i \prod_{j=i+1}^{n}(1-a_j)\right) , \qquad (60)$$

where $\prod_{j=n+1}^{n} (1 - a_j) = 1$, by convention.

Evidently, if all $a_n = p$, $n \geq 2$, then $\mathbf{A}_{\mathcal{A}} = \mathbf{A}^*$ as given in (20) with $\mathbf{A}_{(2)}(x, y) = (1-p)x + py$. Recall that the decomposable aggregation operator $\mathbf{A}_{\mathcal{A}}$ is a weighted mean discussed in Section 4.1, compare with (95). For more details on decomposable aggregation operators we recommend [34].

3 Some construction methods for aggregation operators

There is a wide known demand for an ample variety of aggregation operators having predictable and tailored properties to be used in modelling processes. Recently, several construction methods have been introduced and developed for extending the known classes of aggregation operators (acting either on $[0, 1]$ or, possibly, on some other domains). Obviously, new construction methods should be a central issue in the rapidly developing field of aggregation operators. In this chapter we will put in perspective some of well established construction methods as well as some new ones.

3.1 Idempotization

Recall that the basic (standard, classical) aggregation operators acting on some specific subsets of the real line are operators related to addition, multiplication, minimization and maximization. All these operators are continuous, symmetric, associative and jointly strictly monotone. These operators are the background for many other (continuous, symmetric) aggregation operators. For example, the arithmetic mean $\mathbf{M}$ and the geometric mean $\mathbf{G}$ are results of idempotization of the sum and product, respectively.

Proposition 1. *Let* $\mathbf{B} : \bigcup_{n \in \mathbb{N}} I^n \to I$ *be a jointly strictly monotone and continuous aggregation operator acting on* $I = [a, b] \subseteq [-\infty, \infty]$. *Define the operator*

$$\mathbf{I_B} : \bigcup_{n \in \mathbb{N}} I^n \to I\,, \quad \mathbf{I_B}(x_1, \ldots, x_n) = x, \tag{61}$$

where x *is an element for which*

$$\mathbf{B}_{(n)}(x, \ldots, x) = \mathbf{B}_{(n)}(x_1, \ldots, x_n)\,.$$

Then $\mathbf{I_B}$ *is a continuous, jointly strictly monotone and idempotent aggregation operator on* I.

Proof. The continuity of $\mathbf{B}$ ensures the existence of x occurring in (61) and the joint strict monotonicity of $\mathbf{B}$ ensures the uniqueness of x. Therefore the operator $\mathbf{I_B}$ is well defined. Obviously, the monotonicity of $\mathbf{B}$ ensures the monotonicity of $\mathbf{I_B}$. As $\mathbf{B}$ fulfills the boundary conditions on I, so does the operator $\mathbf{I_B}$, which means that $\mathbf{I_B}$ is an aggregation operator on I.

Its continuity and joint strict monotonicity follow from the continuity and joint strict monotonicity of $\mathbf{B}$. The idempotency of $\mathbf{I_B}$ is a direct consequence of (61). $\square$

Observe that if we have an idempotent aggregation operator $\mathbf{A}$ acting on some interval $I = [a,b]$, and $[c,d] \subset [a,b]$, then $\mathbf{A}_{[c,d]} = \mathbf{A}|_{\bigcup_{n\in\mathbb{N}} [c,d]^n}$ is an idempotent aggregation operator acting on $[c,d]$.

Applying Proposition 1 to the sum–operator $\sum$ acting on $[0,\infty]$ (or on $[-\infty,0]$, $[-\infty,\infty]$) we obtain the standard arithmetic mean $\mathbf{M} = \mathbf{I}_{\sum}$. Indeed, applying (61) we obtain

$$\mathbf{M}(x_1,\ldots,x_n) = x \quad \text{if} \quad \sum_{i=1}^{n} x_i = \sum_{i=1}^{n} x = nx\,,$$

i.e.,

$$\mathbf{M}(x_1,\ldots,x_n) = \frac{1}{n}\sum_{i=1}^{n} x_i\,.$$

Note that due to the previous remark we do not distinguish the interval on which $\mathbf{M}$ is acting. $\mathbf{M}$ can act on any interval $[a,b] \subseteq [-\infty,\infty]$. Similarly, starting from the product operator $\mathbf{\Pi}$ acting either on $[0,1]$ or $[0,\infty]$, by the described procedure we get the geometric mean $\mathbf{G}$. Further, the idempotization of the operator $\mathbf{Q}$ given in (30) by

$$\mathbf{Q}(x_1,\ldots,x_n) = \prod_{i=1}^{n} x_i^i\,,$$

leads to the weighted geometric mean

$$\mathbf{G}_\Delta : \bigcup_{n\in\mathbb{N}} [0,1]^n \to [0,1], \quad \mathbf{G}_\Delta(x_1,\ldots,x_n) = \prod_{i=1}^{n} x_i^{w_{in}}\,,$$

with weights $w_{in} = \frac{2i}{n(n+1)}$, $i = 1,\ldots,n$, $n \in \mathbb{N}$, see also (112).

Obviously, a jointly strictly monotone aggregation operator $\mathbf{A}$ is idempotent if and only if $\mathbf{A} = \mathbf{I_A}$, see, e.g., $\mathbf{max} = \mathbf{I_{max}}$ and $\mathbf{min} = \mathbf{I_{min}}$.

From the nonstandard examples take, e.g., the Einstein sum $\mathbf{E}_{[1]}$ from (41), whose binary form is

$$\mathbf{E}_{[1]}(x,y) = \frac{x+y}{1+xy}\,.$$

Applying (61), we obtain an idempotent aggregation operator $\mathbf{A} = \mathbf{I}_{\mathbf{E}_{[1]}}$, whose binary form is

$$\mathbf{A}(x,y) = \frac{1 + xy - \sqrt{(1-x^2)(1-y^2)}}{x+y}, \quad (x,y) \neq (0,0). \tag{62}$$

Observe that this operator is a quasi–arithmetic mean discussed in Section 4.3.

3.2 Transformation, duality, symmetric sums, invariantness

Linear transformations of aggregation operators acting on an (bounded) interval $[a,b]$ into aggregation operators acting on another (bounded) interval were already applied in Sections 2.8 and 2.9. In general, for any two intervals $[a,b]$ and $[c,d]$ (even unbounded) transformations of relevant aggregation operators are treated by means of some monotone bijections $\varphi : [a,b] \to [c,d]$.

Proposition 2. *Let* $\mathbf{A} : \bigcup_{n\in\mathbb{N}} [c,d]^n \to [c,d]$ *be an aggregation operator on* $[c,d]$ *and let* $\varphi : [a,b] \to [c,d]$ *be a monotone bijection. Then the operator* $\mathbf{A}_\varphi : \bigcup_{n\in\mathbb{N}} [a,b]^n \to [a,b]$ *defined by*

$$\mathbf{A}_\varphi(x_1,\dots,x_n) = \varphi^{-1}\left(\mathbf{A}\left(\varphi(x_1),\dots,\varphi(x_n)\right)\right) \tag{63}$$

is an aggregation operator on $[a,b]$.

Proof. It is enough to show that $\mathbf{A}_\varphi$ satisfies the properties (9), (10) and (11) from Definition 1. However, this is a direct consequence of (63) and the fact that $\mathbf{A}$ is an aggregation operator. □

Note that aggregation operators $\mathbf{A}_\varphi$ inherit most of the properties from the aggregation operator $\mathbf{A}$. Some exceptions can be, e.g., the Lipschitz property, or the properties introduced in Definition 13, i.e., shift–invariantness, homogeneity, linearity, additivity and comonotone additivity. Observe that the transformations can be applied consecutively, i.e., if $\mathbf{A} : \bigcup_{n\in\mathbb{N}} [c,d]^n \to [c,d]$ is a given aggregation operator, $\varphi : [a,b] \to [c,d]$ and $\psi : [u,v] \to [a,b]$ are given monotone bijections, then $\mathbf{A}_\varphi$ is an aggregation operator on $[a,b]$ and hence $(\mathbf{A}_\varphi)_\psi$ is an aggregation operator on $[u,v]$. Evidently,

$$(\mathbf{A}_\varphi)_\psi = \mathbf{A}_{\varphi\circ\psi}. \tag{64}$$

There are many examples of aggregation operators among already introduced in the previous sections, which can be seen as transformations of some other aggregation operators. For example, the geometric mean $\mathbf{G}$ is the transformation of the arithmetic mean $\mathbf{M}$ acting on $[0,\infty]$. Indeed, take $\varphi : [0,1] \to [0,\infty]$, $\varphi(x) = -\log x$. Then

$$\begin{aligned}\mathbf{M}_\varphi(x_1,\ldots,x_n) &= \exp\left(-\mathbf{M}(-\log x_1,\ldots,-\log x_n)\right)\\ &= \exp\left(\frac{1}{n}\sum_{i=1}^{n}\log x_i\right) = \left(\prod_{i=1}^{n} x_i\right)^{\frac{1}{n}}\end{aligned}$$

for all n–tuples $(x_1,\ldots,x_n) \in [0,1]^n$, $n \in \mathbb{N}$. Recall that $-\log 0 = \infty$ gives the standard extension of the real valued function log to an extended real valued function. Also observe that $\mathbf{G}$ has annihilator $a = 0$ which corresponds to the annihilator $\varphi(a) = -\log 0 = \infty$ of $\mathbf{M}$ on $[0,\infty]$.

In the rest of this section we will deal with aggregation operators acting on the unit interval $[0,1]$ only. One of the most applied transformations $\varphi : [0,1] \to [0,1]$ is the duality transformation $\varphi_d(x) = 1 - x$. Applying this transformation to any aggregation operator $\mathbf{A}$ acting on $[0,1]$ as described in (63), we obtain the dual aggregation operator $\mathbf{A}^d$.

Definition 15. Let $\mathbf{A} : \bigcup_{n\in\mathbb{N}} [0,1]^n \to [0,1]$ be an aggregation operator. Then the operator $\mathbf{A}^d = \mathbf{A}_{\varphi_d} : \bigcup_{n\in\mathbb{N}} [0,1]^n \to [0,1]$ given by

$$\mathbf{A}^d(x_1,\ldots,x_n) = 1 - \mathbf{A}(1-x_1,\ldots,1-x_n) \tag{65}$$

is called a dual aggregation operator of $\mathbf{A}$.

Because of (64), $\left(\mathbf{A}^d\right)^d = \mathbf{A}$ for any aggregation operator $\mathbf{A}$, i.e., duality acts on the class of all aggregation operators as an involutive operator. Couples of dual operators are, for example, **min** and **max**, $\mathbf{A}_w$ and $\mathbf{A}_s$, $\mathbf{E}_{[0]}$ from (40) and $\mathbf{E}_{[1]}$ from (41), $\mathbf{med}_a$ and $\mathbf{med}_{1-a}$ for $a \in [0,1]$, see (43), etc. Some aggregation operators are *self–dual*, i.e., with the property $\mathbf{A}^d = \mathbf{A}$. This holds, e.g., for the arithmetic mean $\mathbf{M}$, $\mathbf{med}_{0.5}$, **med** and all weighted arithmetic means $\mathbf{W}$ which will be in detail discussed in Section 4.1. Note that duality will be important especially in Chapter 6, where t–norms and t–conorms are discussed, which are just dual operators to each other, while the classes of uninorms and nullnorms are closed under duality.

Observe that duality preserves most of the introduced properties of aggregation operators up to some modifications, for instance, if $e \in [0,1]$ ($a \in [0,1]$) is the neutral element (annihilator) of an aggregation operator $\mathbf{A}$, then its dual operator $\mathbf{A}^d$ has neutral element $1-e$ (annihilator $1-a$).

Already mentioned self–dual aggregation operators are also called symmetric sums, see [117] or [27]. These operators are characterized by the equality

$$\mathbf{A}(x_1,\ldots,x_n) + \mathbf{A}(1-x_1,\ldots,1-x_n) = 1\,, \tag{66}$$

which is fulfilled by all n–tuples $(x_1, \ldots, x_n) \in [0,1]^n$, $n \in \mathbb{N}$.

Modifying Silvert's result from [117], we obtain the following characterization.

Proposition 3. *An aggregation operator* $\mathbf{A} : \bigcup_{n \in \mathbb{N}} [0,1]^n \to [0,1]$ *is a symmetric sum if and only if there is an aggregation operator* $\mathbf{B} : \bigcup_{n \in \mathbb{N}} [0,1]^n \to [0,1]$ *such that* $\mathbf{A} = \mathbf{B}^\sharp$ *where*

$$\mathbf{B}^\sharp(x_1, \ldots, x_n) = \frac{\mathbf{B}(x_1, \ldots, x_n)}{\mathbf{B}(x_1, \ldots, x_n) + \mathbf{B}(1 - x_1, \ldots, 1 - x_n)}, \tag{67}$$

with convention $\frac{0}{0} = 0.5$.

Observe that symmetric sums can be characterized either by the equality $\mathbf{A} = \mathbf{A}^d$ or by the equality $\mathbf{A} = \mathbf{A}^\sharp$. However, the $\sharp$–operator introduced in (67) gives a hint how to construct symmetric sums. Moreover, we can introduce an equivalence relation $\sim$ on the class of all aggregation operators, namely, $\mathbf{A} \sim \mathbf{B}$ if and only if $\mathbf{A}^\sharp = \mathbf{B}^\sharp$. The factorization of the class of all aggregation operators with respect to this equivalence relation can be represented exactly by the class of all symmetric sums.

Take, e.g., the product $\mathbf{\Pi}$. The corresponding symmetric sum $\mathbf{\Pi}^\sharp$ coincides with the $3 - \Pi$–operator $\mathbf{E}$ introduced in (39) up to the convention $\frac{0}{0}$, which is equal 0.5 for $\mathbf{\Pi}^\sharp$ but 0 for $\mathbf{E}$.

Observe that $\mathbf{A}_w^\sharp = \mathbf{A}_s^\sharp = \frac{1}{2}$ whenever possible, i.e., only boundary conditions (9) and (10) force exceptions.

As a peculiar example define the aggregation operator $\mathbf{A} = \mathbf{M}.\mathbf{A}_w^\sharp$, which is for all $n > 1$, $(x_1, \ldots, x_n) \neq (1, \ldots, 1)$ given by

$$\mathbf{A}(x_1, \ldots, x_n) = \frac{\mathbf{M}(x_1, \ldots, x_n)}{2}.$$

Then $\mathbf{A} \neq \mathbf{M}$ but $\mathbf{A}^\sharp = \mathbf{M}$, though $\mathbf{A}$ is a non–continuous aggregation operator and $\mathbf{M}$ is a continuous aggregation operator.

Note that the symmetric sums need not be symmetric aggregation operators. Examples of non–symmetric aggregation operators are weighted arithmetic means, e.g., $\mathbf{M}^*$ and $\mathbf{M}_*$ given in (22) and (23), respectively.

The self–duality $\mathbf{A} = \mathbf{A}^d$ of symmetric sums is a specific invariantness property. Recall that for a bijection $\varphi : [0,1] \to [0,1]$ an operator $\mathbf{A} : \bigcup_{n \in \mathbb{N}} [0,1]^n \to [0,1]$ is called φ–invariant if $\mathbf{A}_\varphi = \mathbf{A}$. The only aggregation operators invariant under any monotone bijection φ are the projections, for example $\mathbf{P}_F$, $\mathbf{P}_L$. The class of continuous aggregation operators invariant under any increasing bijection φ was recently characterized by Ovchinnikov and Dukhovny [108] by means of the Choquet integral with respect to $\{0,1\}$–valued fuzzy measures, see also [82]. For more details about these operators we recommend Section 5.2. Note only that applying minimal fuzzy measures we get just the operator **min**, while the application of maximal fuzzy measures leads to the **max** operator, i.e., **min** and **max** operators are invariant under any increasing transformation φ. Non–continuous aggregation operators invariant under any increasing transformation φ are characterized in [112]. Examples of such aggregation operators are $\mathbf{A}_w$ and $\mathbf{A}_s$.

3.3 Weighted aggregation operators

In several applications some weights (importances) are assigned to single inputs. This is, e.g., in the case of multi–criteria decision making when aggregating score given by some experienced and some unexperienced jurymen.

One possible approach how to incorporate the (integer) weights into a symmetric aggregation is based on the next idea which is given for a fixed number n of inputs to be aggregated.

Let $m_1, \ldots, m_n \in \mathbb{N} \cup \{0\}$ be given integer weights and let $\mathbf{m} = (m_1, \ldots, m_n)$ be the corresponding non–zero weighting vector. For a given symmetric aggregation operator $\mathbf{A} : \bigcup_{n \in \mathbb{N}} [0,1]^n \to [0,1]$ we can incorporate weights m_i as follows:

$$\mathbf{A}_{\mathbf{m}}(x_1, \ldots, x_n) = \mathbf{A}(\underbrace{x_1, \ldots, x_1}_{m_1-times}, \underbrace{x_2, \ldots, x_2}_{m_2-times}, \ldots, \underbrace{x_n, \ldots, x_n}_{m_n-times}) \,. \tag{68}$$

It is easy to check that $\mathbf{A}_{\mathbf{m}}$ acts as an n–ary aggregation operator on $[0,1]$. Moreover, if $m_i = 0$ for some $i \in \{1, \ldots, n\}$, then the value x_i has no influence on the global weighted evaluation output. Applying (68) to the arithmetic mean $\mathbf{M}$, we obtain the class of weighted means (with rational weights $\frac{m_i}{\sum_{j=1}^{n} m_j}$). Similarly, (68) applied to $\mathbf{G}$ leads to the weighted geometric mean. However, applying (68) to the **max** operator, we get operator also discussed in [13], $\mathbf{max}_{\mathbf{m}}(x_1, \ldots, x_n) = \max\{x_i; i \in \{1, \ldots, n\}, m_i > 0\}$. Hence, if all weights m_i are positive, $\mathbf{max}_{\mathbf{m}} = \mathbf{max}$.

Because of the rich variety of aggregation operators, there is no general approach leading to a satisfactory incorporation of weights into a symmetric aggregation. Therefore we sketch only some of the most important approaches.

First note that if the weights are successfully incorporated into an aggregation operator $\mathbf{A}$, then we can do the same in the case of any transformed aggregation operator $\mathbf{A}_\varphi$. Based on (68), we can define the weighted arithmetic mean $\mathbf{M}_{\mathbf{v}}$ for any non–zero weighting vector $\mathbf{v} = (v_1, \ldots, v_n)$ with $v_i \in [0, \infty[$, $i = 1, \ldots, n$, as an n–ary operator

$$\mathbf{M}_{\mathbf{v}}(x_1, \ldots, x_n) = \sum_{i=1}^{n} w_i x_i \,, \quad \text{where } w_i = \frac{v_i}{\sum_{j=1}^{n} v_j} \,. \tag{69}$$

Then $w_i \in [0,1]$, $\sum_{i=1}^{n} w_i = 1$ and for $\mathbf{w} = (w_1, \ldots, w_n)$ we get $\mathbf{M_v} = \mathbf{M_w}$.

Observing that $\mathbf{G} = \mathbf{M}_\varphi$ with $\varphi(x) = -\log x$, we can introduce

$$\mathbf{G_w}(x_1, \ldots, x_n) = \varphi^{-1}\left(\mathbf{M_w}\left(\varphi(x_1), \ldots, \varphi(x_n)\right)\right) = \prod_{i=1}^{n} x_i^{w_i}, \tag{70}$$

which is a weighted geometric mean related to (68) applied to $\mathbf{G}$.

In the case of aggregation operators with neutral element $e \in [0,1]$, following ideas given in [27,135,136,13] we can apply another approach. For a normalized weighting vector $\mathbf{u} = (u_1, \ldots, u_n) \in [0,1]^n$, $\max(u_1, \ldots, u_n) = 1$, let $\mathbf{A}$ be an aggregation operator with neutral element $e \in [0,1]$. Let $h : [0,1]^2 \to [0,1]$ be a mapping which is monotone in the first component and non–decreasing in the second component such that

$$h(0,x) = e \quad \text{and} \quad h(1,x) = x \quad \text{for all } x \in [0,1].$$

Moreover, suppose $\mathbf{A}\left(h(u_1,0), \ldots, h(u_n,0)\right) \neq \mathbf{A}\left(h(u_1,1), \ldots, h(u_n,1)\right)$. Then we can define the n–ary operator $\mathbf{A_u} : [0,1]^n \to [0,1]$ by

$$\mathbf{A_u}(x_1, \ldots, x_n) = g\left(\mathbf{A}\left(h(u_1,x_1), \ldots, h(u_n,x_n)\right)\right), \tag{71}$$

where g is a linear transformation given by

$$g(x) = \frac{x - \mathbf{A}\left(h(u_1,0), \ldots, h(u_n,0)\right)}{\mathbf{A}\left(h(u_1,1), \ldots, h(u_n,1)\right) - \mathbf{A}\left(h(u_1,0), \ldots, h(u_n,0)\right)} \tag{72}$$

Observe that if $\mathbf{u} = (1, \ldots, 1)$ then $\mathbf{A_u} = \mathbf{A}$ for any h. So, for example, take the product $\mathbf{\Pi}$ with neutral element $e = 1$. As a convenient mapping we can take the mapping h, $h(u,x) = x^u$ with convention $0^0 = 1$, see [13]. Then

$$\mathbf{\Pi_u}(x_1, \ldots, x_n) = \prod_{i=1}^{n} x_i^{u_i}.$$

Similarly, for **min** with neutral element $e = 1$, put $h(u,x) = \max(1-u, x)$ and then

$$\mathbf{min_u}(x_1, \ldots, x_n) = \min\left(\max(1-u_i, x_i) \mid i \in \{1, \ldots, n\}\right).$$

For the aggregation operator **max** with neutral element $e = 0$, $h(u,x) = \min(u,x)$ was proposed, see [27,135], and then

$$\mathbf{max_u}(x_1, \ldots, x_n) = \max\left(\min(u_i, x_i) \mid i \in \{1, \ldots, n\}\right).$$

Finally, take the aggregation operator $\mathbf{A}_{\{c\}}$ with neutral element $e = c$ given in (25). As an appropriate mapping h we can choose the mapping $h(u,x) = ux + (1-u)c$, compare [136,121]. Then

$$\mathbf{A}_{\{c\}}(x_1, \ldots, x_n) = \max\left(0, \min\left(1, c + \sum_{i=1}^{n} u_i(x_i - c)\right)\right).$$

3.4 Ordinal sums

The idea of ordinal sums of operators has its roots in the semigroup theory [19,21]. In the framework of special aggregation operators, namely t–norms discussed in Section 6.1, this method is studied in detail in [61]. In general, ordinal sums extend operators acting on some subdomains $[a_i, b_i]$ of $[0,1]$ to a global operator acting on $[0,1]$. Depending on the properties which have to be preserved, several types of ordinal sums of operators can be introduced. From well–known types of ordinal sums recall ordinal sums of t–norms and t–conorms [114,61] or copulas [114,107]. Our aim is to discuss the construction methods for constructing new aggregation operators from given ones, hence the crucial point is the monotonicity (11) and the boundary conditions (9) and (10) of resulting operators. From given aggregation operators $\mathbf{A}_i$ acting on non–overlapping domains $[a_i, b_i] \subset [0,1]$, $i = 1, \ldots, k$, several possible extensions of $\mathbf{A}_i's$ to an aggregation operator $\mathbf{A}$ acting on $[0,1]$ can be introduced, in general. We introduce only some distinguished (and unique in some sense) possible extensions. The next results are in detail shown in [93].

Proposition 4. *Let* $(\mathbf{A}_i : \bigcup_{n \in \mathbf{N}} [a_i, b_i]^n \to [a_i, b_i],\ i = 1, \ldots, k)$, *be a system of aggregation operators acting on non–overlapping domains* $[a_i, b_i]$, $i = 1, \ldots, k$, $0 \le a_1 < b_1 \le a_2 < b_2 \le \ldots < b_k \le 1$. *Define two operators*

$$\mathbf{A}^{(w)} = (< a_i, b_i, \mathbf{A}_i > \mid i = 1, \ldots, k)_w : \bigcup_{n \in \mathbf{N}} [0,1]^n \to [0,1]$$

and

$$\mathbf{A}^{(s)} = (< a_i, b_i, \mathbf{A}_i > \mid i = 1, \ldots, k)_s : \bigcup_{n \in \mathbf{N}} [0,1]^n \to [0,1]$$

by

$$\mathbf{A}^{(w)}(x_1, \ldots, x_n) = \begin{cases} 0 & \text{if } a < a_1, \\ \mathbf{A}_i\left(\min(x_1, b_i), \ldots, \min(x_n, b_i)\right) & \text{if } a_i \le a < a_{i+1}, \\ 1 & \text{if } a = 1, \end{cases} \tag{73}$$

where $a_{k+1} = 1$ *and* $a = \min(x_1, \ldots, x_n)$,
and

$$\mathbf{A}^{(s)}(x_1, \ldots, x_n) = \begin{cases} 0 & \text{if } b = 0, \\ \mathbf{A}_i\left(\max(x_1, a_i), \ldots, \max(x_n, a_i)\right) & \text{if } b_{i-1} < b \le b_i, \\ 1 & \text{if } b_k < b, \end{cases} \tag{74}$$

where $b_0 = 0$ *and* $b = \max(x_1, \ldots, x_n)$.
Then $\mathbf{A}^{(w)}$ *and* $\mathbf{A}^{(s)}$ *are the weakest and the strongest aggregation operators, respectively, such that the corresponding restrictions to the inputs from* $[a_i, b_i]$

coincide with aggregation operators $\mathbf{A}_i$, $i = 1, \ldots, k$. *The operator* $\mathbf{A}^{(w)}$ *is called the lower ordinal sum (of aggregation operators* $\mathbf{A}_i$*), while* $\mathbf{A}^{(s)}$ *is called the upper ordinal sum (of aggregation operators* $\mathbf{A}_i$*).*

Note that formally $k = 0$ can also be accepted, and then (73) gives $\mathbf{A}_w$, see (17), and (74) leads to $\mathbf{A}_s$, see (18).

Example 3 To illustrate Proposition 3, suppose that $\mathbf{A}_1 = \mathbf{G}$ acts on $[a_1, b_1] = [0, 0.5]$ while $\mathbf{A}_2 = \mathbf{M}$ acts on $[a_2, b_2] = [0.7, 1]$. Then

$$\mathbf{A}^{(w)}(x,y) = \begin{cases} \mathbf{G}(x,y) = \sqrt{xy} & \text{if } (x,y) \in [0,\, 0.5]^2, \\ \mathbf{M}(x,y) = \frac{x+y}{2} & \text{if } (x,y) \in [0.7,\, 1]^2, \\ \mathbf{G}(x,0.5) = \sqrt{0.5x} & \text{if } (x,y) \in [0,\, 0.5]\times]0.5, 1], \\ \mathbf{G}(0.5,y) = \sqrt{0.5y} & \text{if } (x,y) \in]0.5,\, 1] \times [0, 0.5], \\ \mathbf{G}(0.5,0.5) = 0.5 & \text{else}. \end{cases}$$

Further,

$$\mathbf{A}^{(s)}(x,y) = \begin{cases} \mathbf{G}(x,y) = \sqrt{xy} & \text{if } (x,y) \in [0,\, 0.5]^2, \\ \mathbf{M}(x,y) = \frac{x+y}{2} & \text{if } (x,y) \in [0.7,\, 1]^2, \\ \mathbf{M}(x,0.7) = \frac{x+0.7}{2} & \text{if } (x,y) \in [0.7,\, 1] \times [0, 0.7[, \\ \mathbf{M}(0.7,y) = \frac{0.7+y}{2} & \text{if } (x,y) \in [0,\, 0.7] \times [0.7, 1], \\ 0.7 & \text{else}. \end{cases}$$

The lower (upper) ordinal sums preserve the symmetry of incoming summands $\mathbf{A}_i$, however, they may violate other properties, in general. To preserve the idempotency of the incoming summands $\mathbf{A}_i$ also by the constructed extension, we have the next result introduced in [93].

Proposition 5. *Let all aggregation operators* $\mathbf{A}_i$ *from Proposition 3 be idempotent. Then the weakest idempotent aggregation operator* $\mathbf{A}^{(iw)}$ *and the strongest idempotent aggregation operator* $\mathbf{A}^{(is)}$ *acting on* $[0,1]$ *and extending all operators* $\mathbf{A}_i$, $i = 1, \ldots, k$, *are given by*

$$\mathbf{A}^{(iw)}(x_1, \ldots, x_n) = \begin{cases} \mathbf{A}_i\left(\min(x_1, b_i), \ldots, \min(x_n, b_i)\right) & \textit{if } a_i \le a < b_i\,, \\ a & \textit{else}, \end{cases} \tag{75}$$

and

$$\mathbf{A}^{(is)}(x_1, \ldots, x_n) = \begin{cases} \mathbf{A}_i\left(\max(x_1, a_i), \ldots, \min(x_n, a_i)\right) & \textit{if } a_i < b \le b_i\,, \\ b & \textit{else}, \end{cases} \tag{76}$$

where $a = \min(x_1, \ldots, x_n)$ *and* $b = \max(x_1, \ldots, x_n)$.

Observe that admitting formally $k = 0$, we obtain $\mathbf{A}^{(iw)} = \mathbf{min}$ and $\mathbf{A}^{(is)} = \mathbf{max}$.

Example 4 Under requirements of Example 3 we have

$$\mathbf{A}^{(iw)}(x,y) = \begin{cases} \min(x,y) & \text{if } \min(x,y) \in [0.5, 0.7[, \\ \mathbf{A}^{(w)}(x,y) & \text{else,} \end{cases}$$

and

$$\mathbf{A}^{(is)}(x,y) = \begin{cases} \max(x,y) & \text{if } \max(x,y) \in]0.5, 0.7], \\ \mathbf{A}^{(s)}(x,y) & \text{else.} \end{cases}$$

Observe that the standard ordinal sums of t–norms and t–conorms or copulas (though neither t–norms nor copulas are idempotent operators, in general), as introduced in [114,61], are just our $\mathbf{A}^{(iw)}$ ordinal sum, while $\mathbf{A}^{(is)}$ describes exactly the standard ordinal sum of t–conorms or of dual copulas, see [114,107,61].

Note that $\mathbf{A}^{(w)} = \mathbf{A}^{(iw)}$ and $\mathbf{A}^{(is)} = \mathbf{A}^{(s)}$ if and only if $\bigcup_{i=1}^{k} [a_i, b_i] = [0,1]$.

Further, each extension $\mathbf{A} : \bigcup_{n \in \mathbb{N}} [0,1]^n \to [0,1]$ of aggregation operators $\mathbf{A}_i$ acting on non–overlapping intervals $[a_i, b_i] \subset [0,1]$, $i = 1, \ldots, k$, is characterized by the inequalities

$$\mathbf{A}^{(w)} \le \mathbf{A} \le \mathbf{A}^{(s)}.$$

Moreover, $\mathbf{A}^w = \mathbf{A}^s$, i.e., there is a unique extension only if $k = 2$, $\mathbf{A}_1$ acts on $[0,a]$ and $\mathbf{A}_2$ acts on $[a,1]$ where a is the annihilator of both operators $\mathbf{A}_1$ and $\mathbf{A}_2$. Then this unique extension is given by

$$\mathbf{A}(x_1, \ldots, x_n) = \begin{cases} \mathbf{A}_1(x_1, \ldots, x_n) & \text{if all } x_i \in [0,a], \\ \mathbf{A}_2(x_1, \ldots, x_n) & \text{if all } x_i \in [a,1], \\ a & \text{else.} \end{cases}$$

As an example recall the a–median $\mathbf{med}_a$ introduced in (43) with $\mathbf{A}_1 = \mathbf{max}$ on $[0,a]$ and $\mathbf{A}_2 = \mathbf{min}$ on $[a,1]$.

An important property for technical applications is the continuity of an aggregation operator to be applied. But neither $\mathbf{A}^{(w)}$, $\mathbf{A}^{(s)}$ nor $\mathbf{A}^{(iw)}$, $\mathbf{A}^{(is)}$ need not be continuous though all summands are continuous operators. Therefore we propose a method resulting in a continuous extension. For more details we recommend [95].

Suppose that all operators $\mathbf{A}_i$ in Proposition 3 are continuous and $\bigcup_{i=1}^{k} [a_i, b_i] = [0,1]$, $a_1 = 0, b_k = 1$. If the latter condition is not satisfied we can add any continuous summands such as **min**, **max**, **M**, **G**, etc., to act on occurring gaps. Let $f : [0,1] \to [-\infty, \infty]$ be any continuous strictly monotone function. Then the mapping $\mathbf{A}^{(f)} : \bigcup_{n\in\mathbb{N}} [0,1]^n \to [0,1]$ given by

$$\mathbf{A}^{(f)}(x_1,\dots,x_n) = f^{-1}\left(\sum_{i=1}^{k} f\left(\mathbf{A}_i(x_1^{(i)},\dots,x_n^{(i)})\right) - \sum_{i=2}^{k} f(a_i)\right), \quad (77)$$

where $x^{(i)} = \max(a_i, \min(b_i, x))$, is a continuous aggregation operator such that

$$\forall i = 1,\dots,k: \quad \mathbf{A}^{(f)}_{|\bigcup_{n\in\mathbb{N}}[a_i,b_i]^n} = \mathbf{A}_i\,.$$

Note that $x^{(i)}$ is a point from $[a_i, b_i]$ closest to x. So, e.g., if $f(x) = id(x) = x$, we get

$$\mathbf{A}^{(id)}(x_1,\dots,x_n) = \sum_{i=1}^{k}\left(\mathbf{A}_i(x_1^{(i)},\dots,x_n^{(i)}) - a_i\right),$$

while for $f(x) = \log x$ we have

$$\mathbf{A}^{(\log)}(x_1,\dots,x_n) = \prod_{i=1}^{k} \frac{\mathbf{A}_i(x_1^{(i)},\dots,x_n^{(i)})}{b_i}\,.$$

Remarkably, if all operators $\mathbf{A}_i$ are t–norms (t–conorms, copulas, dual copulas) then for any f, $\mathbf{A}^{(f)}$ is the standard ordinal sum of t–norms (t–conorms, copulas, dual copulas).

Observe that (77) is a special case of the $\mathbf{D}$–based ordinal sum construction [95] where $\mathbf{D} : \bigcup_{n\in\mathbb{N}} [0,1]^n \to [0,1]$ is a symmetric continuous idempotent aggregation operator which is strictly monotone (cancellative) on $\bigcup_{n\in\mathbb{N}}]0,1[^n$. Then the $\mathbf{D}$–ordinal sum $\mathbf{A}^{\mathbf{D}} : \bigcup_{n\in\mathbb{N}} [0,1]^n \to [0,1]$ of aggregation operators $\mathbf{A}_i : \bigcup_{n\in\mathbb{N}} [a_{i-1}, a_i]^n \to [a_{i-1}, a_i]$, $i = 1,\dots,k$, $0 = a_0 < a_1 \dots < a_k = 1$, is given as a (unique) solution of the equation

$$\mathbf{D}\left(\mathbf{A}^{\mathbf{D}}(x_1,\dots,x_n), a_1,\dots,a_{k-1}\right) =$$
$$\mathbf{D}\left(\mathbf{A}_1(x_1^{(1)},\dots,x_n^{(1)}),\dots,\mathbf{A}_k(x_1^{(k)},\dots,x_n^{(k)})\right), \quad (78)$$

where $x^{(i)} = \max(a_{i-1}, \min(a_i, x))$, $i = 1,\dots,k$.

Evidently, if $\mathbf{D}(x_1,\dots,x_n) = f^{-1}\left(\frac{1}{n}\sum_{i=1}^{n} f(x_i)\right)$ for some continuous strictly monotone function $f : [0,1] \to [-\infty, \infty]$, that is, if $\mathbf{D}$ is a quasi–arithmetic mean, see Section 4.3, then (78) results in (77).

3.5 Composed aggregation

The composition of aggregation operators can be derived from the standard composition of real functions.

Proposition 6. *Let* $\mathbf{A}, \mathbf{B}_1, \ldots, \mathbf{B}_k : \bigcup_{n \in \mathbb{N}} [0,1]^n \to [0,1]$ *be aggregation operators. Then the operator* $\mathbf{C} = \mathbf{A}(\mathbf{B}_1, \ldots, \mathbf{B}_k) : \bigcup_{n \in \mathbb{N}} [0,1]^n \to [0,1]$, *defined by*

$$\mathbf{C}(x_1, \ldots, x_n) = \mathbf{A}(\mathbf{B}_1(x_1, \ldots, x_n), \ldots, \mathbf{B}_k(x_1, \ldots, x_n)) \tag{79}$$

is an aggregation operator if and only if $\mathbf{A}_{(k)}$ *is an idempotent aggregation operator.*

Proof. We can directly verify the monotonicity of $\mathbf{C}$, i.e., the property (11), as well as the boundary condition (9). Both hold independently of the properties of $\mathbf{A}$. Finally, the boundary condition (10) of $\mathbf{C}$ is equivalent to

$$x = \mathbf{C}(x) = \mathbf{A}(\mathbf{B}_1(x), \ldots, \mathbf{B}_k(x)) = \mathbf{A}_{(k)}(x, \ldots, x)$$

for any $x \in [0,1]$, i.e., to the idempotency of $\mathbf{A}_{(k)}$. □

Evidently, for each idempotent aggregation operator $\mathbf{A}$ we have

$$\mathbf{A}(\mathbf{A}, \ldots, \mathbf{A}) = \mathbf{A}.$$

However, if we compose different aggregation operators from the same class (such as, e.g., weighted means, Choquet–integral based operators, t–norms, etc.) we can obtain qualitatively different operators, in general. This is not the case of weighted sum, because the composition of weighted sums results always in a weighted sum, possibly with different weights.

Typical examples of composed aggregation are the following operators:

$$\mathbf{max}(\mathbf{B}_1, \ldots, \mathbf{B}_k), \quad \mathbf{min}(\mathbf{B}_1, \ldots, \mathbf{B}_k),$$

$$\left(\prod_{i=1}^{k} \mathbf{B}_i\right)^{1/k} = \mathbf{G}(\mathbf{B}_1, \ldots, \mathbf{B}_k), \quad \frac{1}{k}\sum_{i=1}^{k} \mathbf{B}_i = \mathbf{M}(\mathbf{B}_1, \ldots, \mathbf{B}_k),$$

etc.

An interesting composite aggregation is a Cartesian product–based aggregation introduced in [111].

Definition 16. Let $\mathbf{A}, \mathbf{B} : \bigcup_{n \in \mathbb{N}} [0,1]^n \to [0,1]$ be two idempotent aggregation operators and let $\mathbf{A}$ be symmetric. For any $k \in \mathbb{N}$, the aggregation operator $\mathbf{A}_k^{\mathbf{B}} : \bigcup_{n \in \mathbb{N}} [0,1]^n \to [0,1]$ given by

$$\mathbf{A}_k^{\mathbf{B}}(x_1, \ldots, x_n) =$$

$$\mathbf{A}_{(n^k)}\left(\mathbf{B}_{(k)}(x_1,\ldots,x_1),\mathbf{B}_{(k)}(x_1,\ldots,x_1,x_2),\ldots,\mathbf{B}_{(k)}(x_n,\ldots,x_n)\right) \quad (80)$$

is called a k–Cartesian $\mathbf{A}-\mathbf{B}$ product.

Note that the right–hand side runs over all points $(x_{\alpha(1)},\ldots,x_{\alpha(k)})$, where α is an element of the Cartesian product $\{1,\ldots,n\}^k$.

Observe that while in the composition of aggregation operators given in (79) the number of inner aggregation operators $\mathbf{B}_1,\ldots,\mathbf{B}_k$ is fixed, in (80) we apply $\mathbf{B}$ n^k–times. Note that $\mathbf{max}^{\mathbf{B}}=\mathbf{max}$ and $\mathbf{min}^{\mathbf{B}}=\mathbf{min}$ for any idempotent aggregation operator $\mathbf{B}$ and $k\in\mathbb{N}$. Moreover, for any symmetric idempotent aggregation operator $\mathbf{A}$ we have $\mathbf{A}_1^{\mathbf{B}}=\mathbf{A}$, independently of $\mathbf{B}$.

Example 5 (i) Put $\mathbf{A}=\mathbf{M}$ and $\mathbf{B}=\mathbf{max}$. Then $\mathbf{M}_k^{\mathbf{max}}=\mathbf{C}_k$ is an OWA operator which is generated by a non–decreasing quantifier $q(x)=x^k$, see Section 4.2,

$$\mathbf{C}_k(x_1,\ldots,x_n)=\sum_{i=1}^{n}\left|\left(\frac{i}{n}\right)^k-\left(\frac{i-1}{n}\right)^k\right|x_i', \quad (81)$$

where $(x_1',\ldots,x_n')$ is a non–decreasing permutation of $(x_1,\ldots,x_n)$.

(ii) Similarly to (i), $\mathbf{D}_k=\mathbf{M}_k^{\mathbf{min}}$ is an OWA operator related to a decreasing quantifier $q(x)=(1-x)^k$,

$$\mathbf{D}_k(x_1,\ldots,x_n)=\sum_{i=1}^{n}\left|\left(1-\frac{i}{n}\right)^k-\left(1-\frac{i-1}{n}\right)^k\right|x_i', \quad (82)$$

(iii) Denote $\mathbf{M}_{1/k}=\mathbf{M}_k^{\mathbf{G}}$. Then

$$\mathbf{M}_{1/k}(x_1,\ldots,x_n)=\left(\frac{1}{n}\sum_{i=1}^{n}x_i^{\frac{1}{k}}\right)^k, \quad (83)$$

i.e., $\mathbf{M}_{1/k}$ is a root–power operator (special quasi–arithmetic mean), see Section 4.3.

(iv) The operator $\mathbf{G}_k^{\mathbf{M}}$, $k>1$, is a new operator not belonging to any of classes discussed in next sections. For example, for $n=2$ we obtain

$$\mathbf{G}_2^{\mathbf{M}}(x,y)=\left(xy\left(\frac{x+y}{2}\right)^2\right)^{1/4},$$

while

$$\mathbf{G}_3^{\mathbf{M}}(x,y)=\left(xy\left(\frac{2x+y}{3}\right)^3\left(\frac{x+2y}{3}\right)^3\right)^{1/8}.$$

In general,

$$\mathbf{G}_k^{\mathbf{M}}(x,y)=\left(\prod_{i=0}^{k}\left(\frac{(k-i)x+iy}{k}\right)^{\binom{k}{i}}\right)^{1/2^k}.$$

Proposition 6 describes two-step aggregation. First, the inputs are aggregated by means of operators $\mathbf{B}_1, \ldots, \mathbf{B}_k$, and then the obtained outputs are aggregated by means of $\mathbf{A}$ to get the final output. In a real situation operators $\mathbf{B}_1, \ldots, \mathbf{B}_k$ can correspond to the jurymen $J_1, \ldots, J_k$ and the operator $\mathbf{A}$ to the head of a jury.

By induction, easily more step aggregation can be introduced. This type of aggregation is philosophically related to the Bayesian computations of the global apriori probabilities and is discussed in the framework of the Choquet integral-based aggregation operators in [37,38,98].

3.6 Some other construction methods

There are several construction methods in some special classes of aggregation operators. For example, the widely used additive continuous generators in the field of t–norms and t–conorms can be generalized to special non–continuous generators. Then several surprising results, never appearing in the continuous case, may emerge, e.g., the existence of non–trivial idempotent elements. For interested readers we recommend the works of Viceník [127,126,128]. Similarly, transformations discussed in Section 3.2 can be further generalized applying non–continuous or non–strictly monotone functions. However, then instead of standard inverse functions, so-called quasi–inverses and pseudo–inverses should be applied, see e.g., [114,60,61]. Among other specific construction methods we recall the following three ones.

Constraining

This method was proposed in [66]. For a given aggregation operator $\mathbf{A} : \bigcup_{n\in\mathbb{N}} [0,1]^n \to [0,1]$ and a given constant $u \in [0,1]$ satisfying the condition

$$\forall n \geq 2 : \quad \mathbf{A}_{(n)}(u, 1, \ldots, 1) > \mathbf{A}_{(n)}(u, 0, \ldots, 0)$$

a new aggregation operator ${}_u\mathbf{A} : \bigcup_{n\in\mathbb{N}} [0,1]^n \to [0,1]$ is defined by

$${}_u\mathbf{A}(x_1, \ldots, x_n) = \frac{\mathbf{A}_{(n+1)}(u, x_1, \ldots, x_n) - \mathbf{A}_{(n+1)}(u, 0, \ldots, 0)}{\mathbf{A}_{(n+1)}(u, 1, \ldots, 1) - \mathbf{A}_{(n+1)}(u, 0, \ldots, 0)} . \tag{84}$$

Symmetric operators invariant under constraining (84), i.e., with the property ${}_u\mathbf{A} = \mathbf{A}$ for all considered u, are appropriate operators for evaluating information with possible collapse of some of the inputs, for instance if the i–th sensor is broken (with unknown i). By [66], such operators form a parameterized class $(\mathbf{K}_\alpha)_{\alpha\in]0,1[}$ where

$$\mathbf{K}_\alpha(x_1, \ldots, x_n) = \frac{\prod_{i=1}^{n} (\alpha + (1-2\alpha)x_i) - \alpha^n}{(1-\alpha)^n - \alpha^n} , \quad \text{for } \alpha \neq 0.5 , \tag{85}$$

and $\mathbf{K}_{0.5} = \mathbf{M}$. The limit members of this class are operators $\mathbf{K}_0 = \mathbf{\Pi}$ and $\mathbf{K}_1 = \mathbf{\Pi}^d$, where $\mathbf{\Pi}^d$ is the probabilistic sum, i.e., the dual operator to $\mathbf{\Pi}$, and they are constrained invariant for all $u \in [0,1]$ up to 0 or 1, respectively. Observe that for each $\alpha \in [0,1]$, $\alpha = \mathbf{K}_\alpha(0,1) = \mathbf{K}_\alpha(1,0)$, that is, the parameter α corresponds to the aggregation output when aggregating the contradictory inputs 0 and 1.

Augmenting and decreasing

There are several reasons for augmenting (or decreasing) the output given by chosen aggregation operators. Then two basic approaches can be used. In the first one, see [138,25], we try to get the desired effect by means of some additional aggregation operators and some arithmetic operations. So, e.g., for given aggregation operators $\mathbf{A}, \mathbf{B} : \bigcup\limits_{n \in \mathbb{N}} [0,1]^n \to [0,1]$ and a constant $\beta \in]0, \infty[$ we can define a new operator $\mathbf{A}_{\beta\mathbf{B}} = \max(0, \mathbf{A} - \beta(1 - \mathbf{B}))$.

Evidently, $\mathbf{A}_{\beta\mathbf{B}} \le \mathbf{A}$ and $\mathbf{A}_{\alpha\mathbf{B}} \le \mathbf{A}_{\beta\mathbf{B}}$ whenever $\beta \le \alpha$, hence we get a decreasing family of aggregation operators $(\mathbf{A}_{\beta\mathbf{B}})_{\beta \in [0,\infty[}$.

Similarly, the augmentation of $\mathbf{A}$ can be defined by

$$\mathbf{A}^{\beta\mathbf{B}} = \min(1, \mathbf{A} + \beta\mathbf{B}),$$

see [138].

Example 6 (i) Take the aggregation operator $\mathbf{\Pi}$ and decrease it by means of its dual operator $\mathbf{\Pi}^d$. In the binary form we have for $\mathbf{A}_\beta = \mathbf{\Pi}_{\beta\mathbf{\Pi}^d}$:

$$\begin{aligned} \mathbf{A}_\beta(x,y) &= \max(0, xy - \beta(1 - x - y + xy)) \\ &= \max(0, \beta(x + y - 1) + (1 - \beta)xy). \end{aligned} \tag{86}$$

Note that $\mathbf{A}_{\beta(2)}$, is just a Weber t–norm [130], see also Section 6.1.

(ii) When augmenting $\mathbf{\Pi}^d$ by means of $\mathbf{\Pi}$, for $\mathbf{B}_\beta = \left(\mathbf{\Pi}^d\right)^{\beta\mathbf{\Pi}}$ we obtain:

$$\mathbf{B}_\beta(x,y) = \min(1, x + y - xy + \beta xy) = \max(1, x + y + (\beta - 1)xy), \tag{87}$$

i.e., $\mathbf{B}_\beta$ is just the Sugeno λ– addition introduced in [119] for $\lambda = \beta - 1$. Note that this means that $\mathbf{B}_{\beta(2)}$ is a continuous Archimedean t–conorm , see Section 6.1. For more examples and applications of these aggregation operators we refer the reader to [138,25].

An alternative approach was proposed in [69]. In this approach, first the inputs are augmented (decreased) by means of some transformations and then the transformed inputs are aggregated by an aggregation operator $\mathbf{A}$, i.e.,

$$\mathbf{A}_{\mathcal{F}}(x_1, \dots, x_n) = \mathbf{A}\left(f_1^{(n)}(x_1), \dots, f_n^{(n)}(x_n)\right),$$

where $\mathcal{F} = \left(f_i^{(n)} \,|\, n \in \mathbb{N},\, i \in \{1, \dots, n\}\right)$ is a system of $[0,1] \to [0,1]$ non–decreasing mappings with fixed points 0 and 1.

Example 7 Put $f_i^{(n)} : [0,1] \to [0,1]$, $f_i^{(n)}(x) = x^{1/n}$, $n \in \mathbb{N}$, $i \in \{1,\dots,n\}$. As $f_i^{(n)}(x) \geq x$, we augment the input values. For example, the product will be transformed into the geometric mean, i.e., $\mathbf{\Pi}_{\mathcal{F}} = \mathbf{G}$. For more details and examples see [69].

Flying parameter

Having a parameterized family $(\mathbf{A}_\alpha)_{\alpha\in[0,1]}$ of aggregation operators, the crucial task for application is the choice of a parameter α. If this choice depends on input values to be aggregated, we get just the method of flying parameter. This idea has appeared already in [139] and can be formalized as follows.

Proposition 7. *Let* $(\mathbf{A}_\alpha)_{\alpha\in[0,1]}$ *be a non–decreasing family of aggregation operators and* $\mathbf{B}$ *be any aggregation operator. Then* $\mathbf{A_B} : \bigcup_{n\in\mathbb{N}} [0,1]^n \to [0,1]$ *given by*

$$\mathbf{A_B}(x_1,\dots,x_n) = \mathbf{A}_{\mathbf{B}(x_1,\dots,x_n)}(x_1,\dots,x_n) \tag{88}$$

is an aggregation operator.

Proof. The boundary conditions (9) and (10) can be verified directly. To see the monotonicity (11), observe that $x_1 \leq y_1, \dots, x_n \leq y_n$ implies $\mathbf{B}(x_1,\dots,x_n) \leq \mathbf{B}(y_1,\dots,y_n)$ and due to the non–decreasingness of the family $(\mathbf{A}_\alpha)_{\alpha\in[0,1]}$ we have

$$\begin{aligned}\mathbf{A_B}(x_1,\dots,x_n) &= \mathbf{A}_{\mathbf{B}(x_1,\dots,x_n)}(x_1,\dots,x_n) \leq \mathbf{A}_{\mathbf{B}(y_1,\dots,y_n)}(x_1,\dots,x_n)\\ &\leq \mathbf{A}_{\mathbf{B}(y_1,\dots,y_n)}(y_1,\dots,y_n) = \mathbf{A_B}(y_1,\dots,y_n)\,. \quad \square\end{aligned}$$

A nice example of an aggregation operator constructed by means of (88) was given by Yager and Filev [139]. For $\alpha \in [0,1]$, put

$$\mathbf{A}_\alpha = (1-\alpha)\mathbf{\Pi} + \alpha\mathbf{\Pi}^d\,.$$

As $\mathbf{\Pi} \leq \mathbf{\Pi}^d$, $(\mathbf{A}_\alpha)_{\alpha\in[0,1]}$ is a non–decreasing family of aggregation operators. Now, for $\mathbf{B} = \mathbf{E}$, where $\mathbf{E}$ is the $3-\Pi$–operator given in (39), we obtain

$$\mathbf{A_E}(x_1,\dots,x_n) =$$

$$\left(1 - \frac{\prod_{i=1}^n x_i}{\prod_{i=1}^n x_i + \prod_{i=1}^n (1-x_i)}\right)\prod_{i=1}^n x_i + \frac{\prod_{i=1}^n x_i}{\prod_{i=1}^n x_i + \prod_{i=1}^n (1-x_i)}\left(1 - \prod_{i=1}^n (1-x_i)\right)$$

$$= \frac{\prod_{i=1}^n x_i}{\prod_{i=1}^n x_i + \prod_{i=1}^n (1-x_i)} = \mathbf{E}(x_1,\dots,x_n).$$

4 Aggregation operators based on arithmetic mean

The arithmetic mean is a prototype of a strictly monotone (i.e., cancellative) continuous symmetric idempotent aggregation operator. **M** fulfills several other properties, such as self-identity property, linearity, additivity, etc. In fact, the arithmetic mean is the most applied aggregation operator, and in the natural language it is the operator modelling the average (average income, average temperature, average note, etc.). If for some reasons some of the properties of the arithmetic mean does not fit the real situation we have to model, we try to modify the arithmetic mean with respect to that violated property, but still preserving as much other properties as possible. Roughly speaking, this is the philosophical background of several classes of aggregation operators related to arithmetic mean, i.e., coming from the arithmetic mean applying some construction method. This chapter is devoted to the presentation of some of these classes of aggregation operators related to the arithmetic mean.

4.1 Weighted arithmetic means

The symmetry of the arithmetic mean does not reflect the possibly different importances of single criteria in multi-criteria decision making. The need of incorporating the importances into arithmetic mean resulted in the introducing the class of weighted arithmetic means as a result of the method presented in Section 3.3, see (69).

For n-ary operators, the weights $w_1, \ldots, w_n$ form an n-dimensional weighting vector $\mathbf{w} = (w_1, \ldots, w_n) \in [0,1]^n$ with $\sum_{i=1}^{n} w_i = 1$. However, if we have to think on a weighted arithmetic mean $\mathbf{W} : \bigcup_{n \in \mathbb{N}} [0,1]^n \to [0,1]$ as an operator on any input tuples, we need to know the relevant weights for all possible input cardinalities n. Therefore, a weighting triangle $\triangle = (w_{in} \mid n \in \mathbb{N}, i \in \{1, \ldots, n\})$, such that all $w_{in} \in [0,1]$ and $\sum_{i=1}^{n} w_{in} = 1$ for all $n \in \mathbb{N}$, is needed. For more details about weighting triangles we recommend [11,12].

Definition 17. Let $\triangle$ be a given weighting triangle. An aggregation operator $\mathbf{W}_\triangle : \bigcup_{n \in \mathbb{N}} [0,1]^n \to [0,1]$ defined by

$$\mathbf{W}_\triangle(x_1, \ldots, x_n) = \sum_{i=1}^{n} w_{in} x_i \tag{89}$$

is called a weighted arithmetic mean associated with $\triangle$. If no confusion appears, the notation **W** and name weighted arithmetic mean can be used.

Weighted arithmetic means are continuous, idempotent, linear, additive and self-dual aggregation operators. Note that each additive aggregation operator $\mathbf{W} : \bigcup_{n\in\mathbb{N}} [0,1]^n \to [0,1]$ is necessarily a weighted arithmetic mean with weighting triangle $\Delta = (w_{in}), w_{in} = \mathbf{W}_{(n)}(0,\ldots,1,\ldots,0)$, where all inputs up to the ith one are equal to zero. Evidently, the arithmetic mean $\mathbf{M}$ is the only symmetric weighted mean associated with the weighting triangle $\Delta = (w_{in}) = (\frac{1}{n})$.

We have already introduced weighted arithmetic mean $\mathbf{M}^*$ in (22) and $\mathbf{M}_*$ in (23). The relevant weighting triangles Δ^* and Δ_* are given by

$$\Delta^* = (w_{in}), \quad w_{nn} = \frac{1}{2^{n-1}}, \quad \text{and} \quad w_{in} = \frac{1}{2^i} \quad \text{for} \quad i < n,\, n \in \mathbb{N}, \tag{90}$$

and

$$\Delta_* = (w_{in}{}'), \quad w_{1n}{}' = \frac{1}{2^{n-1}}, \quad \text{and } w_{in}{}' = \frac{1}{2^{n-i+1}} \quad \text{for } i > 1,\, n \in \mathbb{N}. \tag{91}$$

Note that Δ^* and Δ_* are an example of reversed weighting triangles (w_{in}) and $(w_{in}{}')$ for which $w_{in} = w_{jn}{}'$ whenever $i+j = n+1$, i.e., $w_{jn}{}' = w_{(n+1-j)n}$.

An appropriate choice of a weighted arithmetic mean $\mathbf{W}$, or, equivalently, of a weighting triangle Δ, is a crucial task when applying these operators. There are some methods of generating weighting triangles. First of all, we can apply methods described in (20) and (21) for extending a binary weighted arithmetic mean to a general weighted arithmetic mean. Indeed, let $p \in [0,1]$ be any given fixed constant and let

$$\mathbf{W}(x,y) = px + (1-p)y. \tag{92}$$

Applying (20) to the binary operator from (92), we obtain an aggregation operator $\mathbf{W}^*$ which is the weighted arithmetic mean associated with the weighting triangle

$$\Delta^* = (w_{in}), \quad w_{in} = p(1-p)^{i-1} \quad \text{for} \quad i < n \quad \text{and} \quad w_{nn} = (1-p)^{n-1}. \tag{93}$$

Observe that the weighting triangle Δ^* described in (90) is a special case of (93) for $p = \frac{1}{2}$. Analogously, applying (21) to the binary operator from (92), we obtain an aggregation operator $\mathbf{W}_*$ which is a weighted arithmetic mean associated with the weighting triangle

$$\Delta_* = (w_{in}{}'), \quad w_{1n}{}' = p^{n-1} \quad \text{and} \quad w_{in}{}' = (1-p)\cdot p^{n-i} \quad \text{for} \quad i > 1. \tag{94}$$

Observe that the weighting triangle Δ_* from (94) related to p is a reversed weighting triangle Δ^* given in (93) related to $(1-p)$.

Recall that the weighting triangle Δ^* in (93) (and similarly Δ_* from (94)) is also called the Sierpiński carpet [11,12]. Both two extremal cases, that is, $p = 1$ and $p = 0$, lead to $\mathbf{W}^* = \mathbf{W}_* = \mathbf{P}_F$ and $\mathbf{W}^* = \mathbf{W}_* = \mathbf{P}_L$, respectively. Further generalizations of the Sierpiński carpets are weighting triangles of weighted arithmetic means possessing the self-identity property (48) (in $\triangle_\star$ - case) and weighting triangles of invariant constrained weighted arithmetic means, see (84) (in $\triangle^\star$ - case). Since these weighting triangles are linked by reversion (and by duality of p and $1-p$), we only describe generalizations of $\triangle_\star$ - case.

Let $\mathcal{A} = (a_n) \in [0,1]^{\mathbb{N}}$ be a system of real constants such that $a_1 = 1$. Define a weighting triangle $\triangle_{\mathcal{A}} = (w_{in})$ as follows:

$$w_{in} = a_i \cdot \prod_{j=i+1}^{n} (1 - a_j), \tag{95}$$

where convention $\prod\limits_{j=n+1}^{n} (1 - a_j) = 1$ is applied, see also [12].

Note that weighted means with weights given by (95) are decomposable aggregation operators, see Section 2.11. Note also that no other weighted means possess the decomposability property.

Evidently, if $\forall n \geq 2, a_n = 1 - p$, then $\triangle_{\mathcal{A}} = \triangle_\star$ as given in (94). Recall also that a weighted arithmetic mean $\mathbf{W}$ possesses the self-identity property (48) if and only if the relevant weighting triangle is equal to $\triangle_{\mathcal{A}}$ for some $\mathcal{A}$ as given in (95).

More details about $\triangle_{\mathcal{A}}$ weighting triangles can be found in [11,12], where such weighting triangles are call left–balancing. Among several examples of $\triangle_{\mathcal{A}}$ different from Sierpiński carpet recall the normalized Fibonacci triangle, $w_{in} = \frac{f_i}{f_{n+2}-1}$, where (f_n) is the Fibonacci sequence. Note that all weighted means related to a weighting triangle $\triangle_{\mathcal{A}}$ are monotonic with respect to the α–order and to the β–order, see Section 2.2, but the opposite need not be true.

Completely different approach to generating of weighting triangles was proposed in [139]. This approach is based on so called quantifiers $q : [0,1] \to [0,1]$, which are monotone real functions such that $\{0,1\} \subseteq \operatorname{Ran} q$. Consequently, either a quantifier q is non-decreasing and then necessarily $q(0) = 0$ and $q(1) = 1$, or q is non-increasing and then $q(0) = 1$ and $q(1) = 0$.

Weighting triangle $\triangle_q$ is defined for non-decreasing quantifiers q as follows:

$$w_{in} = q\left(\frac{i}{n}\right) - q\left(\frac{i-1}{n}\right). \tag{96}$$

In the case of non-increasing quantifiers, the weighting triangle $\triangle^q$ is given by

$$w_{in} = q\left(\frac{i-1}{n}\right) - q\left(\frac{i}{n}\right). \tag{97}$$

Observe that the quantifiers approach to generating of weighting triangles is, in fact, related to the Lebesgue integral-based aggregation described in Section 5.1. Observe also that if q is a non-decreasing quantifier, then

$1-q$ is a non-increasing quantifier, and $\Delta_q = \Delta^{1-q}$. So, for example, the arithmetic mean $\mathbf{M}$ is related to the quantifier generated weighting triangle $\Delta_{id} = \Delta^{1-id}$, where $id(x) = x$. On the other hand, two weighting triangles Δ_q and Δ^h are a couple of reversed weighting triangles if and only if $q(x) = h(1-x)$. Combining both mentioned properties, a quantifier generating weighting triangle Δ_q is self-reversed, that is, $w_{in} = w_{(n+1-i)n}$ for all $n \in \mathbb{N}$, $i \in \{1,\ldots,n\}$, if and only if $1 - q(x) = q(1-x)$, that is, $q(x) + q(1-x) = 1$ (for all $x \in [0,1]$). Geometrically, this means that the graph of the quantifier q is symmetric with respect to the point $(0.5, 0.5)$.

Example 8 (i) Define $\mathcal{A} = (a_n)_{n\in\mathbb{N}}$ by $a_n = \frac{1}{n}$ and apply (95). Then $w_{in} = \frac{1}{n}$ for all $n \in \mathbb{N}$, $i \in \{1,\ldots,n\}$, that is, the corresponding weighted arithmetic mean is just the arithmetic mean $\mathbf{M}$.

(ii) Define the quantifier $q : [0,1] \to [0,1]$ by $q(x) = x^k$, $k \in]0,\infty[$. Then, applying (96) and (89), for the corresponding weighted arithmetic mean $\mathbf{W}$ we get

$$\mathbf{W}(x_1,\ldots,x_n) = \sum_{i=1}^{n} \left(\left(\frac{i}{n}\right)^k - \left(\frac{i-1}{n}\right)^k \right) \cdot x_i, \tag{98}$$

compare also (81).

(iii) Two limit cases in (98) we obtain for $k = 0$ and $k = \infty$. In the first case,

$$q(x) = \begin{cases} 0 \text{ if } x = 0, \\ 1 \text{ if } x > 1 \end{cases}$$

and the corresponding weighted mean is $\mathbf{W} = \mathbf{P}_F$. In the second case,

$$q(x) = \begin{cases} 0 \text{ if } x < 1, \\ 1 \text{ if } x = 1 \end{cases}$$

and the corresponding mean is $\mathbf{W} = \mathbf{P}_L$.

(iv) An interesting median–like operator is related to the quantifier

$$q(x) = \begin{cases} 0 & \text{if } x < 0.5, \\ 0.5 & \text{if } x = 0.5, \\ 1 & \text{if } x > 0.5. \end{cases}$$

For the corresponding weighted arithmetic mean $\mathbf{W}$ we have

$$\mathbf{W}(x_1,\ldots,x_n) = \begin{cases} \frac{x_j + x_{j+1}}{2} & \text{if } n = 2j, \\ \\ x_{j+1} & \text{if } n = 2j+1. \end{cases} \tag{99}$$

Among several others properties of weighted means note that a weighted mean $\mathbf{W}$ is both α– and β–monotone, see Section 2.2, if and only if the corresponding weighting triangle Δ verifies for for all $n \in \mathbb{N}$, $i \in \{1,\ldots,n\}$, $w_{i(n+1)} \le w_{in}$.

4.2 OWA operators

Ordered weighted average operators (OWA operators for short) were introduced by Yager [133]. Enormous interest in OWA operators resulted in many papers and even an edited volume [140]. A very simple characterization of OWA operators is their representation as symmetrized weighted means when the symmetrization method described in Section 2.6 is applied.

Definition 18. Let $\mathbf{W} : \bigcup_{n\in\mathbb{N}} [0,1]^n \to [0,1]$ be a weighted mean associated with the weighting triangle $\triangle = (w_{in})$. The operator $\mathbf{W}' : \bigcup_{n\in\mathbb{N}} [0,1]^n \to [0,1]$ given by

$$\mathbf{W}'(x_1,\ldots,x_n) = \sum_{i=1}^{n} w_{in} \cdot x_i', \tag{100}$$

where $(x_1',\ldots,x_n')$ is a non-decreasing permutation of the input n-tuple $(x_1,\ldots,x_n)$ is called an OWA operator associated with $\triangle$ (sometimes also notation $\mathbf{W}'_\triangle$ can be used).

Evidently, each OWA operator is a symmetric continuous idempotent aggregation operator which is linear and comonotone additive. There is a close relationship between OWA operators and the Choquet integral-based operators, see Section 5.2, which were studied, for example, in [46,32,80]. Indeed, each OWA operator can be represented by means of the Choquet integral with respect to an appropriate system of symmetric fuzzy measures. For more details we refer to Section 5.2. Following [5], this means that each symmetric comonotone additive aggregation operator is just an OWA operator.

The only OWA operator $\mathbf{W}'_\triangle$ coinciding with the original weighted mean $\mathbf{W}_\triangle$, i. e., $\mathbf{W}'_\triangle = \mathbf{W}_\triangle$, is the arithmetic mean $\mathbf{M}$. The OWA operator related to the projection $\mathbf{P}_F$ is just **min**, $\mathbf{P}'_F = \mathbf{min}$. Similarly, the symmetrization of the projection $\mathbf{P}_L$ leads to the OWA operator **max**, $\mathbf{P}'_L = \mathbf{max}$. These two last observations stress a close relationship between OWA operators and the projection operators (as an alternative look at OWA operators). Indeed, define a system $(\mathbf{P}_{in} \,|\, n \in \mathbb{N}, i \in \{1,\ldots,n\})$ of projection operators $\mathbf{P}_{in} : [0,1]^n \to [0,1]$, $\mathbf{P}_{in}(x_1,\ldots,x_n) = x_i$. For a given weighting triangle $\triangle = (w_{in})$, the corresponding weighted mean operator $\mathbf{W}_\triangle$ can be viewed as a weighted projection operator,

$$(\mathbf{W}_\triangle)_{(n)} = \sum_{i=1}^{n} w_{in} \cdot \mathbf{P}_{in}. \tag{101}$$

Then the corresponding OWA operator $\mathbf{W}'_\triangle$ can be viewed as a weighted order statistic operator,

$$(\mathbf{W}'_\triangle)_{(n)} = \sum_{i=1}^{n} w_{in} \cdot \mathbf{P}'_{in}, \tag{102}$$

where $\mathbf{P}'_{in} : [0,1]^n \to [0,1]$ is the i th order statistic, i.e., $\mathbf{P}'_{in}(x_1, \ldots, x_n) = x_i'$.

As already observed, $\mathbf{P}'_F = \mathbf{min}$ and $\mathbf{P}'_L = \mathbf{max}$, i. e., **min** is an OWA operator related to the weighting triangle $\triangle_m = (w_{in}), w_{1n} = 1$ and $w_{in} = 0$ for $n \in \mathbb{N}, i \neq 1$; similarly, $\mathbf{max} = \mathbf{W}'_{\triangle_M}$ with $\triangle_M = (w_{in}), w_{nn} = 1$ and $w_{in} = 0$ for $n \in \mathbb{N}$, $i < n$. Evidently, the weighting triangles $\triangle_m$ and $\triangle_M$ form a reversed couple, reflecting the duality of operators **min** and **max**. In general, if $\triangle$ is a weighting triangle and $\triangle^r$ is the corresponding reversed weighting triangle, then the relevant OWA operators $\mathbf{W}'_\triangle$ and $\mathbf{W}'_{\triangle^r}$ form a dual couple of aggregation operators $(\mathbf{W}'_\triangle)^d = \mathbf{W}'_{\triangle^r}$. Though all weighted means are self-dual, $\mathbf{W} = \mathbf{W}^d$, an OWA operator $\mathbf{W}'_\triangle$ is self-dual if and only if the corresponding weighting triangle $\triangle$ is symmetric, i. e., $\triangle = \triangle^r$. An important example of a self-dual OWA operator is the median operator **med**, defined in Section 2.10 by (44), which is assigned to the weighting triangle $\tilde{\triangle} = (w_{in})$,

$$w_{in} = \begin{cases} 1 & \text{if } i = k+1,\ n = 2k+1, \\ 0.5 & \text{if } i = k \ \ or \ \ i = k+1,\ n = 2k, \\ 0 & \text{else.} \end{cases} \tag{103}$$

Observe that the median weighting triangle $\tilde{\triangle}$ can be derived by means of the quantifier q given in Example 8 (iv).

In (93) and (94) we have introduced examples of a recursive construction of weighted means starting from a binary operator (92), or equivalently, of weighting triangles leading to the Sierpiǹski carpets. These types of constructions cannot be applied in the framework of OWA operators if we expect to get as a final result an OWA operator. That is, starting from a binary OWA operator, we cannot construct an n-ary OWA operator, in general. Note that the composition of OWA operators need not be an OWA operator in general [98]. Therefore, up to some exception (as $\mathbf{min}, \mathbf{max}, \mathbf{M}$), when applying an OWA operator $\mathbf{W}'_\triangle$, we need to know the corresponding weighting triangle $\triangle$ (at least the weights $w_{1n}, \ldots, w_{nn}$ related to the number n of input values to be aggregated).

Example 9 Define the weighting triangle $\triangle = (w_{in})$ by $w_{1n} = w_{nn} = 0.5$ and $w_{in} = 0$ for all $n \geq 2$. Then the corresponding OWA operator $\mathbf{W}'_\triangle : \bigcup_{n \in \mathbb{N}} [0,1]^n \to [0,1]$ is self-dual and it is given by

$$\mathbf{W}'_\triangle(x_1, \ldots, x_n) = \frac{1}{2}(min(x_1, \ldots, x_n) + max(x_1, \ldots, x_n)),$$

that is, $\mathbf{W}'_\triangle = \mathbf{M}(\mathbf{min}, \mathbf{max})$, see (79).

However, applying the composition construction (79) to $\mathbf{A} = \mathbf{min}$, $\mathbf{B}_1 = \mathbf{W}'_\triangle$ and $\mathbf{B}_2 = \mathbf{med}$, that is, $\mathbf{C} = \min(\mathbf{W}'_\triangle, \mathbf{med})$, we will not obtain an OWA operator. Indeed, the ternary operator $\mathbf{C}_{(3)} : [0,1]^3 \to [0,1]$ is given by

$$\mathbf{C}_{(3)}(x,y,z) = min\left(\frac{1}{2}(min(x,y,z) + max(x,y,z)), med(x,y,z)\right),$$

which in the case $x \leq y \leq z$ means that

$$\mathbf{C}_{(3)}(x,y,z) = min\left(\frac{1}{2}(x+z),y\right).$$

If $\mathbf{C}$ was an OWA operator, then $\mathbf{C}_{(3)}$ would necessarily be a linear combination of $\mathbf{min}_{(3)}, \mathbf{med}_{(3)}$ and $\mathbf{max}_{(3)}$, which evidently is not true in our case.

Note that so far several parametric characterizations of OWA operators have been introduced, which help the user to make his best choice.

The basic characterization is the "orness" measure m_{or} of weighting vectors $\mathbf{w}_n = (w_{1n}, \ldots, w_{nn}), n \geq 2$ [133],

$$m_{or}(w_{1n}, \ldots, w_{nn}) = \sum_{i=1}^{n} w_{in} \cdot \frac{i-1}{n-1} \tag{104}$$

Since the knowledge of a weighting triangle is equivalent to the knowledge of an OWA operator, the "orness" measure of an OWA operator is denoted by $m_{or}\left(\mathbf{W}'_{(n)}\right)$.

Observe that $m_{or}(\mathbf{W}'_{(n)}) = 0$ if and only if $\mathbf{W}'_{(n)} = \mathbf{min}_{(n)}$ and $m_{or}(\mathbf{W}'_{(n)}) = 1$ if and only if $\mathbf{W}'_{(n)} = \mathbf{max}_{(n)}$. This motivated the name *measure of orness*, coming from modelling the operator OR by **max** in fuzzy logic. Note also that $m_{or}(\mathbf{W}'_{(n)}) = \frac{1}{2}$ for all $n \geq 2$ whenever the operator $\mathbf{W}'$ is self-dual, that is, when the corresponding weighting triangle $\triangle$ is self-reversible. However, this means that the measure of "orness" is not enough to determine an OWA operator $\mathbf{W}'$ (or its n-ary restriction $\mathbf{W}'_{(n)}$), in general. Therefore other parameters for OWA operators (or for the relevant weighting triangles) were also proposed. Having in mind that each weighting vector $\mathbf{w}_n = (w_{1n}, \ldots, w_{nn})$ can be viewed also as a discrete probability distribution, the Shannon entropy H [115] was proposed for the characterization of OWA operators

$$H(w_{1n}, \ldots, w_{nn}) = -\sum_{i=1}^{n} w_{in} \cdot \log w_{in}, \tag{105}$$

with convention $0 \cdot \log 0 = 0$. Recall that for a fixed n, $H(w_{1n}, \ldots, w_{nn}) = H(\mathbf{W}'_{(n)}) \in [0, \log n]$, and that $H(\mathbf{W}'_{(n)}) = \log n$ if and only if $\mathbf{W}'_{(n)} = \mathbf{M}_{(n)}$. Similarly, $H(\mathbf{W}'_{(n)}) = 0$ if and only if $\mathbf{W}'_{(n)} = \mathbf{P}'_{in}$ for some $i \in \{1, \ldots, n\}$, i. e., if $\mathbf{W}'_{(n)}$ is some order statistic.

For practical purposes, the (n-ary) OWA operators with given "orness" measure $m_{or}(\mathbf{W}'_{(n)}) = \alpha \in]0,1[$ and with maximal (minimal) entropy can be found, mostly by means of non-linear programming. For illustration observe that if we fix $m_{or} = \alpha = 0.5$, then the maximal entropy requirement leads to the unique solution $\mathbf{W}' = \mathbf{M}$ (the arithmetic mean), while the minimal entropy forces the use of the unique OWA operator $\mathbf{W}' = \mathbf{med}$ (for odd number of inputs $n = 2k+1$).

Recall again the specific α– and β–monotonicity of aggregation operators introduced in Section 2.2. Following [86], an OWA operator $\mathbf{W}'$ related to a weighting triangle $\triangle = (w_{in})$ is both α– and β–monotone if and only if for all $n \in \mathbb{N}$, $p \in \{1, \ldots, n\}$,

$$\sum_{i=1}^{p} w_{i(n+1)} \leq \sum_{i=1}^{p} w_{in} \leq \sum_{i=1}^{p+1} w_{i(n+1)}. \tag{106}$$

As an interesting example of a weighting triangle $\triangle = (w_{in})$ fulfilling (106) recall the normalized Pascal triangle, i.e., $w_{in} = \frac{\binom{n}{i}}{2^n}$.

Note that recently Torra [123] introduced so called WOWA operators, which are a common generalization of OWA operators and weighted means. More details can also be found in Section 4.2 or in the chapter of this book written by Torra and Godo [124].

4.3 Quasi-arithmetic means

An important class of aggregation operators is derived from the arithmetic mean $\mathbf{M}$ (acting on an interval $[a, b] \subset [-\infty, \infty]$) by transformation given in Proposition 2, see (63).

Definition 19. Let $f : [0,1] \to [-\infty, \infty]$ be a continuous strictly monotone function. The aggregation operator $\mathbf{M}_f : \bigcup_{n \in \mathbb{N}} [0,1]^n \to [0,1]$ given by

$$\mathbf{M}_f(x_1, \ldots, x_n) = f^{-1}(\mathbf{M}(f(x_1), \ldots, f(x_n)) = f^{-1}\left(\frac{1}{n} \sum_{i=1}^{n} f(x_i)\right) \tag{107}$$

is called a quasi-arithmetic mean.

Note that if necessary, the convention $+\infty + (-\infty) = -\infty$ is applied. Observe that each quasi-arithmetic mean is symmetric, idempotent, decomposable and bisymmetric. It is continuous and strictly monotone (cancellative) on $\bigcup_{n \in \mathbb{N}}]0,1[^n$. The continuity of a quasi-arithmetic mean $\mathbf{M}_f$ is equivalent with $Ran\, f \neq [-\infty, \infty]$, that is, $\mathbf{M}_f$ is not continuous only if $Ran\, f = [-\infty, \infty]$ as the consequence of the non-continuity of the arithmetic mean $\mathbf{M}$ acting on $[-\infty, \infty]$. Obviously, the non-continuity may appear only for inputs simultaneously containing the values 0 and 1. Quasi-arithmetic means do not have neutral element. However, they may possess an annihilator as well as a weak annihilator.

Example 10 (i) A distinguished class of quasi-arithmetic means are so called root–power operators $\mathbf{M}_p : \bigcup_{n\in\mathbb{N}} [0,1]^n \to [0,1]$ generated for $p \in]-\infty, 0[\cup]0, \infty[$ by the function $f_p : [0,1] \to [-\infty, \infty]$, $f_p(x) = x^p$, that is,

$$\mathbf{M}_p(x_1, \ldots, x_n) = \left(\frac{1}{n} \sum_{i=1}^{n} x_i{}^p \right)^{\frac{1}{p}}. \tag{108}$$

Distinguished root–power operators are $\mathbf{M}_1 = \mathbf{M}$, $\mathbf{M}_2$ (the quadratic mean), $\mathbf{M}_{-1} = \mathbf{H}$ (the harmonic mean). Family $(\mathbf{M}_p)_{p\in]-\infty,0[\cup]0,\infty[}$ is non-decreasing family of aggregation operators with limit members $\mathbf{M}_0 = \mathbf{G} = \mathbf{M}_{\log x}$, i.e., $\mathbf{M}_0$ is the geometric mean, and $\mathbf{M}_\infty = \mathbf{max}$ and $\mathbf{M}_{-\infty} = \mathbf{min}$. Observe that **max** and **min** are not quasi-arithmetic means.

(ii) Define $f : [0,1] \to [-\infty, \infty]$ by $f(x) = \log \frac{x}{1-x}$. Then the corresponding quasi–arithmetic mean $\mathbf{M}_f$ is given by

$$\mathbf{M}_f(x_1, \ldots, x_n) = \begin{cases} \dfrac{\sqrt[n]{\prod_{i=1}^{n} x_i}}{\sqrt[n]{\prod_{i=1}^{n} x_i} + \sqrt[n]{\prod_{i=1}^{n} (1-x_i)}} & \text{if } \{0,1\} \not\subset \{x_1, \ldots, x_n\}, \\ 0 & \text{else,} \end{cases} \tag{109}$$

that is, $\mathbf{M}_f = \dfrac{\mathbf{G}}{\mathbf{G} - \mathbf{G}^d + 1}$ with convention $\frac{0}{0} = 0$.

Observe that the $3 - \Pi$–operator $\mathbf{E}$ given in (39) has a form similar to (109) because of its relationship with the generating function f, see Section 6.2 or 7.1. Moreover, the idempotization method described in Proposition 1 applied to $\mathbf{E}$ leads just to $\mathbf{M}_f$, i.e., $\mathbf{M}_f = \mathbf{I}_\mathbf{E}$, see (61).

Note that root–power operators were intensively studied in [29]. For $p \in]0, \infty[$, they are strictly monotone and continuous, with no annihilator. For $p \in]-\infty, 0[$, including the limit member $p = 0$, they are continuous but strictly monotone only on $\bigcup_{n\in\mathbb{N}}]0,1]^n$, and they possess an annihilator $a = 0$ (but no weak annihilator). The quasi-arithmetic mean $\mathbf{M}_f$ given by (109) is not continuous, it is strictly monotone on $\bigcup_{n\in\mathbb{N}}]0,1[^n$ and it has annihilator $a = 0$ and weak annihilator $b = 1$.

Recall some general properties of quasi-arithmetic means. First of all, $\mathbf{M}_f = \mathbf{M}_{uf+v}$ for any $u, v \in \mathbb{R}, u \neq 0$ (the only exception is if $Ran\, f = [-\infty, \infty]$ and $u < 0$, when the operators $\mathbf{M}_f$ and $\mathbf{M}_{uf+v}$ differs when aggregating inputs $(x_1, \ldots, x_n)$ containing simultaneously 0 and 1). So, e. g., root–power operators can also be generated by $g_p : [0,1] \to [-\infty, \infty]$, $g_p(x) = \frac{x^p - 1}{p}$. Observe here the relationship of root–power operators and the

Schweizer–Sklar family of t-norms [61,114] possessing additive generators g_p, see Section 6.1. Recall that

$$\lim_{p\to 0} g_p(x) = \lim_{p\to 0} \frac{x^p - 1}{p} = \log x\,,$$

and that

$$\lim_{p\to 0} \mathbf{M}_p = \mathbf{M}_0 = \mathbf{M}_{\log x},$$

as a consequence of another general property of quasi-arithmetic means. Namely, having a family $(\mathbf{M}_{f_n})_{n\in\mathbb{N}}$ of quasi-arithmetic means such that

$$\lim_{n\to\infty} f_n = f$$

and $\mathbf{M}_f$ is a quasi-arithmetic mean, then

$$\lim_{n\to\infty} \mathbf{M}_{f_n} = \mathbf{M}_f.$$

Vice–versa, if

$$\lim_{n\to\infty} \mathbf{M}_{f_n} = \mathbf{M}_f$$

is a quasi-arithmetic mean generated by f, then there are real constants $u_n, v_n, n \in \mathbb{N}$ such that

$$\lim_{n\to\infty} (u_n f_n + v_n) = f.$$

Several related limit properties of quasi-arithmetic means can be found in [67]. Interesting is also the comparison of quasi-arithmetic means. Following [65] for two quasi-arithmetic means $\mathbf{M}_f$ and $\mathbf{M}_g$ we have $\mathbf{M}_f \le \mathbf{M}_g$ if and only if either the composite function $f \circ g^{-1}$ is convex and f is decreasing, or $f \circ g^{-1}$ is concave and f is increasing.

Among several characterization results leading to the quasi-arithmetic means, recall the older result of Kolmogorov [71] and Nagumo [106], see also [34,3].

Proposition 8. *An aggregation operator* $\mathbf{A} : \bigcup_{n\in\mathbb{N}} [0,1]^n \to [0,1]$ *is a continuous, symmetric, idempotent, decomposable and strictly monotone operator if and only if there is a monotone bijection* $f : [0,1] \to [0,1]$ *such that* $\mathbf{A} = \mathbf{M}_f$.

For some other characterization results related to quasi-arithmetic means we refer to [3,34]. Observe that the decomposability in the previous proposition can be replaced by the symmetric, see [2]. Moreover, if we omit the strict monotonicity of $\mathbf{A}$, the the class of non-strict means characterized by Fodor and Marichal [33] is recovered. Omitting the details given in [33], note that non–strict means are special ordinal sums of quasi–arithmetic means with annihilator and min and max in the form (77), see Section 3.4.

The class of quasi-arithmetic means is closed under duality. Indeed, for any quasi-arithmetic mean $\mathbf{M}_f$, the corresponding dual aggregation operator $\mathbf{M}_f^d$ is generated by the function $h : [0,1] \to [-\infty, \infty]$, $h(x) = f(1-x)$, that is, $\mathbf{M}_f^d = \mathbf{M}_h$. So, e. g., the dual to the geometric mean $\mathbf{G}^d$ is generated by $\log(1-x)$, $\mathbf{G}^d = \mathbf{M}_{\log(1-x)}$, and obviously, $\mathbf{G}^d$ has annihilator $a = 1$ and

$$\mathbf{G}^d(x_1, \ldots, x_n) = 1 - \left(\prod_{i=1}^{n} (1 - x_i) \right)^{\frac{1}{n}} .$$

Also in the class of quasi-arithmetic means some parameters helping the user to choose an appropriate operator were introduced. We only recall here the "orness" measure m_{or} introduced by Dujmovic [29]. We give m_{or} measure in an explicit form which can be derived from the Dujmovic proposal (which originally concerns root–power operators only), and which in the context of idempotent operators was proposed and studied by Kolesárová [68].

Let $\mathbf{M}_f : \bigcup_{n \in \mathbb{N}} [0,1]^n \to [0,1]$ be a quasi-arithmetic mean. For its n-ary restriction $(\mathbf{M}_f)_{(n)}, n \geq 2$, we put

$$m_{or}\left((\mathbf{M}_f)_{(n)}\right) = \frac{(n+1) \int_0^1 \ldots \int_0^1 \mathbf{M}_f(x_1, \ldots, x_n)\, dx_1 \ldots dx_n - 1}{n-1}. \qquad (110)$$

Observe that (110) can also be applied to aggregation operators **max** and **min**, and that

$$m_{or}(\mathbf{max}_{(n)}) = 1 , \qquad \text{while} \qquad m_{or}(\mathbf{min}_{(n)}) = 0,$$

compare with m_{or} introduced for OWA operators in (104). Consequently, for any quasi-arithmetic mean $\mathbf{M}_f$ and $n \geq 2$ we have $m_{or}\left((\mathbf{M}_f)_{(n)}\right) \in]0,1[$. Note also that $m_{or}(\mathbf{M}_{(n)}) = 0.5$ for all $n \geq 2$, i.e., the "orness" measures m_{or} introduced in (104) and in (110) coincide (observe that the arithmetic mean $\mathbf{M}$ is unique OWA operator which belongs also to the class of quasi-arithmetic means, while OWA operators **min** and **max** are limit members of the class of quasi-arithmetic means). Moreover, formula (110) can be applied to any (continuous, symmetric) idempotent aggregation operator. By direct computation of $m_{or}(\mathbf{W}'_{(n)})$ by means of (110) we obtain (104), hence proving the compatibility of the Dujmovic approach to "orness" measure for root–power operators and the Yager approach to m_{or} for OWA operators. The "orness" degree of the geometric mean $\mathbf{G}$ (depending on the number n of inputs) is

$$m_{or}(\mathbf{G}_{(n)}) = \frac{n^n - (n+1)^{n-1}}{(n-1)\,(n+1)^{n-1}},$$

that is, for large n, $m_{or}(\mathbf{G}_{(n)}) \approx e^{-1} = 0.36788$.

Note that if we fix $m_{or} = \alpha \in [0,1]$ and $n \geq 2$, then there is a unique root–power operator $\mathbf{M}_p$ such that $m_{or}\left((\mathbf{M}_p)_{(n)}\right) = \alpha$, taking into account also limit members.

4.4 Weighted quasi-arithmetic means

In Section 4.1, we have discussed weighted means as a generalization of the arithmetic mean expressing possibly different weights (importances) of single input values. Another type of generalization of the arithmetic mean based on its transformation leads to the class of quasi-arithmetic means, see Section 4.4.

Now, we can combine both approaches (independently of the order) to obtain the class of weighted quasi-arithmetic means. Clearly, the classes of weighted means and of quasi–arithmetic means are then proper subsets of the class of weighted quasi–arithmetic mean. On the other hand, an appropriate choice of a weighted quasi–arithmetic mean to aggregate the data in some real situation can be rather complicated task.

Definition 20. Let $\triangle = (w_{in})$ be a given weighting triangle and let $f : [0,1] \to [-\infty, \infty]$ be a continuous strictly monotone function. An aggregation operator $\mathbf{W}_{\triangle,f} : \bigcup_{n\in\mathbb{N}} [0,1]^n \to [0,1]$ (or $\mathbf{W}_f$ for short) given by

$$\mathbf{W}_{\triangle,f}(x_1,\ldots,x_n) = f^{-1}\left(\sum_{i=1}^{n} w_{in}\, f(x_i)\right) \tag{111}$$

is called a weighted quasi-arithmetic mean.

Each weighted quasi–arithmetic mean is an idempotent bisymmetric aggregation operator. It is continuous whenever *Ran* $f \neq [-\infty, \infty]$. If all weights w_{in} are positive and *Ran* $f \subset \mathbb{R}$ then $\mathbf{W}_f$ is also strictly monotone (cancellative). Similarly to decomposability, which is a genuine property of quasi–arithmetic means, the bisymmetry of weighted quasi–arithmetic means is an important feature allowing to introduce several characterization results. The next one is due to Aczél [2,34].

Proposition 9. *An aggregation operator* $\mathbf{A} : \bigcup_{n\in\mathbb{N}} [0,1]^n \to [0,1]$ *is a continuous, strictly monotone idempotent and bisymmetric operator if and only if there is a weighting triangle* $\triangle = (w_{in})$ *with all positive weights* $w_{in}, n \in \mathbb{N}, i \in \{1,\ldots,n\}$ *and a monotone bijection* $f : [0,1] \to [0,1]$ *so that* $\mathbf{A}$ *is the weighted quasi-arithmetic mean,* $\mathbf{A} = \mathbf{W}_{\triangle,f}$.

In several applications, especially in economics, the weighted geometric mean $\mathbf{G}_\triangle$ is applied,

$$\mathbf{G}_\triangle(x_1,\ldots,x_n) = \prod_{i=1}^{n} x_i^{w_{in}} \tag{112}$$

Weighted geometric mean $\mathbf{G}_{\triangle}$ has annihilator $a = 0$ whenever all weights w_{in} are positive. Also other properties of quasi-arithmetic means are inherited by weighted quasi-arithmetic means, including the limit properties, comparison, etc. Moreover, we can introduce also the "orness" measure m_{or} in the class of weighted quasi-arithmetic means by means of (110). Take, e. g., the weighted geometric mean $\mathbf{G}_{\triangle}$ obtained by the idempotization of the aggregation operator $\mathbf{Q}$ from (30), which was discussed in Section 3.1,

$$\mathbf{G}_{\triangle}(x_1, \ldots, x_n) = \prod_{i=1}^{n} x_i^{\frac{2i}{n\,(n+1)}},$$

the corresponding "orness" measure for $n \geq 2$ is

$$m_{or}\left((\mathbf{G}_{\triangle})_{(n)}\right) = \frac{\frac{n+1}{\prod_{i=1}^{n}\left(1+\frac{2i}{n\,(n+1)}\right)} - 1}{n-1},$$

i. e., $m_{or}\left((\mathbf{G}_{\triangle})_{(2)}\right) = \frac{7}{20} = 0.35$, $m_{or}\left((\mathbf{G}_{\triangle})_{(3)}\right) = \frac{5}{14} = 0.357$, etc. For $n \to \infty$, $m_{or}\left((G_{\triangle})\right)_{(n)} \to e^{-1} = 0.36788$.

4.5 Ordered weighted quasi-arithmetic means (OWQA operators)

Similarly to the previous section, where by means of weights and transformations applied to the arithmetic mean, the class of weighted arithmetic means was obtained, this section combines the OWA operators and the transformation method to obtain OWQA operators (ordered weighted quasi-averages), or, equivalently, ordered weighted quasi-arithmetic means. The same class can be obtained applying the symmetrization method, see Section 2.6, to the weighted quasi-arithmetic means.

Definition 21. Let $\triangle = (w_{in})$ be a given weighting triangle and let $f : [0,1] \to [-\infty, \infty]$ be a continuous strictly monotone function. An aggregation operator $\mathbf{W}'_{\triangle,f} : \bigcup_{n\in\mathbb{N}} [0,1]^n \to [0,1]$ (or $\mathbf{W}'_f$ for short) given by

$$\mathbf{W}'_{\triangle,f}(x_1, \ldots, x_n) = f^{-1}\left(\sum_{i=1}^{n} w_{in}\, f(x'_i)\right), \qquad (113)$$

where $(x'_1, \ldots, x'_n)$ is a non-decreasing permutation of the input n-tuple $(x_1, \ldots, x_n)$ is called an ordered weighted quasi-arithmetic mean.

OWQA operators are symmetric, idempotent, and continuous whenever $\mathit{Ran}\, f \neq [-\infty, \infty]$. Similarly to the case of weighted quasi-arithmetic means, it seems to be a difficult task to choose appropriately a convenient operator from this huge class for a real application without any specific information.

However, in some fields, some special types of OWQA operators related to a special function f are usually applied. This is, for example, the case of economical rate problems related to the geometric mean, that is, to the function $f(x) = \log x$.

Ordered weighted geometric mean was discussed and applied for example in [18]. Recall that for a given weighting triangle $\Delta = (w_{in})$, the corresponding ordered weighted geometric mean $\mathbf{G}'_\Delta : \bigcup_{n\in\mathbb{N}} [0,1]^n \to [0,1]$ is given by

$$\mathbf{G}'_\Delta(x_1,\ldots,x_n) = \prod_{i=1}^{n} ({x_i}')^{w_{in}} \tag{114}$$

To help the user, again the "orness" measure m_{or} can also be introduced in the class of ordered weighted quasi–arithmetic means, applying the formula (110). So, e. g., let $\Delta = (w_{in})$ be given, i. e.,

$$\mathbf{G}'_\Delta(x,y) = (min(x,y))^{w_{12}} \cdot (max(x,y))^{w_{22}}\,.$$

Then

$$m_{or}\left((\mathbf{G}'_\Delta)_{(2)}\right) = \frac{w_{22}}{1+w_{12}} = \frac{1-w_{12}}{1+w_{12}}.$$

Hence if $w_{12} = \frac{1}{3}$ (and therefore $w_{22} = \frac{2}{3}$), $m_{or}\left(({\mathbf{G}'}_\Delta)_{(2)}\right) = 0.5$. Oppositely, if we have chosen $m_{or}\left((\mathbf{G}'_\Delta)_{(2)}\right) = \alpha \in [0,1]$, then we know that the relevant weights should be $w_{12} = \frac{1-\alpha}{1+\alpha}$ and $w_{22} = \frac{2\alpha}{1+\alpha}$.

5 Aggregation operators based on integrals

Integration is one of the oldest classical aggregation methods. For a fixed number of inputs corresponding to an abstract set $\mathbb{X}$, the input vectors correspond to real functions defined on $\mathbb{X}$. If the number of inputs is $n \in \mathbb{N}$, we will always suppose, without any loss of generality, that $\mathbb{X} = \mathbb{X}_n = \{1,\ldots,n\}$. Then the input vector $(x_1,\ldots,x_n) \in [0,1]^n$ (or generally, from $[a,b]^n$) can be understood as a function $\mathbf{x} : \mathbb{X}_n \to [0,1]$, $\mathbf{x}(i) = x_i$, $i \in \mathbb{X}$ (or $\mathbf{x} : \mathbb{X}_n \to [a,b]$). Since this overview deals with aggregation of finitely many inputs only, we do not need discuss any measurability aspects. Each integration method on $\mathbb{X}_n$ is based on some set function $\mathbf{m} : \mathcal{P}(\mathbb{X}_n) \to [0,1]$ (or $\mathbf{m} : \mathcal{P}(\mathbb{X}_n) \to [a,b]$). This set function $\mathbf{m}$ can be understood as a global weight of the relevant set of criteria. Therefore, $\mathbf{m}(\emptyset) = 0$ and $\mathbf{m}(\mathbb{X}_n) = 1$ are supposed to be granted for any such set function $\mathbf{m}$. Moreover, each function $\mathbf{m}$ should satisfy a genuine requirement: $\mathbf{m}(I) \le \mathbf{m}(J)$ whenever $I \subset J \subseteq \mathbb{X}_n$, which means that a greater set of criteria J cannot have the weight which is less than the weight of a smaller set of criteria I. Set functions with these properties are known under several names. For example, these functions are called premeasures in [122], but the most popular seems to be the name *fuzzy measures* introduced by Sugeno [119], and therefore in what follows we adopt this terminology.

5.1 Lebesgue integral–based aggregation

Lebesgue integral on an abstract set $\mathbb{X}$ is always related to some (σ)–additive measure $\mathbf{m}$ defined on a measurable space $(\mathbb{X}, \mathcal{A})$, where $\mathcal{A}$ is a σ–algebra of subsets of $\mathbb{X}$. More details about the Lebesgue integral can be found in [55]. Recall that each additive measure $\mathbf{m}$ defined on $(\mathbb{X}_n, \mathcal{P}(\mathbb{X}_n))$ which is also a fuzzy measure, is necessarily a discrete probability measure on $\mathbb{X}_n$. This means that each additive fuzzy measure $\mathbf{m}$ defined on $\mathbb{X}_n$ has a unique representation in the form

$$\mathbf{m}(I) = \sum_{i \in I} w_i \,, \tag{115}$$

where $w_i = \mathbf{m}(\{i\})$, $i \in \mathbb{X}_n$. For each function $\mathbf{x} : \mathbb{X}_n \to [0,1]$ the Lebesgue integral with respect to $\mathbf{m}$ is given by

$$\int\limits_{\mathbb{X}_n} \mathbf{x} d\mathbf{m} = \sum_{i=1}^{n} w_i x_i \,. \tag{116}$$

Obviously, this means that the Lebesgue integral–based aggregation coincides with the weighted means aggregation, see Section 4.1, and hence it has also the same properties. Recall that the additivity of the applied fuzzy measure $\mathbf{m}$ reflects in multi–criteria decision making the non–interactivity of single criteria. Consequently, the Lebesgue integral–based aggregation (or equivalently, the weighted mean aggregation), models the global evaluation of single objects using non–interactive criteria with possibly different weights (importances).

What is the Lebesgue integral–based aggregation good for, when the weighted means can do the same job? First of all, it offers another point of view how to obtain an appropriate aggregation thinking on the properties of underlying probability measures on $\mathbb{X}_n$ only. An interesting example of such approach is related to the balanced measures introduced in [8]. Note that a balanced fuzzy measure $\mathbf{m}$ on $\mathbb{X}_n$ is characterized by the inequality

$$\mathbf{m}(I) \leq \mathbf{m}(J) \quad \text{whenever} \quad \operatorname{card} I < \operatorname{card} J \,. \tag{117}$$

For $n = 3$ (117) means that all weights w_1, w_2, $w_3 \in [0, \frac{1}{2}]$ and the corresponding ternary aggregation operator $\mathbf{A}$ fulfills

$$\frac{\mathbf{min} + \mathbf{med}}{2} \leq \mathbf{A} \leq \frac{\mathbf{med} + \mathbf{max}}{2} \,.$$

For a global weighted mean $\mathbf{W} : \bigcup\limits_{n \in \mathbb{N}} [0,1]^n \to [0,1]$, the crucial point is the choice of the relevant weighting triangle. This choice can be simplified by the Lebesgue integral.

Proposition 10. *Let* $\mathbf{p}$ *be a given probability measure on the space* $([0,1[, \mathcal{B}([0,1[))$*, where* $\mathcal{B}([0,1[)$ *is the* σ*-algebra of all Borel subsets of* $[0,1[$*. Define the operator* $\mathbf{W_p} : \bigcup_{n\in\mathbb{N}} [0,1]^n \to [0,1]$ *by*

$$\mathbf{W_p}(x_1,\ldots,x_n) = \int_{[0,1[} h d\mathbf{p}\,, \tag{118}$$

where $h : [0,1[\to [0,1]$ *is given by* $h(u) = x_i$ *whenever* $i-1 \le nu < i$*,* $i \in \{1,\ldots,n\}$*, and the right-hand side of (118) is the Lebesgue integral. Then* $\mathbf{W_p}$ *is a weighted mean assigned to a weighting triangle* Δ_q *defined by a non-decreasing quantifier* $q : [0,1] \to [0,1]$*, see (96), where* $q(u) = \mathbf{p}([0,u[)$ *is the distribution function related to* $\mathbf{p}$.

Proof. By a direct application of (118) we obtain

$$\mathbf{W_p}(x_1,\ldots,x_n) = \sum_{i=1}^{n} \int_{[\frac{i-1}{n},\frac{i}{n}[} x_i d\mathbf{p} = \sum_{i=1}^{n} x_i \mathbf{p}\left([\frac{i-1}{n},\frac{i}{n}[\right)$$

$$= \sum_{i=1}^{n} \left(q\left(\frac{i}{n}\right) - q\left(\frac{i-1}{n}\right)\right) x_i\,. \quad \square$$

Proposition 10 shows that the quantifier approach to the construction of weighting triangles $\Delta_q = (w_{in})$ as proposed by Yager [133], see also Section 4.1, is a straightforward application of the Lebesgue integral–based approach to aggregation.

Remark 1 Observe that the approach shown in Proposition 10 allows to define aggregation operators also for infinitely many inputs, see also [44,43]. For example, the standard mean value

$$\mathbf{W_p}(h) = \int_{[0,1[} h d\mathbf{p}$$

can be seen as an aggregation of inputs $(h(x))_{x\in[0,1[}$ (supposing the Borel measurability of h). Thus $\mathbf{W_p}$ can be seen as a generalized weighted mean related to a given fixed probability measure $\mathbf{p}$. The role of the arithmetic mean is here played by the uniform probability $\mathbf{p} = \lambda$, where λ is the Lebesgue measure, and for a piecewise continuous h we have

$$\mathbf{W}_\lambda(h) = \int_0^1 h(x)dx\,,$$

where the integral on right–hand side is the Riemann integral. Observe that applying (118) to $\mathbf{p} = \lambda$, we really obtain $\mathbf{W}_\lambda = \mathbf{M}$.

An interesting and till now not completely examined approach to the aggregation based on the Lebesgue integral is related to the Cartesian product–based aggregation introduced in Definition 1 Observe that for any aggregation operator $\mathbf{A} : \bigcup_{n\in\mathbb{N}} [0,1]^n \to [0,1]$, fixing integers $n, k \in \mathbb{N}$, we can introduce an n–ary aggregation operator $\mathbf{K} : [0,1]^n \to [0,1]$ by

$$\mathbf{K}(x_1,\dots,x_n) = \int_{\mathbb{X}_n^k} h\, d\mathbf{m}_k\,, \tag{119}$$

where $\mathbf{m}_k$ is a probability measure on the product space $\mathbb{X}_n^k$ and $h : \mathbb{X}_n^k \to [0,1]$ is given by $h(i_1,\dots,i_k) = \mathbf{A}(x_{i_1},\dots,x_{i_k})$. Obviously, if $\mathbf{A}$ is continuous (idempotent) then also $\mathbf{K}$ is continuous (idempotent). For more details we refer to [92].

Example 11 (i) Let $\mathbf{A} = \mathbf{min}$ and let, e.g., $\mathbf{m}_k$ be a uniform probability measure on $\mathbb{X}_n^k$, that is,

$$\mathbf{m}_k(I) = \frac{\operatorname{card} I}{\operatorname{card} \mathbb{X}_n^k} = \frac{\operatorname{card} I}{n^k} \quad \text{for any} \quad I \subseteq \mathbb{X}_n^k\,.$$

Then (119) gives

$$\mathbf{K}(x_1,\dots,x_n) = \sum_{i=1}^{n} \left(\left(\frac{i}{n}\right)^k - \left(\frac{i-1}{n}\right)^k \right) x_i'\,,$$

where $(x_1',\dots,x_n')$ is a non–decreasing permutation of $(x_1,\dots,x_n)$. Observe that $\mathbf{K}$ is an OWA operator, in its global form related to the weighting triangle Δ_q generated by the quantifier $q : [0,1] \to [0,1]$, $q(u) = u^k$, compare also (81).

(ii) Repeating (i) with $\mathbf{A} = \mathbf{max}$, we obtain

$$\mathbf{K}(x_1,\dots,x_n) = \sum_{i=1}^{n} \left(\left(\frac{n+1-i}{n}\right)^k - \left(\frac{n-i}{n}\right)^k \right) x_i'\,,$$

that is, $\mathbf{K}$ is an OWA operator related to the weighting triangle Δ_q, $q(u) = (1-u)^k$, see also (82).

Observe that operators $\mathbf{K}$ described in Example 11, in (i) and (ii) respectively, are dual operators. In general, if for the operator $\mathbf{A}$ we obtain by (119) an operator $\mathbf{K}$, then for the dual operator $\mathbf{A}^d$ we obtain by (119) the dual operator $\mathbf{K}^d$. Also note that the construction (119) based on the operator $\mathbf{A} = \mathbf{min}$ always leads to the Choquet integral aggregation with respect to the fuzzy measure $\mathbf{m}$, $\mathbf{m}(I) = \mathbf{m}_k(I^k)$, see [92] or Section 5.2.

5.2 Choquet integral–based aggregation

A general integral with respect to fuzzy measures was introduced by G. Choquet [20]. Originally, a special subclass of fuzzy measures (so called capacities) was considered. The basic idea of the Choquet integral is very simple. Each function h to be integrated can be identified with a chain of subsets $\mathcal{H} = (\{x \in \mathbb{X} \mid h(x) \geq u\})_{u \in \mathbb{R}}$ and for a (σ–additive) measure $\mathbf{m}$ the Lebesgue integral $\int_{\mathbb{X}} h \, d\mathbf{m}$ is then computed by means of the real chain $\mathcal{M}_h = (\mathbf{m}(\{x \in \mathbb{X} \mid h(x) \geq u\}))_{u \in \mathbb{R}}$ only. However, the additivity of $\mathbf{m}$ cannot be seen from the chain $\mathcal{M}_h$. Taking into account any (continuous from below) measure $\mathbf{m}$, the chain $\mathcal{M}_h$ can be applied to compute the Lebesgue integral $\int_{\mathbb{X}} h \, d\mathbf{m}_h$, where $\mathbf{m}_h$ is a σ–additive measure coinciding on the chain of subsets $\mathcal{H}$ with the original fuzzy measure $\mathbf{m}$. For more details we refer to [20,23,109,129]. A closely related integration concept, independent of the Lebesgue integral, was developed by Šipoš [122], see also [23,109]. Both concepts coincide for inputs from $[0,1]$ (generally for inputs from $[0,\infty]$). Further, already mentioned relationship of the Choquet and Lebesgue integrals means that aggregation based on the Choquet integral leads to a weighted means aggregation in which the weights depend on the order of single inputs. For the convenience of the reader we repeat the definition of the Choquet integral on an abstract space $\mathbb{X}$.

Definition 22. Let $(\mathbb{X}, \mathcal{A})$ be a measurable space and let $\mathbf{m} : \mathcal{A} \to [0,1]$ be a fuzzy measure. Let $h : \mathbb{X} \to [0,1]$ be an $\mathcal{A}$–measurable function. Then the Choquet integral of h with respect to $\mathbf{m}$, with the notation $(C) - \int_{\mathbb{X}} h \, d\mathbf{m}$, is defined by

$$(C) - \int_{\mathbb{X}} h \, d\mathbf{m} = \int_0^1 \mathbf{m}(\{x \in \mathbb{X} \mid h(x) \geq u\}) \, du \,, \tag{120}$$

where the right–hand side integral is the standard Riemann integral.

Recent results related to the Choquet integral can be found in [52]. Now, we recall the Choquet integral–based aggregation of n–inputs.

Proposition 11. *Let* $\mathbf{m}$ *be a fuzzy measure defined on* $\mathbb{X}_n$*, that is* $\mathbf{m} : \mathcal{P}(\mathbb{X}_n) \to [0,1]$. *Define the* n*–ary operator* $\mathbf{Ch} : [0,1]^n \to [0,1]$ *by*

$$\mathbf{Ch}(x_1, \ldots, x_n) = (C) - \int_{\mathbb{X}_n} h \, d\mathbf{m} \,, \tag{121}$$

where $h : \mathbb{X}_n \to [0,1]$ *is given by* $h(i) = x_i$. *Then* $\mathbf{Ch}$ *is an idempotent continuous linear and comonotone additive aggregation operator.*

For the proof we refer the reader to [5]. Note that by [5] each comonotone additive aggregation operator $\mathbf{A}_{(n)}$ can be always represented in the form of the Choquet integral (121) with respect to the fuzzy measure $\mathbf{m}:\mathcal{P}(\mathbb{X}_n)\to[0,1]$,

$$\mathbf{m}(I)=\mathbf{A}_{(n)}(\mathbf{1}_I(1),\ldots,\mathbf{1}_I(n)),$$

where

$$\mathbf{1}_I(i)=\begin{cases}1 \text{ if } i\in I,\\ 0 \text{ else}.\end{cases}$$

Another axiomatic characterization of the Choquet integral–based aggregation was shown by Marichal in [79,80].

There are several equivalent formulae describing the operator **Ch** given in (121), see, e.g., [53,24,105] or the chapter of Marichal in this volume [81].

$$\mathbf{Ch}(x_1,\ldots,x_n)=\sum_{i=1}^{n}(x_i'-x_{i-1}')\mathbf{m}(\{j\,|\,x_j\geq x_i'\}) \tag{122}$$

$$=\sum_{i=1}^{n}x_i'\left(\mathbf{m}(\{j\,|\,x_j\geq x_i'\})-\mathbf{m}(\{j\,|\,x_j\geq x_{i+1}'\})\right) \tag{123}$$

$$=\sum_{i=1}^{n}x_i'\left(\mathbf{m}(I_i')-\mathbf{m}(I_{i+1}')\right) \tag{124}$$

where $(x_1',\ldots,x_n')$ is a non–decreasing permutation of the input n–tuple $(x_1,\ldots,x_n)$ and $x_0'=0$, $x_{n+1}'=\infty$ by convention. Further, if $(\alpha(1),\ldots,\alpha(n))$ is a permutation of indexes $(1,\ldots,n)$, such that $x_i'=x_{\alpha(i)}$ then $I_i'=\{\alpha(i),\ldots,\alpha(n)\}$, $i=1,\ldots,n$, and $I_{n+1}'=\emptyset$. Observe that if we fix the order of input arguments, e.g., $x_1\leq x_2\leq\ldots\leq x_n$, then (124) becomes a weighted mean $\sum_{i+1}^{n}w_ix_i$ with weights

$$w_i=\mathbf{m}(I_i')-\mathbf{m}(I_{i+1}'),\quad i=1,\ldots,n,\ \text{and}\quad \sum_{i=1}^{n}w_i=1.$$

Example 12 For $n=2$, let $\mathbf{m}:\mathcal{P}(\mathbb{X}_2)\to[0,1]$ be a fuzzy measure given by $\mathbf{m}(\{1\})=a\in[0,1]$, $\mathbf{m}(\{2\})=b\in[0,1]$. Then

$$\mathbf{Ch}(x,y)=\begin{cases}(1-b)x+by \text{ if } x\leq y,\\ ax+(1-a)y \text{ if } x\geq y.\end{cases} \tag{125}$$

Observe that $\mathbf{Ch}=\mathbf{W}$ is a weighted mean if and only if $(1-b)x+by=ax+(1-a)y$ for all $x,y\in[0,1]$, that is iff $a+b=1$. This means that $\mathbf{m}$ is a probability measure on $\mathbb{X}_2$.

Further, $\mathbf{Ch}=\mathbf{W}'$ is an OWA operator if and only if $\mathbf{Ch}(x,y)=\mathbf{Ch}(y,x)$, i.e., when $(1-b)x+by=ay+(1-a)x$ for all $0\leq x\leq y\leq 1$. This holds iff $a = b$, that is iff $\mathbf{m}(I) = h(\operatorname{card} I)$ which means that $\mathbf{m}$ depends only on the cardinality of the underlying set I. Fuzzy measures with this property are called *symmetric fuzzy measures.*

As just observed, OWA operators are special Choquet integral–based aggregation operators related to the symmetric fuzzy measures [32,46], or in other words, symmetric Choquet integral–based aggregation operators. Similarly to the case of weighted means related to the weighting triangles generated by quantifiers, see Proposition 10 also global OWA operators related to quantifiers can be represented as Choquet integrals on $[0,1[$. As it was already shown in Section 5.1, it is enough to deal with non–decreasing quantifiers only.

Proposition 12. *Let $q : [0,1] \to [0,1]$ be a non–decreasing quantifier. The corresponding OWA operator $\mathbf{W}'_{\Delta_q} = \mathbf{W}' : \bigcup_{n \in \mathbb{N}} [0,1]^n \to [0,1]$, see Section 4.2, is given by*

$$\mathbf{W}'(x_1, \ldots, x_n) = (C) - \int_{[0,1[} h \, d\mathbf{m}, \tag{126}$$

where $h : [0,1[\to [0,1]$ is given by $h(u) = x_i$ whenever $u \in \left[\frac{i-1}{n}, \frac{i}{n}\right[$ and $\mathbf{m}$ is a fuzzy measure on $([0,1[\,, \mathcal{B}([0,1[))$ given by $\mathbf{m}(I) = q(\lambda(I))$, where λ is the standard Lebesgue measure on $\mathcal{B}([0,1[)$.

Proof. Applying (120) we immediately obtain (122) and hence also (124), where $\mathbf{m}(I'_i) = q\left(\frac{n+1-i}{n}\right)$, that is, (126) is of the form

$$\sum_{i=1}^{n} \left(q\left(\frac{n+1-i}{n}\right) - q\left(\frac{n-i}{n}\right)\right) x'_i \, .$$

Comparing this result to (96) and (100), we can see that (126) is just an OWA operator $\mathbf{W}'$ related to the quantifier q^*, $q^*(u) = 1 - q(1-u)$. □

As a special case recall the border OWA operators **min** and **max**. They can be represented by the Choquet integral (126) related to the quantifiers

$$q(x) = \begin{cases} 1 \text{ if } x = 1, \\ 0 \text{ if } x < 1, \end{cases} \qquad \text{and} \qquad q(x) = \begin{cases} 1 \text{ if } x > 0, \\ 0 \text{ if } x = 0, \end{cases}$$

respectively, that is to the fuzzy measures on $\mathbb{X}_n$ given by

$$\mathbf{m}(I) = \begin{cases} 1 \text{ if } I = \mathbb{X}_n, \\ 0 \text{ else}, \end{cases} \qquad \text{and} \qquad \mathbf{m}(I) = \begin{cases} 1 \text{ if } I \neq \emptyset, \\ 0 \text{ else}, \end{cases}$$

respectively.

Remark 2 Similarly to Remark 1, we can introduce an extension of OWA operators for infinitely many inputs $(h(x))_{x\in[0,1[}$,

$$\mathbf{W}'(h) = (C) - \int_{[0,1[} h\, d\mathbf{m}\,,$$

where $\mathbf{m}$ is described in Proposition 12. Observe that if $\mathbf{m} = \lambda$, that is, the quantifier is the identity $q(x) = x$, $x \in [0,1]$, then $\mathbf{W}' = \mathbf{M}$ is the arithmetic mean if (126) is applied and in general, $\mathbf{W}' = \mathbf{W}_\lambda$.

For extremal quantifiers corresponding to **min** and **max** in the case of uncountably many inputs we obtain the ess inf and ess sup, respectively.

A general fuzzy measure $\mathbf{m}$ on $\mathbb{X}_n$ models all possible interactions between the criteria. However, to determine such fuzzy measure, we need to find $2^n - 2$ values related to $\mathbf{m}(I)$ in general, only values $\mathbf{m}(\emptyset)$ and $\mathbf{m}(\mathbb{X}_n)$ are always equal to 0 and 1, respectively. To avoid the problems with computational complexity and practical estimations , Grabisch introduced the concept of k–order additivity, see [48,47]. Note that in the case of k–order additive fuzzy measure we have to find (estimate) only $n+\binom{n}{2}+\ldots+\binom{n}{k}-1$ values. For more details on k–order additivity and its consequences for aggregation problems we refer the reader to [49,78,51,92]. Here we recall only aggregation related to $k = 2$, that is, the case of 2–order additive fuzzy measures.

Example 13 For a fixed $n \in \mathbb{N}, n \geq 2$, let c_i, d_{ij}, $i,j \in \{1,\ldots,n\}$, $i < j$, be real constants such that $\forall c_i \geq 0$, $\forall I \subseteq \mathbb{X}_n$, $\forall j \in \mathbb{X}_n \setminus I$, $\sum\limits_{i\in I} d_{ij} \geq c_j$ and $\sum\limits_i c_i + \sum\limits_{i<j} d_{ij} = 1$. Then

$$\mathbf{m} : \mathcal{P}(\mathbb{X}_n) \to [0,1]\,, \qquad \mathbf{m}(I) = \sum_{i\in I} c_i + \sum_{\{i,j\}\subset I} d_{ij}\,,$$

is a 2–order additive fuzzy measure on $\mathbb{X}_n$. The operator $\mathbf{Ch} : [0,1]^n \to [0,1]$ given by (124) related to this fuzzy measure is given by

$$\mathbf{Ch}(x_1,\ldots,x_n) = \sum_i c_i x_i + \sum_{i<j} d_{ij} \min(x_i,x_j)\,. \tag{127}$$

The operator $\mathbf{Ch}$ is an OWA operator if and only if $c_i = c$ and $d_{ij} = d$ for all possible i, j, where

$$c \in \left[0, \frac{2}{n}\right] \quad \text{and} \quad nc + \binom{n}{2} d = 1\,, \quad \text{that is } d = \frac{2-2nc}{n(n-1)}\,.$$

Then

$$\mathbf{Ch}(x_1,\ldots,x_n) = c\sum_{i=1}^{n} x_i + d\sum_{i=1}^{n}(n-i)x'_i$$
$$= \sum_{i=1}^{n} x'_i(c+(n-i)d) = \sum_{i=1}^{n} w_i x'_i\,, \tag{128}$$

where

$$w_i = c + \frac{(n-i)(2-2nc)}{n(n-1)}\,.$$

If $c = \frac{1}{n}$ then $d = 0$ and we get the arithmetic mean $\mathbf{M}_{(n)}$. For $c = 0$ we have $d = \frac{2}{n(n-1)}$ and

$$\mathbf{Ch}(x_1,\ldots,x_n) = \frac{2}{n(n-1)}\sum_{i=1}^{n}(n-i)x'_i\,. \tag{129}$$

Similarly, for $c = \frac{2}{n}$ we obtain $d = -\frac{2}{n(n-1)}$ and

$$\mathbf{Ch}(x_1,\ldots,x_n) = \frac{2}{n(n-1)}\sum_{i=1}^{n}(i-1)x'_i\,. \tag{130}$$

Observe that (129) and (130) describe a pair of OWA operators with reversed weights. Also note that the parametric family of OWA operators introduced in (128) with parameter $c \in [0, \frac{2}{n}]$ has the "orness" measure $m_{or}(\mathbf{Ch}) \in \left[\frac{n-2}{3n-3}, \frac{2n-1}{3n-3}\right]$, with limit values corresponding to operators given in (129) and (130), respectively.

An interesting type of the Choquet integral–based aggregation is the two–step (or more–step) Choquet integral, see [37,38,98]. In the two–step Choquet integral over $\mathbb{X}_n$, first a family $(I_k)_{k\in\mathcal{K}}$ of index subsets $I_k \subset \mathbb{X}_n$, $k \in \mathcal{K}$ is introduced and Choquet integrals of given inputs over I_k with respect to relevant fuzzy measures measures $\mathbf{m}_k$ on I_k are computed,

$$c_k = (C) - \int_{[0,1[} h_k\,\mathbf{dm}_k\,, \quad h_k : I_k \to [0,1], \quad h_k(i) = x_i\,,$$

and then, as the second step, the Choquet integral of inputs c_k over $\mathcal{K}$ is computed to obtain the value

$$\mathbf{A}(x_1,\ldots,x_n) = (C) - \int_{\mathcal{K}} h\,\mathbf{dm} \tag{131}$$

where $h : \mathcal{K} \to [0,1]$, $h(k) = c_k$. Observe that two–step integration is a special case of the composed aggregation, see Section 3.5. More details can be found in [98]. Recall that the two–step Choquet integral leads to an idempotent linear continuous aggregation operator, which is not comonotone additive, in general, and hence, in general, it cannot be described by the standard Choquet integral.

Example 14 For $n = 3$ define the ternary operator $\mathbf{A}_{(3)} : [0,1]^3 \to [0,1]$ by

$$\mathbf{A}_{(3)}(x, y, z) = \frac{1}{2}\left(\min(x, y) + \max(y, z)\right) . \tag{132}$$

This operator is the two–step Choquet integral, where all involved Choquet integrals are related to the OWA operators, that is $\mathbf{A}_{(3)}$ is the two–step OWA operator. Here $\mathcal{K} = \{1,2\}$, $I_1 = \{1,2\}$, $I_2 = \{2,3\}$, $c_1 = \min(x,y)$, $c_2 = \max(y,z)$ and finally, we apply $\mathbf{M}_{(2)}$ on c_1 and c_2. The operator $\mathbf{A}_{(3)}$ is not an OWA operator, but it can be represented by the standard Choquet integral with respect to a fuzzy measure $\mathbf{m}$ on $\mathbb{X}_3$ given by

$$\mathbf{m}(\emptyset) = \mathbf{m}(\{1\}) = 0, \;\; \mathbf{m}(\{1,2\}) = \mathbf{m}(\mathbb{X}_{(3)}) = 1, \;\; \text{and } \mathbf{m}(I) = 0.5 \text{ otherwise.}$$

As already mentioned, fuzzy measures model interaction between single criteria. The aggregation of interacting criteria can be thus described by the Choquet integral with respect to the relevant fuzzy measure. This type of aggregation is extensively studied in the chapter of this book written by Marichal [81]. For other applications of the Choquet integral in the area of aggregation we refer the reader to [52].

Remark 3 Recently, Torra [123] introduced so called WOWA operators (Weighted Ordered Weighted Average), generalizing both weighted means and OWA operators. For more details we recommend the chapter of this book written by Torra and Godo [124]. Note that also WOWA operators can be represented by the Choquet integral, similarly to weighted means in (118) and OWA operators in (126). Indeed, for a given probability measure $\mathbf{p}$ acting on $\mathcal{B}([0,1[)$ and a non–decreasing quantifier $q : [0,1] \to [0,1]$, a WOWA operator $\mathbf{W}^\diamond : \bigcup_{n\in\mathbb{N}} [0,1]^n \to [0,1]$ is given by

$$\mathbf{W}^\diamond(x_1, \ldots, x_n) = (C) - \int_{[0,1[} h \, d\mathbf{m} ,$$

where h is given as in Propositions 10 and 12 and $\mathbf{m}$ is a fuzzy measure on $\mathcal{B}([0,1[)$ given by $\mathbf{m}(I) = q(\mathbf{p}(I))$. Obviously, if $q = id_{|[0,1]}$, then $\mathbf{W}^\diamond = \mathbf{W}_\mathbf{p}$, see Proposition 10. Similarly, if $\mathbf{p} = \lambda$, then $\mathbf{W}^\diamond = \mathbf{W}'$ is the OWA operator from Proposition 12. Further, the above integral approach allows to extend WOWA operators also to infinitely many inputs $(h(x))_{x\in[0,1[}$ putting

$$\mathbf{W}^{\circ}(h) = (C) - \int_{[0,1[} h\,d\mathbf{m}\,.$$

5.3 Sugeno integral–based aggregation

A maxitive fuzzy measure $\mathbf{m} : \mathbb{X} \to [0,1]$ was introduced by Shilkret [116]. For $\mathbb{X} = \mathbb{X}_{(n)}$ it is characterized by

$$\mathbf{m}(I) = \bigvee_{i\in I} w_i\,, \tag{133}$$

where $w_i = \mathbf{m}(\{i\})$, that is, $w_i \in [0,1]$ and $\bigvee_{i\in\mathbf{X}_{(n)}} w_i = \max(w_1,\dots,w_n) = 1$.

Lebesque–like integration related to maxitive measures is based on a pseudo–multiplication $\odot : [0,1]^2 \to [0,1]$, which can be equal to the standard product, that is, $\odot = \cdot$, see [116], or to the minimum, i.e., $\odot = \wedge$ [119], to some strict t–norm, $\odot = \mathbf{T}$ [131], to some left–continuous uninorm, that is $\odot = \mathbf{U}$ [62], etc.

The aggregation operators corresponding to the mentioned multiplications are given as follows:

(i) For $\odot = \cdot$ we obtain

$$\mathbf{A}(x_1,\dots,x_n) = \max(w_1\cdot x_1,\dots,w_n\cdot x_n) = \bigvee_{i=1}^{n} w_i\cdot x_i\,. \tag{134}$$

(ii) For $\odot = \wedge$ we have

$$\mathbf{A}(x_1,\dots,x_n) = \bigvee_{i=1}^{n}(w_i \wedge x_i) \quad \text{that is, } \mathbf{A} \text{ is a weighted maximum.} \tag{135}$$

(iii) If $\odot = \mathbf{T}_2^H$, where $\mathbf{T}_2^H$ is the Hamacher t–norm, see (40), then

$$\mathbf{A}(x_1,\dots,x_n) = \bigvee_{i=1}^{n} \frac{w_i x_i}{2 - w_i - x_i + w_i x_i}\,. \tag{136}$$

(iv) Finally, for $\odot = \mathbf{E}(x,y) = \frac{xy}{xy+(1-x)(1-y)}$ (with convention $\frac{0}{0} = 0$), where $\mathbf{E}$ is the uninorm called $3 - \Pi$–operator, see (39), we obtain

$$\mathbf{A}(x_1,\dots,x_n) = \bigvee_{i=1}^{n} \frac{w_i x_i}{w_i x_i + (1-w_i)(1-x_i)}\,. \tag{137}$$

The most often applied aggregation operator from these operators is the weighted maximum given by (135), see [28]. Note that the dual operator to the weighted maximum is the weighted minimum defined by

$$\mathbf{A}^d(x_1,\dots,x_n) = \bigwedge_{i=1}^{n} ((1-w_i) \vee x_i) . \tag{138}$$

The weighted maximum, and also operators given in (136) and (137), can be understood as the maximum operator applied to criteria with different weights (importances), see [27,135] and Section 3.3, and similarly in the case of the weighted minimum.

For a general fuzzy measure $\mathbf{m}$, a counterpart of the Choquet integral based on the **max** operator and coinciding with the weighted maximum for each fixed order of inputs, was introduced by Sugeno [119].

Definition 23. Let $(\mathbb{X},\mathcal{A})$ be a measurable space and let $\mathbf{m} : \mathcal{A} \to [0,1]$ be a fuzzy measure. Let $h : \mathbb{X} \to [0,1]$ be an $\mathcal{A}$–measurable function. Then the Sugeno integral of h with respect to $\mathbf{m}$ with notation $(S) - \int_{\mathbb{X}} h\, d\mathbf{m}$ is defined by

$$(S) - \int_{\mathbb{X}} h\, d\mathbf{m} = \bigvee_{u\in[0,1]} \min\left(u, \mathbf{m}(\{x \in \mathbb{X} \mid h(x) \geq u\})\right) . \tag{139}$$

For an exhaustive overview of the Sugeno integral theory we refer interested readers to [129], see also [109,46]. For recent results we recommend [52].

Coming back to aggregation operators, we can easy prove the following results, see [5].

Proposition 13. *Let* $\mathbf{m}$ *be a fuzzy measure defined on* $\mathbb{X}_n$. *Define the* n*–ary operator* $\mathbf{Su} : [0,1]^n \to [0,1]$ *by*

$$\mathbf{Su}(x_1,\dots,x_n) = (S) - \int_{\mathbb{X}_n} h\, d\mathbf{m}, \tag{140}$$

where $h : \mathbb{X}_n \to [0,1]$ *is given by* $h(i) = x_i$. *Then* $\mathbf{Su}$ *is an idempotent continuous aggregation operator which is comonotone maxitive, and max- and min–homogeneous.*

Recall that comonotone maxitivity means that

$$\mathbf{Su}(x_1 \vee y_1,\dots,x_n \vee y_n) = \mathbf{Su}(x_1,\dots,x_n) \vee \mathbf{Su}(y_1,\dots,y_n) \tag{141}$$

whenever the input n–tuples are comonotone. Further, max- and min–homogeneity mean that for all $a \in [0,1]$

$$\mathbf{Su}(a \vee x_1,\dots,a \vee x_n) = a \vee \mathbf{Su}(x_1,\dots,x_n) , \tag{142}$$

and

$$\mathbf{Su}(a \wedge x_1, \ldots, a \wedge x_n) = a \wedge \mathbf{Su}(x_1, \ldots, x_n) \tag{143}$$

respectively.

Observe that each continuous max- and min–homogeneous n–ary aggregation operator $\mathbf{A}_{(n)} : [0,1]^n \to [0,1]$ can be represented as a Sugeno integral–based aggregation operator related to the fuzzy measure $\mathbf{m} : \mathcal{P}(\mathbb{X}_n) \to [0,1]$, $\mathbf{m}(I) = \mathbf{A}_{(n)}(\mathbf{1}_I(1), \ldots, \mathbf{1}_I(n))$, see [79] or [5]. Observe that the Sugeno integral given in (139) can be modified replacing the min operator by some other pseudo–multiplication $\odot$, for example, by the product $\cdot$, some strict t–norm $\mathbf{T}$, by a uninorm $\mathbf{U}$, etc. We will discuss the related operators only exceptionally, since they have similar properties to the original Sugeno integral–based aggregation. Exploiting the notation from Section 5.2, see (124), we immediately obtain the next representation of the operator $\mathbf{Su}$ introduced in Proposition 13, namely

$$\mathbf{Su}(x_1, \ldots, x_n) = \bigvee_{i=1}^{n} \min(x'_i, \mathbf{m}(I'_i)) = \bigvee_{i=1}^{n} (w'_i \wedge x'_i), \tag{144}$$

where $w'_i = \mathbf{m}(I'_i) = \mathbf{m}(\{\alpha(i), \ldots, \alpha(n)\})$. For alternative forms of the representation of the operator $\mathbf{Su}$ see [79]. Observe that for any fixed order of inputs $x_1, \ldots, x_n$ the operator $\mathbf{Su}$ is the weighted maximum, see (135). Note also that $w'_1 = 1 \geq w'_2 \geq \ldots \geq w'_n \geq 0$ while $0 \leq x'_1 \leq \ldots \leq x'_n \leq 1$. The operator $\mathbf{Su}$ is symmetric if and only if the related fuzzy measure $\mathbf{m}$ is symmetric, that is, if $\mathbf{m}(I) = h(\operatorname{card} I)$ for some non–decreasing function $h : (\mathbb{X}_n) \cup \{0\} \to [0,1]$, with $h(0) = 0$ and $h(n) = 1$. However, then independently of the order of input values $x_1, \ldots, x_n$, $\mathbf{m}(I'_i) = h(n+1-i)$, $i = 1, \ldots, n$. Putting $h(n+1-i) = w_i$, the operator $\mathbf{Su}$ becomes the ordered weighted maximum (OWM operator, for short), see [28,110],

$$\mathbf{Su}'(x_1, \ldots, x_n) = \bigvee_{i=1}^{n} (w_i \wedge x'_i) = \operatorname{med}(x_1, \ldots, x_n, w_1, \ldots, w_n), \tag{145}$$

and as it was mentioned before, $w_1 = 1 \geq w_2 \geq \ldots \geq w_n \geq 0$.

Note that paraphrasing Proposition 12, we can introduce a global ordered weighted maximum operator $\mathbf{Su}' : \bigcup_{n \in \mathbb{N}} [0,1]^n \to [0,1]$.

Proposition 14. *Let $q : [0,1] \to [0,1]$ be a non–decreasing quantifier. Define the aggregation operator $\mathbf{Su}' : \bigcup_{n \in \mathbb{N}} [0,1]^n \to [0,1]$ by*

$$\mathbf{Su}'(x_1, \ldots, x_n) = (S) - \int_{[0,1[} h \, d\mathbf{m}, \tag{146}$$

where $h : [0,1[\to [0,1]$ is given by $h(u) = x_i$ whenever $u \in \left[\frac{i-1}{n}, \frac{i}{n}\right[$ and $\mathbf{m}$ is a fuzzy measure on $([0,1[, \mathcal{B}([0,1[))$ given by $\mathbf{m}(I) = q(\lambda(I))$. For each fixed $n \in \mathbb{N}$, $\mathbf{Su}'_{(n)} : [0,1]^n \to [0,1]$ is an OWM operator.

Proof. Applying directly (139), we find out that

$$\mathbf{Su}'_{(n)}(x_1,\ldots,x_n) = \bigvee_{i=1}^{n}(w_i \wedge x'_i), \text{ where } w_i = q\left(\frac{n+1-i}{n}\right), \ i=1,\ldots,n \ \square.$$

Example 15 (i) For $w_1 = 0.5$, $w_2 = 1$, define the weighted maximum $\mathbf{A}$ by (135), that is

$$\mathbf{A}(x,y) = \begin{cases} x & \text{if } y \le x \le 0.5, \\ 0.5 & \text{if } y \le 0.5 \le x, \\ y & \text{else.} \end{cases}$$

(ii) For the fuzzy measure $\mathbf{m} : \mathcal{P}(\mathbb{X}_2) \to [0,1]$ given in Example 12, that is, $\mathbf{m}(\{1\}) = a$, $\mathbf{m}(\{2\}) = b$, related to the Sugeno integral–based aggregation operator $\mathbf{Su} : [0,1]^2 \to [0,1]$, see (140), is given by

$$\mathbf{Su}(x,y) = (a \wedge x) \vee (b \wedge y) \vee (x \wedge y) = \begin{cases} x \vee (b \wedge y) & \text{if } x \le y, \\ (a \wedge x) \vee y & \text{if } x \ge y. \end{cases} \quad (147)$$

Observe that $\mathbf{Su}$ is the weighted maximum if and only if $\mathbf{m}$ is a maxitive measure, that is if either $a = 1$ or $b = 1$. Further, $\mathbf{Su} = \mathbf{Su}'$ is an OWM operator if and only if $\mathbf{m}$ is a symmetric measure, that is if $a = b$. Then, following (145),

$$\mathbf{Su}'(x,y) = \text{med}(x,y,a) = \mathbf{med}_a(x,y)\,.$$

(iii) The dual operator $\mathbf{A}^d : [0,1]^2 \to [0,1]$ to the weighted maximum $\mathbf{A}$ given in (i) of this example is given by

$$\mathbf{A}^d(x,y) = (x \vee 0.5) \wedge y = \begin{cases} x \vee (0.5 \wedge y) & \text{if } x \le y, \\ y & \text{if } x \ge y. \end{cases}$$

Evidently, $\mathbf{A}^d$ is a Sugeno integral–based operator, compare to (147) for $a = 0$ and $b = 0.5$.

As already mentioned, weighted maxima and weighted minima and ordered weighted maxima and minima, respectively, are special Sugeno integral–based aggregation operators. For more details on these operators and other Sugeno integral–based aggregation operators we recommend [79]. Also observe that the class of Sugeno integral–based aggregation operators is closed under compositions as discussed in Section 3.5, Proposition 6. Indeed, each two–step (or more–step) Sugeno integral can be represented by the standard Sugeno integral, see [98]. For recent applications of the Sugeno integral–based aggregation we refer the reader to [52].

5.4 Aggregation based on some other types of integrals

Choquet–like, Sugeno–like, t–conorm–based, and other types of integrals can also be applied to obtain new types of aggregation operators. Some of them are only transformations of the previous ones. For example, the Choquet–like integrals introduced in [90], which are based on some non–decreasing transformation $g : [0,1] \to [0,1]$ (or $g : [0,1] \to [0,\infty]$). Applying these integrals in the domain $\mathbb{X}_n$, one gets the aggregation operator of the form:

$$\mathbf{A}(x_1,\ldots,x_n) = g^{-1}\left((C) - \int_{\mathbb{X}_n} g \circ h \, dg \circ \mathbf{m}\right)$$
$$= g^{-1}\left(\sum_{i=1}^{n} g(x_i') \left(g(\mathbf{m}(I_i')) - g(\mathbf{m}(I_{i+1}'))\right)\right), \qquad (148)$$

compare to (147). $\mathbf{A}$ is an idempotent continuous n–ary aggregation operator which is pseudo–linear and comonotone pseudo–additive with respect to the pseudo–addition $\oplus : [0,1]^2 \to [0,1]$, $u \oplus v = g^{-1}(\min(g(1), g(u) + g(v)))$ and pseudo–multiplication $\odot : [0,1]^2 \to [0,1]$, $u \odot v = g^{-1}(g(u)g(v))$. Observe that weighted quasi–arithmetic means related to g are special cases of (148).

Example 16 For $n = 2$, take the fuzzy measure $\mathbf{m} : \mathcal{P}(\mathbb{X}_2) \to [0,1]$ as in Example 12, that is, $\mathbf{m}(\{1\}) = a$, $\mathbf{m}(\{2\}) = b$. Further, let $g : [0,1] \to [0,1]$ be given by $g(x) = x^p$, $p \in]0,\infty[$. Then the Choquet–like integral–based operator $\mathbf{A}$ given by (148) has the form

$$\mathbf{A}(x,y) = \begin{cases} \left((1-b^p)x^p + b^p y^p\right)^{1/p} & \text{if } x \le y, \\ \left(a^p x^p + (1-a^p)y^p\right)^{1/p} & \text{if } x \ge y. \end{cases} \qquad (149)$$

Observe that if $a^p + b^p = 1$, then $\mathbf{A}(x,y) = \left(a^p x^p + (1-a^p)y^p\right)^{1/p}$ is the weighted quasi–arithmetic mean. Moreover, if $\mathbf{m}$ is a symmetric fuzzy measure, that is, if $a = b$, then

$$\mathbf{A}(x,y) = \left((1-a^p)x'^p + a^p y'^p\right)^{1/p},$$

where $x' = \min(x,y)$, $y' = \max(x,y)$, which means that $\mathbf{A}$ is an ordered weighted quasi–arithmetic mean.

In limit cases we obtain the following operators:

(i) For $p \to \infty$:

$$\mathbf{A}(x,y) = \begin{cases} x \vee by & \text{if } x \le y, \\ ax \vee y & \text{if } x \ge y, \end{cases} \qquad (150)$$

(ii) for $p \to 0^+$:

$$\mathbf{A}(x,y) = \begin{cases} \max(x,y) & \text{if } a > 0,\ b > 0, \\ x & \text{if } a > 0,\ b = 0, \\ y & \text{if } a = 0,\ b > 0, \\ \min(x,y) & \text{if } a = 0,\ b = 0. \end{cases} \qquad (151)$$

Sugeno–like integrals are related to the original Sugeno integral (139), and as already mentioned in the previous Section 5.3, min operator in (139) is replaced by some pseudo–multiplication $\odot$. The corresponding n–ary aggregation operator $\mathbf{A} : [0,1]^n \to [0,1]$ related to a fuzzy measure $\mathbf{m} : \mathcal{P}(\mathbb{X}_n) \to [0,1]$ is then given by

$$\mathbf{A}(x_1, \ldots, x_n) = \bigvee_{i=1}^{n} (w'_i \odot x'_i), \tag{152}$$

where $(x'_1, \ldots, x'_n)$ is a non–decreasing permutation of $(x_1, \ldots, x_n)$, $x'_i = x_{\alpha(i)}$, and $w'_i = \mathbf{m}(\{\alpha(i), \ldots, \alpha(n)\})$.

For example, taking $\odot = \cdot$, see also [116], we obtain $\mathbf{A}(x_1, \ldots, x_n) = \bigvee_{i=1}^{n} (w'_i . x'_i)$. An example of such operator was already given in (150). Note that all aggregation operators based on Sugeno–like integrals are continuous idempotent operators which are comonotone maxitive. Moreover, if the pseudo–multiplication $\odot$ is associative then the corresponding aggregation operators are also homogeneous with respect to $\odot$, that is, $\mathbf{A}(a \odot x_1, \ldots, a \odot x_n) = a \odot \mathbf{A}(x_1, \ldots, x_n)$ for all $a \in [0,1]$.

For readers interested in pseudo–multiplications and further generalizations of the Choquet and Sugeno integrals we recommend [6].

(S,T)- and (S,U)- integrals introduced and discussed in [62] offer another generalization of quasi–arithmetic means. We only recall a special type of such operators related to some function $f : [0,1] \to [-\infty, \infty]$ which is continuous and strictly monotone, and to a weighting triangle $\Delta = (w_{in})$, $n \in \mathbb{N}$, $i \in \{1, \ldots, n\}$, where all weights $w_{in} \in [0,1]$ and $\sum_{i=1}^{n} w_{in} \geq 1$ for all $n \in \mathbb{N}$, $w_{11} = 1$. Then the resulting operator $\mathbf{A}_f : \bigcup_{n \in \mathbb{N}} [0,1]^n \to [0,1]$ is given by

$$\mathbf{A}_f(x_1, \ldots, x_n) = f^{-1}\left(/ \sum_{i=1}^{n} w_{in} f(x_i) / \right), \tag{153}$$

where $/x/$ is the point from $Ran\, f$ closest to x. Obviously, if $\sum_{i=1}^{n} w_{in} = 1, n \in \mathbb{N}$, then the operator $\mathbf{A}_f$ is just the quasi–arithmetic mean $\mathbf{M}_f$, see Section 4.3. Note that $\mathbf{A}_f$ is a quasi–arithmetic mean if and only if $\mathbf{A}_f$ is idempotent.

Example 17 (i) For $f : [0,1] \to [0,\infty]$ such that $f(0) = 0$ and $w_{in} = 1$ for all i, n, the operator $\mathbf{A}_f$ defines a continuous Archimedean t–conorm, see Section 6.1. Similarly, for $f : [0,1] \to [0,\infty]$ with $f(1) = 0$ and $w_{in} = 1$ for all i, n, the operator $\mathbf{A}_f$ is just a continuous Archimedean t–norm, see Section 6.1. Finally, if $Ran\, f = [-\infty, \infty]$ and all $w_{in} = 1$, then $\mathbf{A}_f$ is a uninorm, see Section 6.2.

(ii) Consider $f : [0,1] \to [-\infty, \infty]$, $f(x) = x$. Then applying (153), we obtain

$$\mathbf{A}_f(x_1, \ldots, x_n) = \min\left(1, \sum_{i=1}^{n} w_{in} x_i\right). \tag{154}$$

(iii) Consider $f : [0,1] \to [-\infty, \infty]$, $f(x) = 1 - x$. Then applying (153), we obtain

$$\begin{aligned}\mathbf{A}_f(x_1, \ldots, x_n) &= 1 - \min\left(1, \sum_{i=1}^{n} w_{in}(1 - x_i)\right) \\ &= \max\left(0, \sum_{i=1}^{n} w_{in} x_i - a_n\right),\end{aligned} \tag{155}$$

where $a_n = \sum\limits_{i=1}^{n} w_{in} - 1$, $n \in \mathbb{N}$.

Put, for example, $w_{in} = \frac{2}{n}$, $n \geq 2$, $i \in \{1, \ldots, n\}$. Then $a_n = 1$, $n \geq 2$, and

$$\mathbf{A}_f(x_1, \ldots, x_n) = \max\left(0, \frac{2}{n} \sum_{i=1}^{n} x_i - 1\right),$$

that is,

$$\mathbf{A}_f = \max(0, 2\mathbf{M} - 1), \quad n \geq 2 .$$

An integration concept based on copulas, see Section 6.4, was proposed by Imaoka [56]. This concept includes the Choquet and Sugeno integrals. Corresponding aggregation operators based on a fuzzy measure $\mathbf{m} : \mathcal{P}(\mathbb{X}_n) \to [0,1]$ and a copula $\mathbf{C} : [0,1]^2 \to [0,1]$ are given by

$$\mathbf{A}(x_1, \ldots, x_n) = \sum_{i=1}^{n} (\mathbf{C}(x'_i, w'_i) - \mathbf{C}(x'_{i-1}, w'_i)) \tag{156}$$

$$= \sum_{i=1}^{n} (\mathbf{C}(x'_i, w'_i) - \mathbf{C}(x'_i, w'_{i+1})), \tag{157}$$

where $(x'_1, \ldots, x'_n)$ is a non-decreasing permutation of $(x_1, \ldots, x_n)$, $x'_i = x_{\alpha(i)}$, $i = 1, \ldots, n$ and $w'_i = \mathbf{m}(\{\alpha(i), \ldots, \alpha(n)\})$, with convention $x'_0 = 0$ and $w'_{n+1} = 0$. For more details we refer the reader to [57]. Observe that if $\mathbf{C} = \mathbf{\Pi}$, we obtain just the Choquet integral while for $\mathbf{C} = \mathbf{min}$, the Sugeno integral is recognized. All Imaoka's integral-based aggregation operators are continuous and idempotent.

Example 18 The weakest copula is the Łukasiewicz t-norm $\mathbf{T}_L : [0,1]^2 \to [0,1]$, $\mathbf{T}_L(x,y) = \max(0, x + y - 1)$. The corresponding Imaoka's integral is sometimes called the opposite Sugeno integral. For $n = 2$ and our standard fuzzy measure $\mathbf{m} : \mathcal{P}(\mathbb{X}_2) \to [0,1]$, $\mathbf{m}(\{1\}) = a$, $\mathbf{m}(\{2\}) = b$, applying (156) or (157) we obtain

$$\mathbf{A}(x,y) = \begin{cases} x & \text{if } x \le y \le 1-b \text{ or } 1-a \le y \le x, \\ y & \text{if } 1-b \le x \le y \text{ or } y \le x \le 1-a, \\ x+y+b-1 & \text{if } x \le 1-b \le y, \\ x+y+a-1 & \text{if } y \le 1-a \le x. \end{cases} \tag{158}$$

Observe that in the symmetric case, that is, for $a = b$, the operator $\mathbf{A}$ given in (158) is just the ordinal sum of min operator acting on $[0, c]$ and max acting on $[c, 1]$, $c = 1 - a$, given in the formula (77) with $f = id$.

Some other types of integrals, for example, t–conorm–based integrals [130] and fuzzy t–conorm–based integrals [104], are also interesting for aggregation, and interested readers can find the details in the mentioned literature.

6 Aggregation operators based on triangular norms and conorms

Triangular norms were originally introduced by Menger [88] as operators for the fusion of distribution functions needed by triangle inequality generalization of a metric on statistical metric spaces. Menger's triangular norms formed a large, rather heterogeneous class of symmetric binary aggregation operators fulfilling $\mathbf{A}(1, a) > 0$ whenever $a > 0$. Nowadays axioms of triangular norms are due to Schweizer and Sklar [114], requiring associativity and neutral element $e = 1$. Note that associativity allowed to extend the triangle inequality to the polygonal inequality, including the fact, that now triangular norms can be applied to any finite number of inputs, that is, they form global aggregation operators in the sense of Definition 1. Triangular norms have become especially popular as models for fuzzy sets intersection. They are applied also in probabilistic metric spaces, many–valued logic, non–additive measures and integrals, etc. For an exhaustive overview of the state of art in the field of triangular norms we recommend the recent monograph [61]. In this chapter we will discuss not only triangular norms (t-norms for short) and their dual operators t–conorms, but also several related aggregation operators, such as uninorms, nullnorms, copulas, composed operators based on t–norms and t–conorms, etc.

6.1 Triangular norms

Definition 24. Triangular norm is an associative symmetric aggregation operator with neutral element 1. Its dual operator, i. e., an associative symmetric aggregation operator with neutral element 0 is called a triangular conorm.

Obviously, each t–norm $\mathbf{T}$ is an aggregation operator with annihilator 0.

The class of all triangular norms is wide, including continuous, non-continuous, and even Borel non-measurable t–norms [58]. The weakest t–norm is the drastic product $\mathbf{T}_D : \bigcup_{n\in\mathbb{N}} [0,1]^n \to [0,1]$,

$$\mathbf{T}_D(x_1,\ldots,x_n) = \begin{cases} x_i \text{ if } \forall j \neq i, x_j = 1, \\ 0 \ \text{ else.} \end{cases} \tag{159}$$

The strongest (and the only idempotent) t–norm is the standard **min** operator. Hence, for any t–norm $\mathbf{T}$,

$$\mathbf{T}_D \leq \mathbf{T} \leq \mathbf{min}.$$

The drastic product $\mathbf{T}_D$ is an example of a non–continuous but right–continuous (upper semi–continuous) t–norm. An important example of a left–continuous (lower semi–continuous) non–continuous t–norm is the nilpotent minimum operator $\mathbf{T}^{nM} : \bigcup_{n\in\mathbb{N}} [0,1]^n \to [0,1]$, see [30],

$$\mathbf{T}^{nM}(x_1,\ldots,x_n) = \begin{cases} 0 \ \text{ if } x'_1 + x'_2 \leq 0.5, \\ x'_1 \text{ else,} \end{cases} \tag{160}$$

where $(x'_1,\ldots,x'_n)$ is a non–decreasing permutation of $(x_1,\ldots,x_n)$. An example of a t–norm which is neither left–continuous nor right–continuous is the operator $\mathbf{T}_\diamond$, for example, any non–measurable t–norm [58].

Non–continuous t–norms play an important role in several applications. Observe, for example, the key role of left–continuous t–norms in theory of probabilistic metric spaces [114] and in many-valued logic [45,54] or preference modelling [34]. Now we focus on the continuous case only. Interested readers can find more details, including non-continuous case, in [61].

Observe that **min** is a continuous t–norm. Another typical continuous t–norm is the product $\mathbf{\Pi}$. The third basic continuous t–norm is the Łukasiewicz t–norm $\mathbf{T}_L : \bigcup_{n\in\mathbb{N}} [0,1]^n \to [0,1]$,

$$\mathbf{T}_L(x_1,\ldots,x_n) = \max\left(0, \sum_{i=1}^{n} x_i - (n-1)\right). \tag{161}$$

A nice probabilistic characterization of these three basic continuous t–norms is the next one: for events $E_1,\ldots,E_n$, let $x_1,\ldots,x_n$ be their respective probabilities. What can we say about the probability x of the intersection $\bigcap_{i=1}^{n} E_i$?. Evidently,

$$\mathbf{T}_L(x_1,\ldots,x_n) \leq x \leq \min(x_1,\ldots,x_n),$$

that is, $\mathbf{T}_L$ and **min** are the lower and the upper bound, respectively, for operators $\mathbf{A}$ such that $\mathbf{A}(x_1, \ldots, x_n) = x$. Moreover, if the events $E_1, \ldots, E_n$ are jointly independent, then $x = \prod(x_1, \ldots, x_n)$.

An important subclass of continuous t–norms form Archimedean t–norms.

Definition 25. An aggregation operator $\mathbf{A} : \bigcup_{n \in \mathbb{N}} [0,1]^n \to [0,1]$ is called Archimedean whenever for each $x \in [0,1]$,

$$\lim_{n \to \infty} x^{(n)} \in TID_{\mathbf{A}},$$

where $x^{(n)} = \mathbf{A}_{(n)}(x, \ldots, x)$ and $TID_{\mathbf{A}}$ is the set of trivial idempotents of $\mathbf{A}$, that is,

$$TID_{\mathbf{A}} = \{0,1\} \cup E_{\mathbf{A}} \cup A_{\mathbf{A}},$$

where $E_{\mathbf{A}}$ is the set of neutral elements of $\mathbf{A}$, $A_{\mathbf{A}}$ is the set of annihilators of $\mathbf{A}$.

Observe that for each $\mathbf{A}$, $E_{\mathbf{A}}$ is either empty set or a singleton $\{e\}$. Similarly, $A_{\mathbf{A}} = \{a\}$ whenever $A_{\mathbf{A}}$ is non-empty. Note that Definition 25 is a modification of the classical Archimedean property of the addition on $[-\infty, \infty]$ (with trivial idempotents $\{-\infty, 0, \infty\}$, 0 is the neutral element) or on $[0, \infty]$ (with trivial idempotents $\{0, \infty\}$), and of the multiplication on $[0, \infty]$ (with trivial idempotents $\{0, 1, \infty\}$, 1 is the neutral element. Further, card $TID_{\mathbf{A}} \leq 3$ for any aggregation operator $\mathbf{A}$.

Coming back to continuous t–norms, Definition 25 can be relaxed in the sense that Archimedeanity of a continuous t–norm $\mathbf{T}$ is equivalent to the diagonal inequality $\mathbf{T}(x,x) < x$ for all $x \in]0,1[$. From already introduced examples, $\mathbf{\Pi}$, $\mathbf{T}_L$, $\mathbf{T}_D$ are Archimedean t–norms, while **min**, $\mathbf{T}^{nM}$ and $\mathbf{T}_\diamond$ given in (32) are not. Continuous Archimedean t–norms can be characterized (and constructed) by means of the next representation theorem of Ling [76].

Proposition 15. *An aggregation operator* $\mathbf{T} : \bigcup_{n \in \mathbb{N}} [0,1]^n \to [0,1]$ *is a continuous Archimedean t–norm if and only if there is a continuous strictly decreasing mapping* $t : [0,1] \to [0, \infty]$, $t(1) = 0$, *such that*

$$\mathbf{T}(x_1, \ldots, x_n) = t^{-1}\left(\min(t(0), \sum_{i=1}^{n} t(x_i))\right). \tag{162}$$

Note that t is called an additive generator of $\mathbf{T}$ and it is unique up to a positive multiplicative constant. The product $\mathbf{\Pi}$ is related to the additive generator $t_\Pi : [0,1] \to [0, \infty]$, $t_\Pi(u) = -\log u$, while the Łukasiewicz t–norm $\mathbf{T}_L$ is related to the additive generator $t_L : [0,1] \to [0, \infty]$, $t_L(u) = 1 - u$.

A continuous Archimedean t–norm with unbounded additive generator is called a strict t–norm (it is strictly monotone on $\bigcup_{n\in\mathbb{N}}]0,1]^n$). Each strict t–norm is isomorphic to the product $\mathbf{\Pi}$. Non–strict continuous Archimedean t–norms are called nilpotent. They possess bounded additive generators and are isomorphic to the Łukasiewicz t–norm $\mathbf{T}_L$. Also note that non–continuous or non–Archimedean t–norms may be generated by (necessarily) non–continuous additive generators [127,126,128]. Continuous Archimedean t–norms are related to the quasi–arithmetic means, see Section 4.3. Indeed, quasi–arithmetic means transform the arithmetic mean by a (additive) generator, while t–norms (continuous Archimedean) transform the addition (truncated addition), see (162). Therefore, several properties of quasi–arithmetic means are similar to the relevant properties of continuous Archimedean t–norms. For example, for two continuous Archimedean t–norms $\mathbf{T}_1$ and $\mathbf{T}_2$ with additive generators t_1 and t_2, respectively, we have $\mathbf{T}_1 \leq \mathbf{T}_2$ if and only if the composite function $h = t_1 \circ t_2^{-1} : [0, t_2(0)] \to [0, t_1(0)]$ is subadditive, that is,

$$h(u+v) \leq h(u) + h(v) \quad \text{for all } u, v, u+v \in [0, t_2(0)].$$

Put

$$h = t_L \circ t_\Pi^{-1} : [0,\infty] \to [0,1], \qquad h(u) = 1 - \exp(-u).$$

For all $u, v \in [0, \infty]$ we have

$$\begin{aligned} h(u+v) &= 1 - \exp(-u-v) = 1 - \exp(-u)\exp(-v) \\ &\leq 1 - \exp(-u) + 1 - \exp(-v) = h(u) + h(v) \end{aligned}$$

that is, $\mathbf{T}_L \leq \mathbf{\Pi}$.

For practical use, often a parameterized family of t–norms is needed. Then the next result [61] is of importance for this purpose.

Proposition 16. *Let $t : [0,1] \to [0,\infty]$ be an additive generator of some continuous Archimedean t-norm $\mathbf{T}$. Then for all $\lambda \in]0,\infty[$, also t^λ generates a continuous Archimedean t-norm $\mathbf{T}_{(\lambda)}$. The family $\left(\mathbf{T}_{(\lambda)}\right)_{\lambda\in]0,\infty[}$ is increasing and*

$$\lim_{\lambda\to\infty} \mathbf{T}_{(\lambda)} = \mathbf{T}_{(\infty)} = \mathbf{min}$$

uniformly,

$$\lim_{\lambda\to 0^+} \mathbf{T}_{(\lambda)} = \mathbf{T}_{(0)} = \mathbf{T}_D$$

pointwisely.

As we can see, the limit members $\mathbf{T}_{(\infty)}$ and $\mathbf{T}_{(0)}$ of the family $\left(\mathbf{T}_{(\lambda)}\right)_{\lambda\in]0,\infty[}$ are independent of the original additive generator t, or, equivalently, of $\mathbf{T} = \mathbf{T}_{(1)}$. Several well–known families are constructed by means of Proposition 16.

For example, the Yager family $(\mathbf{T}_\lambda^Y)_{\lambda\in]0,\infty[}$ [132] is related to the Łukasiewicz t–norm $\mathbf{T}_L = \mathbf{T}_1^Y$, and $t_\lambda^Y(u) = (1-u)^\lambda = (t_L(u))^\lambda$.

Similarly, starting from the product $\mathbf{\Pi}$, we obtain the Aczél-Alsina family $(\mathbf{T}_\lambda^{AA})_{\lambda\in]0,\infty[}$, with additive generators $t_\lambda^{AA}(u) = (t_\Pi(u))^\lambda = (-\log u)^\lambda$.

For a general continuous t–norm $\mathbf{T}$, we have the next representation [76].

Proposition 17. *An aggregation operator* $\mathbf{T} : \bigcup_{n\in\mathbb{N}} [0,1]^n \to [0,1]$ *is a continuous t–norm if and only if there is a (possibly empty) system* $(]a_k, b_k[)_{k\in K}$ *of pairwise disjoint open subintervals of* $[0,1]$ *and the corresponding system of continuous strictly decreasing mappings* $t_k : [a_k, b_k] \to [0,\infty]$, $t_k(b_k) = 0$, *so that*

$$\mathbf{T}(x_1,\ldots,x_n) = \begin{cases} t_k^{-1}(\min(t_k(a_k), \sum_{i=1}^n t_k(\min(x_i, b_k)))) & \textit{if} \min\ x_i \in]a_k, b_k[, \\ \min(x_1,\ldots,x_n) & \textit{else.} \end{cases} \tag{163}$$

Observe that the mapping $t_{(k)} : [0,1] \to [0,\infty]$, $t_{(k)}(u) = t_k(a_k + (b_k - a_k)\cdot u)$, is an additive generator of a continuous Archimedean t–norm $\mathbf{T}_k$, $k \in K$. The original continuous t–norm $\mathbf{T}$ is called an (t–norm) ordinal sum, with notation $\mathbf{T} = (\langle a_k, b_k, \mathbf{T}_k\rangle)_{k\in K}$. Obviously, $\mathbf{T}$ is also an ordinal sum of aggregation operators as discussed in Section 3.4. Namely, it is a lower idempotent ordinal sum of aggregation operators $\mathbf{A}_k$ acting on $[a_k, b_k]$, $k \in K$, where each $\mathbf{A}_k$ is additively generated by t_k (and it is simply a linear transformation of the corresponding continuous Archimedean t–norm $\mathbf{T}_k$). Note that this type of ordinal sums is called a t–norm ordinal sum and it can be applied to any (not only continuous Archimedean) t–norms $\mathbf{T}_k$, $k \in K$, to get a t–norm.

Recall that t–norms as universal quantifiers (i.e., aggregation operators acting on inputs related to an arbitrary index set) are in [42,43,99]

As already mentioned in Definition 24, t–conorms are dual operators to t–norms. By duality, t–conorms have annihilator $a = 1$. For each t–conorm $\mathbf{S}$ we have

$$\mathbf{max} \le \mathbf{S} \le \mathbf{S}_D,$$

where

$$\mathbf{S}_D(x_1,\ldots,x_n) = \begin{cases} x_i & \text{if for all } j \ne i,\ x_j = 0, \\ 1 & \text{else.} \end{cases} \tag{164}$$

The dual operator to the product $\mathbf{\Pi}$ is called the probabilistic sum and it is denoted by $\mathbf{S}_P$,

$$\mathbf{S}_P(x_1,\ldots,x_n) = 1 - \prod_{i=1}^n (1 - x_i). \tag{165}$$

The Łukasiewicz t–conorm $\mathbf{S}_L$ is often called the bounded sum because of

$$\mathbf{S}_L(x_1,\ldots,x_n)=\min(1,\sum_{i=1}^{n}x_i). \tag{166}$$

A continuous Archimedean t–conorm $\mathbf{S}$ is characterized by the diagonal inequality $\mathbf{S}(x,x)>x$ for all $x\in]0,1[$, and it is always related to some continuous strictly increasing additive generator $s:[0,1]\to[0,\infty]$, $s(0)=0$

$$\mathbf{S}(x_1,\ldots,x_n)=s^{-1}\left(\min(s(1),\sum_{i=1}^{n}s(x_i))\right). \tag{167}$$

Dual t–conorms to strict t–norms are called strict t–conorms. They have unbounded additive generators, and they are isomorphic to $\mathbf{S}_P$. Similarly nilpotent t–conorms are introduced (bounded additive generators, isomorphic to $\mathbf{S}_L$). Observe that the duality of continuous Archimedean t–norms and t–conorms is reflected by the duality $s=t\circ N$ (where N is the negation, $N(x)=1-x$) of the corresponding additive generators, i. e., $s(u)=t(1-u)$, $u\in[0,1]$. Consequently, $s_P(u)=-\log(1-u)$ generates the probabilistic sum $\mathbf{S}_P$, while $s_L(u)=u$ generates the bounded sum $\mathbf{S}_L$. Also representation of continuous t–norms (163) is reflected by the dual representation of continuous t–conorms,

$$\mathbf{S}(x_1,\ldots,x_n)=\begin{cases} s_k^{-1}(\min(s_k(b_k),\sum\limits_{i=1}^{n}s_k(\max(x_i,a_k)))) & \text{if } \max\ x_i\in]a_k,b_k[,\\ \max(x_1,\ldots,x_n) & \text{else,}\end{cases}$$

where $(]a_k,b_k[)_{k\in K}$ is a system of pairwise disjoint subintervals of $[0,1]$, and $s_k:[a_k,b_k]\to[0,\infty]$, $s_k(a_k)=0$, is a corresponding system of continuous strictly increasing mappings. A continuous t–conorm $\mathbf{S}$ is, in general, an upper idempotent ordinal sum of aggregation operators acting on $[a_k,b_k]$, $k\in K$. This type of ordinal sums is called a t–conorm ordinal sum and it results in a t–conorm independently of the type of t–conorms $\mathbf{S}_k$, $k\in K$, applied.

6.2 Uninorms

One of the prominent aggregation operators on $[0,\infty]$ is the product $\mathbf{\Pi}$, which is symmetric, associative, and its neutral element $e=1$ is an inner point of the domain $[0,\infty]$. Observe that this operator is not continuous, independently of the choice of the convention $0\cdot\infty$ (0 or ∞). Further, restriction of the product to $[0,1]$ is a triangular norm, while its restriction to $[1,\infty]$ acts as a t–conorm (i. e., neutral element is the lowest domain element). Coming back to aggregation operators acting on $[0,1]$, operators of just mentioned nature have been introduced by Yager and Rybalov [141].

Definition 26. An aggregation operator $\mathbf{U} : \bigcup_{n\in\mathbb{N}} [0,1]^n \to [0,1]$ which is symmetric, associative and possesses a neutral element $e \in]0,1[$ is called a uninorm.

Note that each uninorm $\mathbf{U}$ possesses an annihilator $a = a_{\mathbf{U}}$. Observe that for any uninorm $\mathbf{U}$ and any input n-tuple $(x_1, \ldots, x_n)$ containing at least one input $x_i = 0$ and one input $x_j = 1$, $\mathbf{U}(x_1, \ldots, x_n) = a_{\mathbf{U}} \in \{0,1\}$, see, for example, Sander's chapter on uninorms in this monograph [113].

The uninorms with $a_{\mathbf{U}} = 0$ are called conjunctive uninorms, the remaining uninorms with $a_{\mathbf{U}} = 1$ are called disjunctive uninorms. Further, there is no continuous uninorm [35]. It is easy to check that for any uninorm $\mathbf{U}$, $\mathbf{U}|_{\bigcup_{n\in\mathbb{N}}[0,e]^n}$ acts on $[0,e]$ as a t–norm, while $\mathbf{U}|_{\bigcup_{n\in\mathbb{N}}[e,1]^n}$ acts on $[e,1]$ as a t–conorm. Moreover,

$$\mathbf{min}(x_1, \ldots, x_n) \leq \mathbf{U}(x_1, \ldots, x_n) \leq \mathbf{max}(x_1, \ldots, x_n)$$

whenever

$$\min(x_1, \ldots, x_n) \leq e \leq \max(x_1, \ldots, x_n).$$

Two prominent classes of uninorms are related to the above observations, and they form in some sense boundaries for uninorms.

Proposition 18. *Let* $\mathbf{T}$ *be a t–norm,* $\mathbf{S}$ *a t–conorm and* $\mathbf{U}$ *a uninorm with neutral element* e*, such that* $\mathbf{U}|_{\bigcup_{n\in\mathbb{N}}[0,e]^n} = (<0,e,\mathbf{T}>)|_{\bigcup_{n\in\mathbb{N}}[0,e]^n}$ *and* $\mathbf{U}|_{\bigcup_{n\in\mathbb{N}}[e,1]^n} = (<e,1,\mathbf{S}>)|_{\bigcup_{n\in\mathbb{N}}[0,e]^n}$*, that is,* $\mathbf{U}$ *acts on* $[0,1]$ *as a linear transformation of* $\mathbf{T}$ *and on* $[e,1]$ *as a linear transformation of* $\mathbf{S}$*. Then*

$$(<0,e,\mathbf{T}>) < \mathbf{U}_{e,\mathbf{T},\mathbf{S}} \leq \mathbf{U} \leq \mathbf{U}_{\mathbf{T},\mathbf{S},e} < (<e,1,\mathbf{S}>), \tag{168}$$

where $(<0,e,\mathbf{T}>)$ *is a t–norm ordinal sum,* $(<e,1,\mathbf{S}>)$ *is a t–conorm ordinal sum, and* $\mathbf{U}_{e,\mathbf{T},\mathbf{S}}$, $\mathbf{U}_{\mathbf{T},\mathbf{S},e} : \bigcup_{n\in\mathbb{N}} [0,1]^n \to [0,1]$ *are given by*

$$\mathbf{U}_{e,\mathbf{T},\mathbf{S}}(x_1, \ldots, x_n) = \begin{cases} e \cdot \mathbf{T}\left(\frac{x_1}{e}, \ldots, \frac{x_n}{e}\right) & \text{if } \max x_i \leq e, \\ e + (1-e) \cdot \mathbf{S}\left(\frac{x_1 - e}{1-e}, \ldots, \frac{x_n - e}{1-e}\right) & \text{if } \min x_i \geq e, \\ \min(x_1, \ldots, x_n) & \text{else,} \end{cases} \tag{169}$$

and

$$\mathbf{U}_{\mathbf{T},\mathbf{S},e}(x_1, \ldots, x_n) = \begin{cases} e \cdot \mathbf{T}\left(\frac{x_1}{e}, \ldots, \frac{x_n}{e}\right) & \text{if } \max x_i \leq e, \\ e + (1-e) \cdot \mathbf{S}\left(\frac{x_1 - e}{1-e}, \ldots, \frac{x_n - e}{1-e}\right) & \text{if } \min x_i \geq e, \\ \max(x_1, \ldots, x_n) & \text{else.} \end{cases} \tag{170}$$

Moreover, for any $\mathbf{T}$, $\mathbf{S}$, e operator $\mathbf{U}_{e,\mathbf{T},\mathbf{S}}$ given by (169) is a conjunctive uninorm, while the operator $\mathbf{U}_{\mathbf{T},\mathbf{S},e}$ given by (170) is a disjunctive uninorm. Also observe that, putting $\mathbf{T}_e = (<0, e, \mathbf{T}>)|_{\bigcup_{n\in\mathbb{N}}[0,e]^n}$ and $\mathbf{S}_e = (<e, 1, \mathbf{S}>)|_{\bigcup_{n\in\mathbb{N}}[e,1]^n}$, $\mathbf{U}_{e,\mathbf{T},\mathbf{S}}$ is just the lower ordinal sum of $\mathbf{T}_e$ and $\mathbf{S}_e$, see (73) in Section 3.4, while $\mathbf{U}_{\mathbf{T},\mathbf{S},e}$ is the upper ordinal sum of $\mathbf{T}_e$ and $\mathbf{S}_e$, see (74). Evidently, for any given neutral element $e \in]0,1[$, each uninorm $\mathbf{U}$ with this prescribed neutral element fulfills

$$\mathbf{U}_{e,\mathbf{T}_D,\mathbf{max}} \leq \mathbf{U} \leq \mathbf{U}_{\mathbf{min},\mathbf{S}_D,e}.$$

Two typical idempotent uninorms related to given neutral element $e \in]0,1[$ are given by

$$\mathbf{U}_{e,\mathbf{min},\mathbf{max}}(x_1,\ldots,x_n) = \begin{cases} \max(x_1,\ldots,x_n) & \text{if } \min\ x_i \geq e, \\ \min(x_1,\ldots,x_n) & \text{else,} \end{cases} \tag{171}$$

and

$$\mathbf{U}_{\mathbf{min},\mathbf{max},e}(x_1,\ldots,x_n) = \begin{cases} \min(x_1,\ldots,x_n) & \text{if } \max\ x_i \leq e, \\ \max(x_1,\ldots,x_n) & \text{else.} \end{cases} \tag{172}$$

Though there is a unique idempotent t–norm (**min**) and a unique idempotent t–conorm (**max**), the class of idempotent uninorms is rather rich [22].

Also observe that the class of all uninorms is closed under duality, that is, the dual $\mathbf{U}^d$ to a uninorm $\mathbf{U}$ is again a uninorm. Of course, if $\mathbf{U}$ is a conjunctive (disjunctive) uninorm with neutral element e, its dual $\mathbf{U}^d$ is a disjunctive (conjunctive) uninorm with neutral element $1-e$. Consequently, no uninorm is self-dual, that is, a symmetric sum.

The original purpose of introducing uninorms in [141] was a need for so called compensatory operators improving the lack of upwards (downwards) compensation by t–norms (t–conorms), observed already by Zimmermann and Zysno in eighties [144]. This desirable effect of compensation in both directions (present, e. g., by quasi-arithmetic means) appears in the next class of Archimedean uninorms continuous up to some points with contradictory inputs investigated by Klement et al. [59], see also [26,35].

Proposition 19. *An operator* $\mathbf{U} : \bigcup_{n\in\mathbb{N}} [0,1]^n \to [0,1]$ *is an Archimedean uninorm continuous in all points of* $\bigcup_{n\in\mathbb{N}} [0,1]^n$ *up to* $(x_1,\ldots,x_n)$, $\{0,1\} \subset \{x_1,\ldots,x_n\}$, *if and only if there is a monotone bijection* $h : [0,1] \to [-\infty,\infty]$ *such that*

$$\mathbf{U}(x_1,\ldots,x_n) = h^{-1}\left(\sum_{i=1}^{n} h(x_i)\right), \tag{173}$$

with convention $+\infty + (-\infty) = -\infty$. *The uninorm* $\mathbf{U}$ *is then called a generated uninorm with additive generator* h.

Evidently, generated uninorms transform the standard summation operator acting on $[-\infty, \infty]$ to the unit interval $[0, 1]$. Observe that the neutral element e of a generated uninorm $\mathbf{U}$ is given by $e = h^{-1}(0)$. The increasigness of an additive generator h of a generated uninorm $\mathbf{U}$ is equivalent to its conjunctive form (i. e., disjunctive generated uninorms are related to the decreasing additive generators). A typical example of conjunctive generated uninorms it the $3 - \Pi$–operator $\mathbf{E}$ given in (39),

$$\mathbf{E}(x_1, \ldots, x_n) = \frac{\prod_{i=1}^{n} x_i}{\prod_{i=1}^{n} x_i + \prod_{i=1}^{n} (1 - x_i)}, \text{ with convention } \frac{0}{0} = 0.$$

Its additive generator $h : [0, 1] \to [-\infty, \infty]$ (necessarily unique up to a positive multiplicative constant) is given by

$$h(x) = \log \frac{x}{1 - x}.$$

Observe that generated uninorms are always related to strict t–norms and strict t–conorms (in the sense of Proposition 18). For corresponding additive generators h, t, s of $\mathbf{U}, \mathbf{T}, \mathbf{S}$ we have the next relationships

$$h(x) = \begin{cases} -t\left(\frac{x}{e}\right) & \text{if } x \in [0, e], \\ s\left(\frac{x-e}{1-e}\right) & \text{if } x \in]e, 1]; \end{cases}$$

$$t(x) = -h(ex), \quad x \in [0, 1];$$
$$s(x) = h(e + (1 - e)x), \quad x \in [0, 1].$$

Here e is the neutral element of the discussed generated uninorm $\mathbf{U}$.

The freedom in the choice of an additive generator of a given strict t–norm $\mathbf{T}$ and a given strict t–conorm $\mathbf{S}$ allows to construct a parameterized class of (conjunctive) generated uninorms related to $\mathbf{T}$ and $\mathbf{S}$ as in Proposition 18, see [59].

Let $t : [0, 1] \to [0, \infty]$ be an (unique) additive generator of a given strict t–norm $\mathbf{T}$ such that $t(0.5) = 1$, and similarly, let $s : [0, 1] \to [0, \infty]$, $s(0.5) = 1$, be an additive generator of a given strict t–conorm $\mathbf{S}$. For a given parameter $p \in]0, \infty[$, define an additive generator $h_p : [0, 1] \to [-\infty, \infty]$ related to a generated uninorm $\mathbf{U}_p$,

$$h_p(x) = \begin{cases} -t\left(\frac{x}{e}\right) & \text{if } x \in [0, e]; \\ p \cdot s\left(\frac{x-e}{1-e}\right) & \text{if } x \in]e, 1]. \end{cases}$$

For each $p \in]0, \infty[$, $\mathbf{U}_p$ is related to $\mathbf{T}$ and $\mathbf{S}$ as stated in Proposition 18. Further, the family $(\mathbf{U}_p)_{p \in]0,\infty[}$ is non-decreasing and its limit member is

$$\mathbf{U}_0 = \lim_{p \to 0+} \mathbf{U}_p = \mathbf{U}_{e,\mathbf{T},\mathbf{S}},$$

the other limit member

$$\mathbf{U}_\infty = \lim_{p\to\infty} \mathbf{U}_p$$

coincides with $\mathbf{U}_{\mathbf{T},\mathbf{S},e}$ on $\bigcup_{n\in\mathbb{N}}]0,1]^n$.

Example 19 The $3-\Pi$–operator $\mathbf{E}$, see (39), is related to the Hamacher t–norm $\mathbf{T}_2^H$ generated by an additive generator $t : [0,1] \to [0,\infty]$, $t(x) = \log_3 \frac{2-x}{x}$, $t(0.5) = 1$. Observe that $\mathbf{T}_2^H(x,y) = \frac{xy}{2-x-y+xy}$, see (40). $\mathbf{E}$ is also related to the Einstein sum (Hamacher t–conorm) $\mathbf{S}_2^H$, see also Section 2.9, $\mathbf{S}_2^H(x,y) = \frac{x+y}{1+xy}$, generated by the additive generator $s : [0,1] \to [0,\infty]$, $s(x) = \log_3 \frac{1+x}{1-x}$, $s(0.5) = 1$.

Observe that $\mathbf{S}_2^H$ is the dual t–conorm to $\mathbf{T}_2^H$. Also note that the neutral element of $\mathbf{E}$ is $e = 0.5$, and that $\mathbf{E}^d = \mathbf{E}$ (self-duality) up to the cases when inputs containing both 0 and 1 are aggregated. For $p \in]0,\infty[$, the corresponding additive generator $h_p : [0,1] \to [-\infty,\infty]$ is given by

$$h_p(x) = \begin{cases} \log_3 \frac{x}{1-x} & \text{if } x \in [0,0.5], \\ p \cdot \log_3 \frac{x}{1-x} & \text{if } x \in]0.5,1]. \end{cases}$$

Aggregate, e.g., for $x = 0.1$ and $y = 0.9$, then

$$\mathbf{U}_p(0.1,0.9) = h_p^{-1}\left(\log_3 \frac{1}{9} + p \cdot \log_3 9\right) = h_p^{-1}(2p-2),$$

that is,

$$\text{for } p \in]0,1[, \quad \mathbf{U}_p(0.1,0.9) = \frac{9^p}{9+9^p} \quad (\to 0.1, \ \text{ when } p \to 0^+);$$

$$\text{for } p = 1, \quad \mathbf{U}_p(0.1,0.9) = 0.5;$$

$$\text{for } p \in]1,\infty[, \quad \mathbf{U}_p(0.1,0.9) = \frac{9}{9+9^{\frac{1}{p}}} \quad (\to 0.9, \ \text{ when } p \to \infty).$$

Limit properties of generated uninorms with respect to the powers of additive generators are discussed in [97]. Here we only note that the limit operators are specific aggregation operators depending on the original additive generator (compare with Proposition 16, from which we know that in the case of generated t–norms the limit operators do not depend on starting additive generators; the same holds, by duality, for generated t–conorms). Note also that the properties of uninorms, t–norms and t–conorms allows to derive for each uninorm U the n–ary operator $U_{(n)}$ from the binary operator $U_{(2)}$. Let T and S be a t–norm and a t–conorm, respectively, which are related to U as given in Proposition 18. Put $T^* = (<0, e, \mathbf{T}>)$ and $S^* = (<e, 1, S>)$ the corresponding ordinal sums. Then

$$U_{(n)}(x_1, \ldots, x_n) =$$

$$U_{(2)}\left(T^*\left(\min(x_1,e),\dots,\min(x_n,e)\right),S^*\left(\max(x_1,e),\dots,\max(x_n,e)\right)\right). \tag{174}$$

Equality (174) allows to extend uninorms to act as universal quantifiers on any input index set, see [43].

As a complementary reading to this section we recommend Sander's chapter in this book [113] dealing with non–symmetric operators related to uninorms.

6.3 Nullnorms

Uninorms are aggregation operators related to ordinal sums of a t–norm acting on $[0,e]$ and a t–conorm acting on $[e,1]$. Among several possible extensions of such operators, the minimal one is the conjunctive uninorm $\mathbf{U}_{e,\mathbf{T},\mathbf{S}}$, and the maximal one is the disjunctive uninorm $\mathbf{U}_{\mathbf{T},\mathbf{S},e}$. However, if we discuss the possible extensions of a t–conorm acting on an interval $[0,a]$, $a\in]0,1[$ and a t–norm acting on $[a,1]$, in spite of Proposition 4, there is a unique such extension, that is, the lower ordinal sum and the upper ordinal sum coincide and a is the annihilator of the resulting aggregation operator. This new type of operators are called nullnorms [10,9]. Nullnorms are associative and symmetric. Due to their associativity, it is enough to define (axiomatically) the relevant binary operator (with the same notation). This is, indeed, the original approach to nullnorms in [10], see also [61].

Definition 27. A symmetric associative aggregation operator $\mathbf{V}:[0,1]^2\to[0,1]$ is called a nullnorm if there is an element $a\in]0,1[$ such that

$$\mathbf{V}(x,0)=x \quad \text{for all} \quad x\le a,\quad \mathbf{V}(x,1)=x \quad \text{for all} \quad x\ge a. \tag{175}$$

The monotonicity of $\mathbf{V}$ ensures that a is the annihilator of $\mathbf{V}$. Further, on $\bigcup_{n\in\mathbb{N}}[0,a]^n$, 0 acts as a neutral element of $\mathbf{V}$ (and a is annihilator), that is, $\mathbf{V}$ acts on $[0,a]$ as a t–conorm. Indeed, define a binary operator $\mathbf{S}:[0,1]^2\to[0,1]$ by

$$\mathbf{S}(x,y)=\frac{\mathbf{V}(ax,ay)}{a}. \tag{176}$$

Then $\mathbf{S}$ is a t–conorm, and

$$\mathbf{V}_{|[0,a]^2}=(<0,a,\mathbf{S}>)_{|[0,a]^2}.$$

Similarly, $\mathbf{V}$ acts on $[a,1]$ as a t–norm,

$$\mathbf{V}_{|[a,1]^2}=(<a,1,\mathbf{T}>)_{|[a,1]^2},$$

where

$$\mathbf{T}(x,y)=\frac{\mathbf{V}(a+(1-a)x,a+(1-a)y)-a}{1-a},\quad (x,y)\in[0,1]^2. \tag{177}$$

Observe that fixing $\mathbf{T}, \mathbf{S}$ and $a \in]0,1[$, we have the unique nullnorm $\mathbf{V}$ satisfying (176) and (177), since

$$\mathbf{V}(x,y) = a \quad \text{whenever} \quad (x,y) \in [0,1]^2 \setminus \left([0,a]^2 \cup [a,1]^2\right).$$

Due to the associativity of $\mathbf{V}$, we can extend these results to arbitrary number of inputs, i. e.,

$$\mathbf{V}(x_1,\ldots,x_n) = \begin{cases} a \cdot \mathbf{S}\left(\frac{x_1}{a},\ldots,\frac{x_n}{a}\right) & \text{if } \max x_i \leq a, \\ a + (1-a)\cdot \mathbf{T}\left(\frac{x_1-a}{1-a},\ldots,\frac{x_n-a}{1-a}\right) & \text{if } \min x_i \geq a, \\ a & \text{else.} \end{cases} \tag{178}$$

For a given annihilator $a \in]0,1[$, there is the unique idempotent nullnorm (related to $\mathbf{S} = \mathbf{max}$ and $\mathbf{T} = \mathbf{min}$), namely $\mathbf{med}_a$ (a-median), discussed already in Section 2.10, see (43), $\mathbf{med}_a(x,y) = \mathrm{med}(x,y,a)$. Recall again that these important operators were introduced by Fung and Fu [39] and further studied in [31]. The next result clarifying the structure of nullnorms is based on a-medians, compare also [75].

Proposition 20. *An aggregation operator* $\mathbf{V} : \bigcup_{n\in\mathbb{N}} [0,1]^n \to [0,1]$ *is a nullnorm if and only if there is a t–norm* $\mathbf{T}$*, a t–conorm* $\mathbf{S}$ *and an element* $a \in]0,1[$ *such that* $\mathbf{V}$ *is a composed aggregation operator (see Proposition 6),*

$$\mathbf{V} = \mathbf{med}_a(\mathbf{T},\mathbf{S}),$$

that is,

$$\mathbf{V}(x_1,\ldots,x_n) = \mathrm{med}\left(\mathbf{T}(x_1,\ldots,x_n), \mathbf{S}(x_1,\ldots,x_n), a\right). \tag{179}$$

Similarly as in the case of uninorms, due to (179) we can extend nullnorms to act as universal quantifiers on any input index set, see [43].

Applying Proposition 20 to the Łukasiewicz t–norm $\mathbf{T}_L$ and t–conorm $\mathbf{S}_L$, we can find an interesting nullnorm $\mathbf{V} : \bigcup_{n\in\mathbb{N}} [0,1]^n \to [0,1]$ given by

$$\mathbf{V}(x_1,\ldots,x_n) = \mathrm{med}\left(\sum_{i=1}^{n} x_i, \sum_{i=1}^{n} x_i - (n-1), a\right). \tag{180}$$

Observe that this nullnorm is Archimedean as all nullnorms based on an Archimedean t–norm $\mathbf{T}$ and an Archimedean t–conorm $\mathbf{S}$. Moreover, it is also continuous as all nullnorms based on a continuous t–norm $\mathbf{T}$ and a continuous t–conorm $\mathbf{S}$.

Continuous nullnorms appear naturally in the framework of associative aggregation operators. Due to [85] we have the next important result.

Proposition 21. *Let* $\mathbf{A} : \bigcup_{n\in\mathbb{N}} [0,1]^n \to [0,1]$ *be a continuous associative symmetric aggregation operator and let* $a = \mathbf{A}(0,1)$. *Then:*

(i) if $a = 0$, $\mathbf{A}$ *is a t–norm;*
(ii) if $a = 1$, $\mathbf{A}$ *is a t–conorm;*

(iii) if $a \in]0,1[$, $\mathbf{A}$ *is a nullnorm with annihilator* a.

Also for nullnorms we have a counterpart of Propositions 15 and 19 concerning additive generators [16].

Proposition 22. *An operator* $\mathbf{V} : \bigcup_{n\in\mathbb{N}} [0,1]^n \to [0,1]$ *is a continuous nilpotent nullnorm with annihilator* $a \in]0,1[$ *if and only if there is an increasing bijection* $q : [0,1] \to [0,1]$ *such that*

$$\mathbf{V}(x_1,\ldots,x_n) = q^{-1}\left(\mathrm{med}(\sum_{i=1}^n q(x_i), \sum_{i=1}^n q(x_i) - (n-1), q(a))\right), \quad (181)$$

where $\mathbf{V}$ *is a nilpotent nullnorm whenever for any* $x \in [0,1]$ *there is* $k \in \mathbb{N}$ *such that*

$$\mathbf{V}_{(k)}(x,\ldots,x) \in \{0, a, 1\}.$$

Each nilpotent nullnorm is related to a nilpotent t–conorm $\mathbf{S}$ (with normed additive generator $s : [0,1] \to [0,\infty]$, $s(1) = 1$) and a nilpotent t–norm $\mathbf{T}$ (with normed additive generator $t : [0,1] \to [0,\infty]$, $t(0) = 1$), and then

$$q(x) = \begin{cases} a \cdot s\left(\frac{x}{a}\right) & \text{if } x \in [0,a], \\ 1 - (1-a)\cdot t\left(\frac{x-a}{1-a}\right) & \text{if } x \in]a,1]. \end{cases}$$

The operator introduced in (180) related to $\mathbf{S}_L$ and $\mathbf{T}_L$ has an additive generator $q : [0,1] \to [0,1]$, $q(x) = x$ (independently of a).

Observe that Archimedean continuous nullnorms related to a strict t–conorm $\mathbf{S}$ and/or to a strict t–norm $\mathbf{T}$ cannot be represented by means of additive generators as in (181). However, then some multiplicative version of (181) is of use. For readers interested in this topic we recommend [25].

6.4 Other aggregation operators related to t–norms

Several other aggregation operators are related to t–norms and t–conorms. As already observed when discussing uninorms, triangular conorms lack upwards compensation while triangular norms lack downwards compensation, both properties being naturally presented in human decision making. Therefore, in last twenties, several alternative attempts to overcome these undesirable properties but still to stay closely by t–norms/t–conorms, especially because of the acceptable computational complexity, have been done.

Recall especially the gamma operators

$$\mathbf{\Gamma}_\gamma : \bigcup_{n\in\mathbb{N}} [0,1]^n \to [0,1]$$

introduced by Zimmermann and Zysno [144] and applied in the car control.

(i) Gamma operators

For a parameter $\gamma \in [0,1]$, the gamma operator $\mathbf{\Gamma}_\gamma$ is given by $\mathbf{\Gamma}_\gamma = \mathbf{\Pi}^{1-\gamma}\mathbf{S}_P^\gamma$, that is,

$$\mathbf{\Gamma}_\gamma(x_1,\ldots,x_n) = \left(\prod_{i=1}^n x_i\right)^{1-\gamma}\left(1-\prod_{i=1}^n(1-x_i)\right)^\gamma. \tag{182}$$

Evidently, gamma operators are composed operators related to the weighted geometric mean (as the outer operator) and to the product $\mathbf{\Pi}$ and probabilistic sum $\mathbf{S}_P$ (as the inner operators). Parameter γ can be viewed as a degree of upwards compensation. Indeed, if $\gamma = 1$ then $\mathbf{\Gamma}_1 = \mathbf{S}_P$ (total upwards compensation) while if $\gamma = 0$ then $\mathbf{\Gamma} = \mathbf{\Pi}$ (absolute lack of upwards compensation). Also observe that for all $\gamma \in [0,1[$, $a = 0$ is the annihilator of $\mathbf{\Gamma}_\gamma$. All gamma operators are symmetric continuous and Archimedean aggregation operators which are neither associative nor possessing a neutral element (up to the boundary cases $\gamma \in \{0,1\}$).

(ii) Exponential convex $\mathbf{T}-\mathbf{S}-$ operators

Gamma operators are a special subclass of so called exponential convex $\mathbf{T}-\mathbf{S}-$ operators [125,144], that is, of weighted geometric means of a t–norm $\mathbf{T}$, and a t–conorm $\mathbf{S}$ (not necessarily a dual pair), $\mathbf{E}_{\mathbf{T},\mathbf{S},\gamma} : \bigcup_{n\in\mathbb{N}} [0,1]^n \to [0,1]$,

$$\mathbf{E}_{\mathbf{T},\mathbf{S},\gamma}(x_1,\ldots,x_n) = (\mathbf{T}(x_1,\ldots,x_n))^{1-\gamma}(\mathbf{S}(x_1,\ldots,x_n))^\gamma. \tag{183}$$

Obviously, $\mathbf{E}_{\mathbf{T},\mathbf{S},0} = \mathbf{T}$ and $\mathbf{E}_{\mathbf{T},\mathbf{S},1} = \mathbf{S}$. Any exponential convex $\mathbf{T}-\mathbf{S}-$ operator is symmetric, and for $\gamma < 1$, $a = 0$ is its annihilator. The continuity of $\mathbf{E}_{\mathbf{T},\mathbf{S},\gamma}$, $\gamma \in]0,1[$, is equivalent to the continuity of $\mathbf{T}$ and $\mathbf{S}$, while its Archimedeanity is equivalent to the Archimedeanity of $\mathbf{T}$. The only idempotent exponential convex $\mathbf{T}-\mathbf{S}-$ operators are related to $\mathbf{T} = \mathbf{min}$ and $\mathbf{S} = \mathbf{max}$, in which case we obtain a special ordered weighted geometric mean,

$$\mathbf{E}_{\mathbf{min},\mathbf{max},\gamma}(x_1,\ldots,x_n) = (x_1')^{1-\gamma}(x_n')^\gamma, \tag{184}$$

$(x_1',\ldots,x_n')$ being as usually, a non-decreasing permutation of $(x_1,\ldots,x_n)$. Observe that for a fixed $n \in \mathbb{N}$, also some other exponential convex $\mathbf{T}-\mathbf{S}-$

operators can be idempotent. For example, let $\mathbf{T}$ be the Hamacher product, $\mathbf{T}_0^H(x,y) = \frac{xy}{x+y-xy}$ (with convention $\frac{0}{0}=0$), and let $\mathbf{S}=\mathbf{S}_P$, $\gamma=0.5$. Then

$$\left(\mathbf{E}_{\mathbf{T}_0^H,\mathbf{S},0.5}\right)_{(2)}(x,y) = \sqrt{xy} = \mathbf{G}_{(2)}(x,y).$$

However, the corresponding ternary operator is not idempotent,

$$\left(\mathbf{E}_{\mathbf{T}_0^H,\mathbf{S},0.5}\right)_{(3)}(x,x,x) = x\cdot\sqrt{\frac{3-3x+x^2}{3-2x}}, \quad x\in[0,1].$$

(iii) Linear convex $\mathbf{T}-\mathbf{S}-$ operators

Another composed aggregation approach based on t–norms and t–conorms proposed in [125,144] is related to the weighted arithmetic mean (as the outer operator). A linear convex $\mathbf{T}-\mathbf{S}-$ operator $\mathbf{L}_{\mathbf{T},\mathbf{S},\gamma} : \bigcup_{n\in\mathbb{N}}[0,1]^n \to [0,1]$ is given by

$$\mathbf{L}_{\mathbf{T},\mathbf{S},\gamma}(x_1,\ldots,x_n) = (1-\gamma)\cdot\mathbf{T}(x_1,\ldots,x_n) + \gamma\cdot\mathbf{S}(x_1,\ldots,x_n). \tag{185}$$

Note that these operators were successfully applied in the fuzzy linear programming [77]. Linear convex $\mathbf{T}-\mathbf{S}-$ operators are symmetric, continuous whenever $\mathbf{T}$ and $\mathbf{S}$ are continuous, neither with annihilator nor with neutral element whenever $\gamma\in]0,1[$. Observe that if $\mathbf{T}$ and $\mathbf{S}$ are Archimedean operators, then for all $x\in]0,1[$,

$$\lim_{n\to\infty}\left(\mathbf{L}_{\mathbf{T},\mathbf{S},\gamma}\right)_{(n)}(x,\ldots,x) = \gamma.$$

Moreover, the only idempotent linear convex $\mathbf{T}-\mathbf{S}-$ operators are special OWA operators

$$\mathbf{L}_{\mathbf{min},\mathbf{max},\gamma}(x_1,\ldots,x_n) = (1-\gamma)\cdot x_1' + \gamma\cdot x_n', \tag{186}$$

where $x_1' = \min(x_1,\ldots,x_n)$, $\quad x_n' = \max(x_1,\ldots,x_n)$.

Also here some specific n-ary linear convex $\mathbf{T}-\mathbf{S}-$ operators can be idempotent and different from (186). Namely, for any Frank t–norm $\mathbf{T}_\lambda^F$, $\lambda\in]0,\infty]$ given in (195), (for more details see [36,61]) and its dual t–conorm $\mathbf{S}_\lambda^F$, $\left(\mathbf{L}_{\mathbf{T}_\lambda^F,\mathbf{S}_\lambda^F,0.5}\right)_{(2)} = \mathbf{M}_{(2)}$. For example, for $\lambda=1$ we have $\mathbf{T}_1^F = \mathbf{\Pi}$, and then

$$\mathbf{L}_{\mathbf{\Pi},\mathbf{S}_P,0.5}(x,y) = 0.5xy + 0.5(x+y-xy) = \frac{x+y}{2},$$

while

$$\mathbf{L}_{\mathbf{\Pi},\mathbf{S}_P,0.5}(x,x,x) = x^3 + \frac{3}{2}(x-x^2).$$

Similarly, for $\lambda = \infty$ we have $\mathbf{T}_\infty^F = \mathbf{T}_L$ and then

$$\mathbf{L}_{\mathbf{T}_L,\mathbf{S}_L,0.5}(x,y) = 0.5\max(0,x+y-1) + 0.5\min(1,x+y) = \frac{x+y}{2},$$

while

$$\mathbf{L}_{\mathbf{T}_L,\mathbf{S}_L,0.5}(x,x,x) = \begin{cases} \frac{3x}{2} & \text{if } x \in [0,\frac{1}{3}[, \\ \frac{1}{2} & \text{if } x \in [\frac{1}{3},\frac{2}{3}], \\ \frac{3x-1}{2} & \text{if } x \in]\frac{2}{3},1[. \end{cases}$$

(iv) Symmetric sums related to t–norms and t–conorms

The parameter $\gamma \in [0,1]$ in previous items (i) - (iii) can be vieved as a kind of orness parameter. Then if we fix γ, it stays constant independently of the input we have to aggregate. However, the human attitude to increase the high inputs and to decrease the low inputs is not reflected in these cases. To overcome this lack, Yager and Filev [139] proposed to choose first the parameter γ in dependence on the actual input values and then provide the final aggregation (see also Section 3.6). As a final result, for t–norms with no zero divisors, the symmetric sum $\mathbf{T}^\sharp$, see (67) in Section 3.2, was proposed,

$$\mathbf{T}^\sharp(x_1,\ldots,x_n) = \frac{\mathbf{T}(x_1,\ldots,x_n)}{\mathbf{T}(x_1,\ldots,x_n) + \mathbf{T}(1-x_1,\ldots,1-x_n)}, \tag{187}$$

where convention $\frac{0}{0} = \frac{1}{2}$ (for symmetric sums) can be replaced by some other convention, e. g., $\frac{0}{0} = 0$. Especially, if $\mathbf{T} = \mathbf{\Pi}$, we obtain the $3 - \Pi$–operator $\mathbf{E} = \mathbf{\Pi}^\sharp$ introduced in (39).

Similarly, t–conorm-based symmetric sums can be introduced by

$$\mathbf{S}^\sharp(x_1,\ldots,x_n) = \frac{\mathbf{S}(x_1,\ldots,x_n)}{\mathbf{S}(x_1,\ldots,x_n) + \mathbf{S}(1-x_1,\ldots,1-x_n)}. \tag{188}$$

As a special idempotent symmetric self-dual continuous aggregation operator we can introduce $\mathbf{min}^\sharp$ and $\mathbf{max}^\sharp$ given by

$$\mathbf{min}^\sharp(x_1,\ldots,x_n) = \frac{x_1'}{x_1' + 1 - x_n'}, \tag{189}$$

and

$$\mathbf{max}^\sharp(x_1,\ldots,x_n) = \frac{x_n'}{x_n' + 1 - x_1'}, \tag{190}$$

where $(x_1',\ldots,x_n')$ is a non-decreasing permutation of $(x_1,\ldots,x_n)$. Inspired by (189) and (190), for any fixed $n \in \mathbb{N}$, $i,j \in \{1,\ldots,n\}$, the operator $\mathbf{A}_{i,j} : [0,1]^n \to [0,1]$ can be introduced,

$$\mathbf{A}_{i,j}(x_1,\ldots,x_n) = \frac{x_i'}{x_i' + 1 - x_j'}. \tag{191}$$

Evidently, $\mathbf{A}_{i,j}$-operator generalizes the order statistics $\mathbf{P}'_i = \mathbf{A}_{i,i}$. Moreover, any $\mathbf{A}_{i,j}$-operator is idempotent, symmetric, continuous, and self-dual whenever $i + j = n + 1$.

(v) Copulas and dual copulas

Copulas are specific aggregation operators with neutral element 1 which are 1-Lipschitz, see Section 2.4 (and hence continuous). Copulas are applied to aggregate marginal distribution functions into an output joint distribution function. For interested readers we recommend the recent monograph [107]. Here we briefly recall so called 2-copulas (aggregating 2 marginal distribution functions).

Definition 28. A binary aggregation operator $\mathbf{C} : [0,1]^2 \to [0,1]$ is called a 2-copula (copula for short) if 1 is its neutral element and $\mathbf{C}$ has the moderate growth property, i. e., for all $x, y, u, v \in [0,1]$, $x \le u$, $y \le v$,

$$\mathbf{C}(x,y) + \mathbf{C}(u,v) \ge \mathbf{C}(x,v) + \mathbf{C}(u,y). \tag{192}$$

In general, copulas are neither symmetric nor associative. Observe that the class of copulas is closed under weighted means composition, that is, if $\mathbf{C}_1, \ldots, \mathbf{C}_k$ are copulas and $w_1, \ldots, w_k \in [0,1]$ are weights, $\sum_{i=1}^{k} w_i = 1$, then also $\mathbf{C} = \sum_{i=1}^{k} w_i \mathbf{C}_i$ is a copula. Remarkably, a copula $\mathbf{C}$ is associative if and only if it is a 1-Lipschitz t–norm. Consequently, an associative copula is necessarily symmetric. Due to Moynihan [102], we have the next result.

Proposition 23. *An Archimedean continuous t–norm $\mathbf{T}$ is a copula if and only if it is generated by a convex additive generator $t : [0,1] \to [0,\infty]$. A copula $\mathbf{C}$ is associative if and only if it is a continuous t–norm represented by (163), where all mappings $t_k : [a_k, b_k] \to [0,\infty], k \in K$, are convex.*

Note that for any copula $\mathbf{C}$ we have

$$\mathbf{T}_L \le \mathbf{C} \le \mathbf{min},$$

and moreover, $\mathbf{T}_L$ is the weakest copula while $\mathbf{min}$ is the strongest copula.

Also the product $\mathbf{\Pi}$ is a copula (observe the convexity of its additive generator $t_\Pi : [0,1] \to [0,\infty]$, $t_\Pi(x) = -\log x$), and obviously, it produces the joint distribution in the case of independent marginal random variables. A linear combination $\mathbf{C} = p \cdot \mathbf{\Pi} + (1-p) \cdot \mathbf{min}$,

$$\mathbf{C}(x,y) = pxy + (1-p)\min(x,y) = x'(1 - p(1-y')), \tag{193}$$

is an example of a non-associative symmetric copula (whenever $p \in]0,1[$).

Dual copulas are not necessarily dual aggregation operators to copulas. For a given copula $\mathbf{C} : [0,1]^2 \to [0,1]$, a dual copula $\tilde{\mathbf{C}} : [0,1]^2 \to [0,1]$ is given by $\tilde{\mathbf{C}}(x,y) = x + y - \mathbf{C}(x,y)$. Dual copulas are 1-Lipschitz aggregation operators with neutral element $e = 0$, and if they are associative, then they are specific continuous t-conorms (related to convex additive generators). An interesting problem when a dual associative copula $\tilde{\mathbf{C}}$ coincides with dual aggregation operator $\mathbf{C}^d$, that is,

$$x + y - \mathbf{C}(x,y) = 1 - \mathbf{C}(1-x, 1-y) \tag{194}$$

was solved by Frank [36]. For Archimedean associative copulas, the only solutions of (194) are the members of the Frank family $\left(\mathbf{T}_\lambda^F\right)_{\lambda \in]0,\infty]}$, $\mathbf{T}_1^F = \mathbf{\Pi}$, $\mathbf{T}_\infty^F = \mathbf{T}_L$, and for $\lambda \in]0,1[\cup]1,\infty[$,

$$\mathbf{T}_\lambda^F(x,y) = \log\left(1 + \frac{(\lambda^x - 1)(\lambda^y - 1)}{\lambda - 1}\right). \tag{195}$$

Observe that

$$\lim_{\lambda \to 0^+} \mathbf{T}_\lambda^F = \mathbf{T}_0^F = \mathbf{min}$$

is also an associative solution of (194), as well as appropriate t–norm ordinal sums related to the Frank family $\left(\mathbf{T}_\lambda^F\right)_{\lambda \in]0,\infty]}$.

7 Generated aggregation operators

Several classes of aggregation operators are directly related to the standard addition of reals. Such a typical example is the bounded sum $\mathbf{S}_L$, $\mathbf{S}_L(x_1, \ldots, x_n) = \min\left(1, \sum_{i=1}^{n} x_i\right)$. Sometimes, the inputs are first modified (by means of some real functions) and the final aggregation is then related to the sum of transformed values, see e. g., (107), (162), (167), (173), etc. Note also that all summation related formulae can be rewritten into a multiplicative form because of the standard logarithm based relationship of multiplication and summation, $\sum_{i=1}^{n} \log x_i = \log \prod_{i=1}^{n} x_i$ for all $x_1, \ldots, x_n \in [0, \infty]$ (and conventions $+\infty + (-\infty) = -\infty$, $0 \cdot \infty = 0$). Throughout this chapter, we will deal with aggregation operators derived from summation only, and these operators will be called generated aggregation operators.

7.1 Generating triples

For the sake of simplicity, in this section only binary aggregation operators will be discussed. However, a generalization to n-ary aggregation operators and also to global aggregation operators is straightforward, compare, e. g., [73] or [69]. The idea of generating triples was firstly introduced by Komorníková in [72].

Definition 29. Let $f, g : [0,1] \to [-\infty, \infty]$ be two non-decreasing continuous functions and let $h : \mathit{Ran}\, f + \mathit{Ran}\, g \to [0,1]$ be a non-decreasing surjection. Then the triple (f, g, h) is called a generating triple.

Proposition 24. *Let* (f, g, h) *be a generating triple and let* $\mathbf{A} : [0,1]^2 \to [0,1]$ *be given by*

$$\mathbf{A}(x, y) = h\left(f(x) + g(y)\right). \tag{196}$$

Then $\mathbf{A}$ *is a binary aggregation operator which is called a generated aggregation operator.*

Observe that a generated aggregation operator is always a continuous aggregation operator, up to the case when $f(0) = -g(1) \in \{-\infty, \infty\}$, in which case the non-continuity in the point $(0,1)$ appears, and when $f(1) = -g(0) \in \{-\infty, \infty\}$ resulting in a non-continuity in the point $(1,0)$. In both cases, $\mathit{Dom}\, h = [-\infty, \infty]$. However, there are generated aggregation operators which are continuous though $\mathit{Dom}\, h = [-\infty, \infty]$. Define, e. g., a generating triple (f, g, h) putting

$$f(x) = \log \frac{x}{1-x}, \quad g(x) = \log(1+x), \quad \text{and} \quad h : [-\infty, \infty] \to [0,1],\ h = f^{-1}.$$

Then we get a continuous generated aggregation operator

$$\mathbf{A} : [0,1]^2 \to [0,1], \qquad \mathbf{A}(x,y) = \frac{x+xy}{1+xy}.$$

Note that a special subclass of all generated aggregation operators generated by linear generating triples (f,g,h), that is, when all involved functions are linear, is just the class of all weighted means discussed in Section 4.1. Similarly, any quasi-arithmetic mean $\mathbf{M}_f$, see Section 4.3, is a special generated aggregation operator generated by a generating triple $\left(\frac{f}{2},\frac{f}{2},f^{-1}\right)$ (or $\left(\frac{-f}{2},\frac{-f}{2},-f^{-1}\right)$ if f is decreasing). Obviously, generated aggregation operators include also any weighted quasi-arithmetic mean.

Based on the results from [73], remark that a generated aggregation operator $\mathbf{A}$ can be derived from different generating triples which differs not only in multiplicative constants. Further, each symmetric generated aggregation operator can be deduced from a generating triple (f,g,h) in which $f=g$. Obviously, a non-symmetric generated aggregation operator $\mathbf{A}$ generated by a generating triple (f,g,h) can be symmetrized to a symmetric aggregation operator $\mathbf{A}' : [0,1]^2 \to [0,1]$ as described in Section 2.6,

$$\mathbf{A}'(x,y) = \mathbf{A}(x',y') = h\left(f(x') + g(y')\right). \tag{197}$$

where $x' = \min(x,y)$ and $y' = \max(x,y)$.

Note also that if $\mathbf{A}$ is generated by a generating triple (f,g,h), then also its dual operator $\mathbf{A}^d$ is a generated aggregation operator which is generated by a generating triple (f^d,g^d,h^d), $f^d(x) = -f(1-x)$ and $g^d(x) = -g(1-x)$ for all $x \in [0,1]$, and $h^d : Ran\ f^d + Ran\ g^d \to [0,1]$, $h^d(x) = 1 - h(-x)$. To get a generated symmetric sum (that is, a self-dual aggregation operator), it is therefore enough to deal with a generating triple (f,g,h) in which the graphs of the functions f and g are symmetric with respect to the point $(0.5,0)$ and the graph of the function h is symmetric with respect to the point $(0,0.5)$. For example, define $f = g : [0,1] \to [-\infty,\infty]$ by $f(x) = x - 0.5$ and $h_k : [-1,1] \to [0,1]$, $h_k(x) = \frac{x^{2k+1}}{2}$, where $k \in \mathbb{N}$. Then the generating triple (f,f,h_k) induces a symmetric self-dual generated aggregation operator

$$\mathbf{A}_k : [0,1]^2 \to [0,1], \quad \mathbf{A}_k(x,y) = \frac{(x+y-1)^{2k+1}+1}{2}.$$

In the next sections we will discuss generated aggregation operators with specific properties.

7.2 Idempotent generated aggregation operators

Recall that any weighted mean is an idempotent generated aggregation operator. The next result from [73] fully characterizes all idempotent generated aggregation operators.

Proposition 25. *A generating triple (f,g,h) defines an idempotent generated aggregation operator $\mathbf{A} : [0,1]^2 \to [0,1]$ if and only if the function $q : [0,1] \to [-\infty,\infty]$, $q = f + g$ is strictly monotone and $h = q^{-1}$.*

As a consequence of Proposition 25, an idempotent generated aggregation operator **A** is symmetric if and only if **A** is a quasi-arithmetic mean.

Example 20 (i) Define $f,g : [0,1] \to [-\infty,\infty]$ by $f(x) = \max(0, 2x-1)$ and $g(x) = \min(2x,1)$. Then

$$q = f + g : [0,1] \to [-\infty,\infty], \quad q(x) = 2x.$$

Put

$$h : [0,2] \to [0,1], \quad h(x) = q^{-1}(x) = \frac{x}{2}.$$

Then the generating triple (f,g,h) induces an idempotent generated aggregation operator $\mathbf{A} : [0,1]^2 \to [0,1]$ given by

$$\mathbf{A}(x,y) = \max(0, x-0.5) + \min(y, 0.5). \tag{198}$$

Observe that **A** is non-symmetric. Remarkably, the corresponding symmetric aggregation operator $\mathbf{A}' : [0,1]^2 \to [0,1]$, see (197), is given by

$$\mathbf{A}'(x,y) = \mathrm{med}(x,y,0.5), \tag{199}$$

that is, $\mathbf{A}' = \mathbf{med}_{0.5}$ is 0.5-median introduced in Section 2.10 in (43).

(ii) Continuing in (i), the generating triple (g,f,h) generates the idempotent generated aggregation operator $\mathbf{B} : [0,1]^2 \to [0,1]$ given by

$$\mathbf{B}(x,y) = \min(x,0.5) + \max(0, y-0.5) = \mathbf{A}(y,x).$$

Now, the symmetrization (197) leads to the operator $\mathbf{B}' : [0,1]^2 \to [0,1]$ given by

$$\mathbf{B}'(x,y) = \begin{cases} \min(x,y) & \text{if } \min(x,y) \leq 0.5, \\ \max(x,y) & \text{if } \max(x,y) > 0.5, \\ x+y-0.5 & \text{else,} \end{cases} \tag{200}$$

that is, $\mathbf{B}'$ is the ordinal sum of the operator **min** acting on $[0,0.5]$ and of the operator **max** acting on $[0.5,1]$, where the relevant ordinal sum is performed by means of $\mathbf{A}^{(id)}$, see Section 3.4.

Observe that similarly to Example 20 (i), each a-median $\mathbf{med}_a$, that is, each idempotent nullnorm (t–norm, t–conorm) can be obtained by symmetrization of a convenient generated aggregation operator.

7.3 Associative generated aggregation operators

Associative generated operators were introduced already by Abel in 1826 in [1]. Generated uninorms, see (173) in Section 6.2, are examples of non-continuous symmetric generated aggregation operators, for example, the famous $3-\Pi$–operator **E** given in (39). Obviously, if a generated uninorm $\mathbf{U} : [0,1]^2 \to [0,1]$ is generated by an additive generator $f : [0,1] \to [-\infty, \infty]$, f being an increasing bijection, then **U** is also generated by the generating triple (f, f, f^{-1}). We conjecture that these operators are the only non-continuous associative generated aggregation operators. However, in the case of continuous associative generated aggregation operators, we have their full characterization in [16].

Proposition 26. *A continuous associative generated aggregation operator* $\mathbf{A} : [0,1]^2 \to [0,1]$ *is:*

(i) non-symmetric if and only if $\mathbf{A} \in \{\mathbf{P}_F, \mathbf{P}_L\}$, *that is, if* $\mathbf{A}$ *is a projection operator;*
(ii) symmetric and $\mathbf{A}(0,1) = \mathbf{A}(1,0) = a \in [0,1]$ *if and only if* a *is annihilator of* $\mathbf{A}$ *and*
 (1) if $a = 0$ *then* $\mathbf{A} = \mathbf{T}$ *is a continuous Archimedean t–norm;*
 (2) if $a = 1$ *then* $\mathbf{A} = \mathbf{S}$ *is a continuous Archimedean t–conorm;*
 (3) if $a \in]0,1[$ *then* $\mathbf{A} = \mathbf{V}$ *is a nilpotent nullnorm, see (181).*

For better understanding of Proposition 26, we now give the relevant generating triples.

(i) For the projection $\mathbf{P}_F$, $\mathbf{P}_F(x,y) = x$, it is enough to take as g any constant function, e. g., $g = 0$. Then f should be strictly monotone and the corresponding generating triple is $(f, 0, f^{-1})$. Similarly, $(0, f, f^{-1})$ generates $\mathbf{P}_L$, $\mathbf{P}_L(x,y) = y$.
(ii) (1) A continuous Archimedean t–norm **T** is related to a generating triple $(f, f, f^{(-1)})$, where $f : [0,1] \to [-\infty, \infty]$ is strictly increasing continuous function such that $f(1) = 0$, and $f^{(-1)} : [2f(0), 0] \to [0,1]$ is given by $f^{(-1)}(x) = f^{-1}(\max(f(0), x))$.
 (2) Similarly to (1), a continuous Archimedean t–conorm **S** is generated by a generating triple $(g, g, g^{(-1)})$, where $g : [0,1] \to [-\infty, \infty]$ is strictly increasing continuous function with $g(0) = 0$, and $g^{(-1)} : [0, 2g(1)] \to [0,1]$ is given by $g^{(-1)}(x) = g^{-1}(\min(g(1), x))$.
 (3) A nilpotent nullnorm **V** is generated by a generating triple (f, f, h), where $f : [0,1] \to [0,1]$ is an increasing bijection, and $h : [0,2] \to [0,1]$ is given by, depending on $a = \mathbf{V}(0,1) \in]0,1[$,

$$h(x) = \begin{cases} f^{-1}(x) & \text{if } x \in [0, f(a)], \\ a & \text{if } x \in]f(a), f(a)+1[, \\ f^{-1}(x-1) & \text{if } x \in [f(a)+1, 2]. \end{cases} \tag{201}$$

Observe that for all $(x,y) \in [0,1]^2$, the formulae (196) applied to the generating triple (f,f,h) and (181) applied to $q = f$ give the same result, that is, $h(x) = f^{-1}\left(\mathbf{med}_{f(a)}(x, x-1)\right)$ ($\mathbf{med}_k$ may act on arbitrary scale, similarly as **min** and **max** operators). Also note that allowing $a = 0$ in (201) we come to a nilpotent t–norm, while $a = 1$ leads to a nilpotent t–conorm.

7.4 Generated aggregation operators with neutral element

The existence of a neutral element has a considerable impact on the corresponding generated aggregation operator. First of all, it forces the symmetry of the corresponding operator. The full characterization of all generated aggregation operators possessing a neutral element is given in the next proposition shown in [73].

Proposition 27. *A generating triple (f,g,h) generates a generated aggregation operator $\mathbf{A} : [0,1]^2 \to [0,1]$ with neutral element $e \in [0,1]$ if and only if f is strictly increasing, $g = f + d$ for some constant $d \in]-\infty, \infty[$, $f(e)$ is finite and $h : [f(0)+g(0), f(1)+g(1)] \to [0,1]$ is given by*

$$h(x) = f^{-1}\left(\min(f(1), \max(f(0), x - g(e)))\right).$$

However, then $\mathbf{A}$ is generated also by a generating triple $(q, q, q^{(-1)})$, where $q = f - f(e)$ and the pseudo-inverse $q^{(-1)} : [2q(0), 2q(1)] \to [0,1]$ is given by

$$q^{(-1)}(x) = \sup\left(u \in [0,1] \mid q(u) < u\right),$$

see [60].

Hence each generated aggregation operator $\mathbf{A}$ with neutral element e can be generated by a generating triple $(q, q, q^{(-1)})$ as given in the above proposition, with $e = q^{(-1)}(0)$. Observe that this class includes also associative aggregation operators, which are:

(i) continuous Archimedean t–norms if $e = 1$; the corresponding additive generator $t = -q$;
(ii) continuous Archimedean t–conorms if $e = 0$; the corresponding additive generator $s = q$;
(iii) generated uninorms whenever $e \in]0,1[$; the corresponding additive generator $h = q$ and necessarily $Ran\, q = [-\infty, \infty]$.

Proposition 27 means that each generated aggregation operator $\mathbf{A}$ with neutral element $e \in \{0,1\}$ is necessarily associative and continuous, and it is a continuous Archimedean t-norm when $e = 1$ and a continuous Archimedean t–conorm when $e = 0$. Also any non-continuous generated aggregation operator $\mathbf{A}$ with neutral element e (which is then necessarily from $]0,1[$) is associative, and namely it is a generated uninorm.

A non-associative generated aggregation operator $\mathbf{A}$ with neutral element e (which is necessarily from $]0,1[$) is always continuous.

Further, each generated aggregation operator $\mathbf{A}$ with neutral element $e \in]0,1[$ is an ordinal sum of a generated t–norm acting on $[0,e]$ and a generated t–conorm acting on $[e,1]$ (of course, an appropriate linear transformation of mentioned operators is taken into account). More details on this topic can be found in [89].

Example 21 Let $q_e : [0,1] \to [-\infty,\infty]$ be given by $q_e(x) = x - e, \quad e \in [0,1]$. Then the generated aggregation operator $\mathbf{A}_e : [0,1]^2 \to [0,1]$ related to the generating triple $(q_e, q_e, q_e^{(-1)})$ is given by

$$\mathbf{A}_e(x,y) = \min\left(1,\ \max(0, x+y-e)\right). \tag{202}$$

Evidently, e is neutral element of $\mathbf{A}_e$. Further, $\mathbf{A}_1 = \mathbf{T}_L$ and $\mathbf{A}_0 = \mathbf{S}_L$ are the only associative members of the family $(\mathbf{A}_e)_{e\in[0,1]}$. Observe that $\mathbf{A}_e$ is the binary form of the operator $\mathbf{A}_{\{e\}}$ introduced in (25).

Conclusions

We have presented some basic properties and classes of aggregation operators. This contribution can serve as a primary handbook on fundamentals of theory of aggregation operators acting on continuous scales. Obviously, this general overview does not contain detailed proofs and ideas, and should be taken as a first look at aggregation. For specific purposes, the original sources should be consulted. So, for example, many deep results related to functional equations, such as associativity, bisymmetry, decomposability, etc., are investigated in details in Aczél's monograph [3]. Several important results related to aggregation in multicriteria decision making and preference modelling can be found in the monograph of Fodor and Roubens [34]. Many interesting ideas and examples offers the Yager and Filev monograph [139] and edited volumes [140,7], as well as monographs [107] and [61]. Nevertheless, this contribution brings the most exhaustive general overview of aggregation operators and construction methods related to them. Moreover, many new results, not yet published, are also included here. We hope that many researchers interested in the theory and applications of aggregation operators will gain profit from this work.

Acknowledgments

The authors are grateful for the support of Department of Mathematics, Faculty of Civil Engineering and Faculty of Chemical Engineering, Slovak University of Technology, Bratislava, and Department of Computer Sciences of University of Alcalá. The partial support of grants VEGA 1/7146/20, 1/8331/21,

1/7076/20, GA ČR 402/99/0032, and of projects BFN2000–1114, TIC2000–1368–C03–01, E007/2001 are also kindly acknowledged. The authors thank also to Prof. J. Fodor and Prof. L. Gonzâlez for many valuable comments and suggestions during the work on this manuscript.

References

1. N. H. Abel: Untersuchungen der Funktionen zweier unabhängigen veränderlichen Grössen x und y wie $f(x,y)$, welche die Eigenschaft haben, dass $f(z, f(x,y))$ eine symmetrische Funktion von x, y und z ist. *J. Reine Angew. Math.* **1** (1826) 11–15.
2. J. Aczél: On mean values. *Bulletin of the American Math. Society* **54** (1948) 392-400.
3. J. Aczél: *Lectures on Functional Equations and their Applications.* Academic Press, New York, 1966.
4. J. Aczél and C. Alsina: Characterization of some classes of quasilinear functions with applications to triangular norms and to synthesizing judgements. *Methods Oper. Res.* **48** (1984) 3–22.
5. P. Benvenuti and R. Mesiar: Integrals with respect to a general fuzzy measure. In: M. Grabisch, T. Murofushi and M. Sugeno, eds. *Fuzzy Measures and Integrals. Theory and Applications.* Physica- Verlag, Heidelberg, 2000, pp. 205–232.
6. P. Benvenuti and R. Mesiar: Pseudo–arithmetical operations as a basis for integration with respect to a general fuzzy measure. *Inform. Sc.*, to appear.
7. B. Bouchon–Meunier, ed.: *Aggregation and Fusion of Imperfect Information.* Physica–Verlag, Heidelberg, 1998.
8. T. Calvo, J. Martin, G. Mayor and J. Torrens: Balanced discrete fuzzy measures. *Int. J. Uncertainty, Fuzziness and Knowledge–Based Systems* **8** (2000) 665–676.
9. T. Calvo and B. De Baets: On a generalization of the absorption equation. *Int. Fuzzy. Math. Publ.* **8** (2000) 141–149.
10. T. Calvo, B. De Baets and J.C. Fodor: The functional equations of Alsina and Frank for uninorms and nullnorms. *Fuzzy Sets and Systems* **120** (2001) 15–24.
11. T. Calvo and G. Mayor: Remarks on two types aggregation functions. *Tatra Mount. Math. Publ.* **16** (1999) 235–254.
12. T. Calvo, G. Mayor, J. Torrens, J. Suñer, M. Mas and M. Carbonell: Generation of weighting triangles associated with aggregation functions. *Int. J. Uncertainty, Fuzziness and Knowledge–Based Systems* **8** (2000) 417–451.
13. T. Calvo and R. Mesiar: Weighted means based on triangular conorms. *Int. J. of Uncertainty, Fuzziness and Knowledge–Based Systems* **9** (2001).
14. T. Calvo and R. Mesiar: Criteria importances in median–like aggregation. *IEEE Transactions on Fuzzy Systems*, to appear.
15. T. Calvo and R. Mesiar: Generalized medians. *Fuzzy Sets and Systems*, to appear.
16. T. Calvo and R. Mesiar: Continuous generated associative aggregation operators. *Fuzzy Sets and Systems*, to appear.
17. T. Calvo and R. Mesiar: Stability of aggregation operators *Proceedings Eusflat'2001*, Leicester, 2001, to appear.
18. F. Chiclana, F. Herrera and F. Herrera-Viedma: The ordered weighted geometric operator. *Proceedings IPMU'2000*, Madrid, 2000, pp. 985–991.

19. A.H. Clifford: Naturally totally ordered commutative semigroups. *Amer. J. Math.* **76** (1954) 631–646.
20. G. Choquet: Theory of capacities. *Ann. Inst. Fourier* **5** (1953-54) 131–295.
21. A.C. Climescu: Sur l'équation fonctionelle de l'associativité. *Bull. École Polytechn. Iassy* **1** (1946) 1–16.
22. B. De Baets: Idempotent uninorms. *Europ. J. Oper. Research* **180** (1999) 631–642.
23. D. Denneberg: *Non–additive Measure and Integral.* Kluwer Academic Publishers, Dordrecht, 1994.
24. D. Denneberg: Non–additive measure and integral, basic concepts and their role for applications. In: M. Grabisch, T. Murofushi and M. Sugeno, eds. *Fuzzy Measures and Integrals. Theory and Applications.* Physica- Verlag, Heidelberg, 2000, pp. 42–69.
25. M. Detyniecki: *Mathematical Aggregation Operators and their Applications to Video Querying.* Ph.D. Thesis, University Paris VI, 2000.
26. J. Dombi: Basic concepts for a theory of evaluation: The aggregative operator. *Europ. J. Oper. Research* **10** (1982) 282–293.
27. D. Dubois and H. Prade: A review of fuzzy set aggregation connectives. *Inform. Sci.* **36** (1985) 85–121.
28. D. Dubois and H. Prade: Weighted minimum and maximum in fuzzy set theory. *Inform. Sci.* **39** (1986) 85–121.
29. J.J. Dujmovic: Weighted conjunctive and disjunctive means and their application in system evaluation. *Univ. Beograd Publ. Elektrotech. Fak.*, 1974, pp. 147–158.
30. J.C. Fodor: Contrapositive symmetry of fuzzy implications. *Fuzzy Sets and Systems* **69** (1995) 141–156.
31. J.C. Fodor: An extension of Fung-Fu's theorem. *Int. J. of Uncertainty, Fuziness and Knowledge–Based Systems* **4** (1996) 235–243.
32. J.C. Fodor, J.–L. Marichal and M. Roubens: Characterization of the ordered weighted averaging operators. *IEEE Transactions on Fuzzy Systems* **3** (1995) 236–240.
33. J.C. Fodor and J.–L. Marichal: On nonstrict means. *Aequationes Mathematicae* **54** (1997) 308–327.
34. J.C. Fodor and M. Roubens: *Fuzzy Preference Modelling and Multicriteria Decision Support.* Kluwer Academic Publishers, Dordrecht, 1994.
35. J.C. Fodor, R.R. Yager and A. Rybalov: Structure of uninorms. *Int. J. of Uncertainty, Fuzziness and Knowledge–Based Systems* **5** (1997) 411–427.
36. M.J Frank: (1979) On the simultaneous associativity of $F(x,y)$ and $x+y-F(x,y)$. *Aequationes Math.* **19** (1979) 194–226.
37. K. Fujimoto, T. Murofushi and M. Sugeno: Canonical hierarchical decomposition of the Choquet integral over a finite set with respect to null–additive fuzzy measure. *Int. J. of Uncertainty, Fuzziness and Knowledge–Based Systems* **6** (1998) 345–363.
38. K. Fujimoto and T. Murofushi: Hierarchical decomposition of the Choquet integral. In: M. Grabisch, T. Murofushi and M. Sugeno, eds. *Fuzzy Measures and Integrals. Theory and Applications.* Physica- Verlag, Heidelberg, 2000, pp. 94–103.
39. L.W. Fung and K.S. Fu: An axiomatic approach to rational decision making in a fuzzy environment. In: L.A. Zadeh, K.S. Fu, K. Tanaka and M. Shimura,

eds., *Fuzzy sets and Their Applications to Cognitive and Decision Processes.* Academic Press, New York, 1975, pp. 227–256.
40. L. Godo and C. Sierra: A new approach to connective generation in the framework of expert systems using fuzzy logic. In: *Proceedings 18th International Symposium on Multiple-Valued Logic.* Palma de Mallorca, IEEE Computer Society Press, 1988, pp. 157–162.
41. L. Godo and V. Torra: Extending Choquet integrals for aggregation of ordinal values. *Proceedings IPMU'2000*, Madrid, 2000, pp. 410–417.
42. L. González: A note on infinitary action of triangular norms and conorms. *Fuzzy Sets and Systems* **101** (1999) 177–180.
43. L. González: Universal aggregation operators. *Proceedings Eusflat'2001*, Leicester, 2001, to appear.
44. L. González: What is arithmetic mean? *Proceedings AGGOP'2001*, Oviedo, 2001, to appear.
45. S. Gottwald: *A Treatise on Many–Valued Logic.* Research Studies Press Ltd., Baldock, Hertforshire, 2001.
46. M. Grabisch: Fuzzy integral in multicriteria decision making. *Fuzzy Sets and Systems* **69** (1995) 279–298.
47. M. Grabisch: k–order additive fuzzy measures. *Proceedings IPMU'96*, Granada, 1996, pp. 1345-1350.
48. M. Grabisch: k–order additive discrete fuzzy measures and their representation. *Fuzzy Sets and Systems* **92** (1997) 167-189.
49. M. Grabisch: The interaction and Möbius representation of fuzzy measures on finite spaces, k–additive measures. In: M. Grabisch, T. Murofushi and M. Sugeno, eds. *Fuzzy Measures and Integrals. Theory and Applications.* Physica-Verlag, Heidelberg, 2000, pp. 70-93.
50. M. Grabisch: Symmetric and asymmetric integrals: the ordinal case. *Proceedings IIZUKA'2000*, Iizuka, 2000, CD–rom.
51. M. Grabisch, J.–L. Marichal and M. Roubens: Equivalent representations of set functions. *Math. Operat. Res.* **25** (2000) 157–178.
52. M. Grabisch, T. Murofushi, M. Sugeno, eds.: *Fuzzy Measures and Integrals. Theory and Applications.* Physica-Verlag, Heidelberg, 2000.
53. M. Grabisch, H.T. Nguyen and E.A. Walker: *Fundamentals of Uncertainty Calculi with Applications to Fuzzy Inference.* Kluwer Academic Publishers, Dordercht, 1995.
54. P. Hájek: *Metamathematics of Fuzzy Logic.* Kluwer Academic Publishers, Dordrecht, 1998.
55. P.R. Halmos: *Measure Theory.* Van Nostrand, New York, 1950.
56. H. Imaoka: On a subjective evaluation model by a generalized fuzzy integral. Int. J. of Uncertainty, Fuzziness and Knowledge–Based Systems **5** (1997) 517–529.
57. H. Imaoka: Comparison between three integrals. In: M. Grabisch, T. Murofushi and M. Sugeno, eds. *Fuzzy Measures and Integrals. Theory and Applications.* Physica- Verlag, Heidelberg, 2000, pp. 273–286.
58. E.P. Klement: Construction of fuzzy σ–algebras using triangular norms. *J. Math. Anal. Appl.* **85** (1982) pp. 543–566.
59. E.P. Klement, R. Mesiar and E. Pap: On the relationship of associative compensatory operators to triangular norms and conorms. Int. J. of Uncertainty, Fuzziness and Knowledge–Based Systems **4** (1996) 129–144.

60. E.P. Klement, R. Mesiar and E. Pap: Quasi– and pseudo–inverses of monotone functions, and the construction of t–norms. *Fuzzy Sets and Systems* **104** (1999) 3–13.
61. E.P. Klement, R. Mesiar and E. Pap: *Triangular Norms*. Kluwer Academic Publishers, Dordrecht, 2000.
62. E.P. Klement, R. Mesiar and E. Pap: Integration with respect to decomposable measures, based on a conditionally distributive semiring on the unit interval. *Int. J. of Uncertainty, Fuzziness and Knowledge–Based Systems* **8** (2000) 701–717.
63. E.P. Klement, R. Mesiar and E. Pap: Geometric approach to aggregation. *Proceedings Eusflat'2001* Leicester, 2001, to appear.
64. G.J. Klir and T.A. Folger: *Fuzzy Sets, Uncertainty and Information*. Prentice Hall, Englewood Cliffs, 1988.
65. A. Kolesárová: On the comparison of quasi–arithmetic means. *Busefal* **80** (1999) 30–34.
66. A. Kolesárová: Collapsed input-based aggregation. *Int. J. of Uncertainty, Fuzziness and Knowledge- Based Systems* **9** (2001).
67. A. Kolesárová: Limit properties of quasi–arithmetic means. *Fuzzy Sets and Systems*, to appear.
68. A. Kolesárová: Parametric evaluation of aggregation operators. Preprint, submitted.
69. A. Kolesárová and M. Komorníková: Triangular norm-based iterative aggregation and compensatory operators. *Fuzzy Sets and Systems* **104** (1999) 109-120.
70. A. Kolesárová and J. Mordelová: 1–Lipschitz and kernel aggregation operators. *Proceedings of AGGOP'2001*, Oviedo, 2001, to appear.
71. A.N. Kolmogoroff: Sur la notion de la moyenne. *Accad. Naz. Lincei Mem. Cl. Sci. Fis. Mat. Natur. Sez.* **12** (1930) 388–391.
72. M. Komorníková: Generated aggregation operators. *Proceedings EUSFLAT'99*, Palma de Mallorca, 1999, pp. 355–358.
73. M. Komorníková: Aggregation operators and additive generators. *Int. J. of Uncertainty, Fuzziness and Knowledge- Based Systems* **9** (2001).
74. J. Lázaro and T. Rückschlossová: Shift invariant binary aggregation operators. *Proceedings AGGOP'2001*, Oviedo, 2001, to appear.
75. Y.-M. Li and Z.-K. Shi: Weak uninorms aggregation operators. *Inform. Sci.* **124** (2000) 317–323.
76. C.M. Ling: Representation of associative functions. *Publ. Math. Debrecen* **12** (1965) 189–212.
77. M.K. Luhandjula: Compensatory operators in fuzzy linear programming with multiple objectives. Fuzzy Sets and Systems **8** (1982) 245–252.
78. J.–L. Marichal: *Aggregations Operators for Multi-Criteria Decision Aid*. Ph.D. Thesis, University of Liége, 1998.
79. J.–L. Marichal: On Choquet and Sugeno integrals as aggregation functions. In: M. Grabisch, T. Murofushi and M. Sugeno, eds. *Fuzzy Measures and Integrals. Theory and Applications*. Physica- Verlag, Heidelberg, 2000, pp. 247–272.
80. J.–L. Marichal: An axiomatic approach of the discrete Choquet integral as a tool to aggregate interacting criteria. *IEEE Transactions on Fuzzy Systems* **8** (2000) 800–807.
81. J.–L. Marichal: Aggregation of interacting criteria by means of the discrete Choquet integral. Chapter in this monograph.
82. J.–L. Marichal: On order invariant synthesizing functions. Preprint, submitted.

83. J.–L. Marichal: On an axiomatization of the quasi–arithmetic mean values without the symmetry axiom. *Aequationes Mathematicae* **59** (2000) 74–83.
84. J.–L. Marichal, P. Mathonet and E. Thousset: Characterization of some aggregations functions stable for positive linear transformations. *Fuzzy Sets and Systems* **102** (1999) 293–314.
85. M. Mas, G. Mayor and J. Torrens: t–operators. *Int. J. Uncertainty, Fuzziness and Knowledge–Based Systems* **7** (1999) 31–50
86. G. Mayor and T. Calvo: On extended aggregation functions. *Proceedings IFSA'97*, Prague, 1997, vol. I, pp. 281–285.
87. G. Mayor and J. Torrens: On a class of operators for expert systems. *Int. J. of Intelligent Systems* **8** (1988) 771–778.
88. K. Menger: Statistical metrics. *Procs. Nat. Acad. Sci. U.S.A.* **37** (1942) 535–537.
89. R. Mesiar: Compensatory operators based on triangular norms. *Proceedings EUFIT'95*, Aachen, 1995, pp. 131–135.
90. R. Mesiar: Choquet–like integrals. *J. Math. Anal. Appl.* **194** (1995) 477–488.
91. R. Mesiar: Generalizations of k–order additive discrete fuzzy measures. *Fuzzy Sets and Systems* **102** (1999) 423-428.
92. R. Mesiar: k–order additive measures. *Int. J. of Uncertainty, Fuzziness and Knowledge-Based Systems* **6** (1999) 561-568.
93. R. Mesiar and B. De Baets: New construction methods for aggregation operators. *Proceedings IPMU'2000*, Madrid, 2000, pp. 701–706.
94. R. Mesiar, T. Calvo and J. Martin: Integral based aggregation of real data. *Proceedings IPMU'2000*, Madrid, 2000, pp. 58–62
95. R. Mesiar and B. De Baets: Continuous ordinal sums of aggregation operators. Manuscript in preparation.
96. R. Mesiar and M. Komorníková: Aggregation operators. In: D. Herceg and K. Surla, eds., *Proceedings PRIM'96, XI. Conference on Applied Mathematics*, 1996, pp. 193–211.
97. R. Mesiar and M. Komorníková: Triangular norm–based aggregation of evidence under fuzziness. In: B. Bouchon–Meunier, ed., *Aggregation and Fusion of Imperfect Information.* Physica–Verlag, Heidelberg, 1998.
98. R. Mesiar and D. Vivona: Two–step integral with respect to fuzzy measure. *Tatra Mount. Math. Publ.* **16** (1999) 359–368.
99. R. Mesiar and H. Thiele: On T–quantifiers and S– quantifiers. In: V. Novák and I. Perfilieva, eds., *Discovering the Word with Fuzzy Logic.* Physica–Verlag, Heidelberg, 2000, pp. 310–326.
100. M. Mizumoto: Pictorial representations of fuzzy connectives, Part I.: Cases of t–norms, t–conorms and averaging operators. *Fuzzy Sets and Systems* **31** (1989) 217–242.
101. M. Mizumoto: Pictorial representations of fuzzy connectives, Part II.: Cases of compensatory operators and self–dual operators. *Fuzzy Sets and Systems* **32** (1989) 45–79.
102. R. Moynihan: On τ_T semigroups of probability distribution functions II. *Aequationes Math.* **17** (1978) 19–40.
103. E. Muel and J. Mordelová: Kernel aggregation operators. *Proceedings AGGOP'2001*, Oviedo, 2001, to appear.
104. T. Murofushi and M. Sugeno: Fuzzy t-conorm integrals with respect to fuzzy measures: generalizations of Sugeno integral and Choquet integral. *Fuzzy Sets and Systems* **42** (1991) 51–57.

105. T. Murofushi and M. Sugeno: Fuzzy measures and fuzzy integrals. In: M. Grabisch, T. Murofushi and M. Sugeno, eds. *Fuzzy Measures and Integrals. Theory and Applications*. Physica- Verlag, Heidelberg, 2000, pp. 3–41.
106. M. Nagumo: Über eine Klasse der Mittelwerte. *Japanese Journal of Mathematics* **6** (1930) 71–79.
107. R.B. Nelsen: *An Introduction to Copulas*. Lecture Notes in Statistic 139, Springer, 1999.
108. S. Ovchinnikov and A. Dukhovny: Integral representation of invariant functionals. *J. Math. Anal. Appl.* **244** (2000) 228–232.
109. E. Pap: *Null-Additive Set Functions*. Kluwer Academic Publishers, Dordrecht, 1995.
110. A.L. Ralescu and D.A. Ralescu: Extensions of fuzzy aggregation. *Fuzzy Sets and Systems* **86** (1997) 321–330.
111. T. Micháliková–Rückschlossová: Some constructions of aggregation operators. *J. Electrical Engin.* **12** (2000) 29–32.
112. T. Rückschlossová: Invariant aggregation operators. Manuscript in preparation.
113. W. Sander: Associative aggregation operators. Chapter in this monograph.
114. B. Schweizer and A. Sklar: *Probabilistic Metric Spaces*. North Holland, New York, 1983.
115. C. Shannon and W. Weaver: *The Mathematical Theory of Communication*. University of Illinois Press, Urbana, 1949.
116. N. Shilkret: Maxitive measures and integration. *Indag. Math.* **33** (1971) 109–116.
117. W. Silvert: Symmetric summation: A class of operations of fuzzy sets. *IEEE Trans. Syst., Man Cybern.* **9** (1979) 657–659.
118. D. Smutná: On a peculiar t–norm. *Busefal* **75** (1998) 60–67.
119. M. Sugeno: *Theory of Fuzzy Integrals and Applications*. Ph.D. Thesis, Tokyo Inst. of Technology, Tokyo, 1974.
120. M. Sugeno and T. Murofushi: Pseudo–additive measures and integrals, *J. Math. Anal. Appl.* **122** (1987) 197-222.
121. M. Šabo, A. Kolesárová and Š. Varga: RET operators generated by triangular norms and copulas. *Int. J. Uncertainty, Fuzziness and Knowledge–Based Systems* **9** (2001).
122. J. Šipoš: Integral with respect to a pre–measure. *Math. Slovaca* **29** (1979) 141–145.
123. V. Torra: The weighted OWA operator. *Int. J. of Intelligent Systems* **12** (1997) 153–166.
124. V. Torra and L. Godo: Continuous WOWA operators with application to defuzzification. Chapter in this monograph.
125. I.B. Türksen: Interval–valued fuzzy sets and "compensatory AND". *Fuzzy Sets and Systems* **51** (1992) 295–307.
126. P. Viceník: A note on generators of t–norms. *Busefal* **75** (1998) 33–38.
127. P. Viceník: Additive generators and discontinuity. *Busefal* **76** (1998) 25–28.
128. P. Viceník: Additive generators of non–continuous triangular norms. In: S. Rodabaugh and P. Klement, eds., *Proceedings of Linz Seminar 1999*, Kluwer Academic Publishers, to appear.
129. Z. Wang and G.J. Klir: *Fuzzy Measure Theory*, Plenum Press, 1992.
130. S. Weber: $\perp$–decomposable measures and integrals for Archimedean t–conorms $\perp$. *J. Math. Anal. Appl.* **101** (1984) 114–138.

131. S. Weber: Two integrals and some modified version–critical remarks. *Fuzzy Sets and Systems* **20** (1986) 97–105.
132. R.R. Yager: On a general class of fuzzy connectives. *Fuzzy Sets and Systems* **4** (1980) 235–242.
133. R.R. Yager: On ordered weighted averaging aggregation operators in multicriteria decisionmaking. *IEEE Trans. Syst.,Man Cybern.* **18** (1988) 183–190.
134. R.R. Yager: Criteria importances in OWA aggregation: An application of fuzzy modeling. *Proceedings IEEE'FUZZ'97*, Barcelona, 1997, pp. 1677–1682.
135. R.R. Yager: Fusion od ordinal information using weighted median aggregation. *Int. J. Approx. Reasoning* **18** (1998) 35–52.
136. R.R. Yager: Uninorms in fuzzy modeling. *Fuzzy Sets and Systems*, to appear.
137. R.R. Yager: Using importances in group preference aggregation to block strategic manipulation. Chapter in this monograph.
138. R.R. Yager, M. Detyniecki and B. Bouchon–Meunier: Specifying t–norms based on the value of $T(1/2, 1/2)$. *Mathware and Soft Computing* **7** (2000) 77-78.
139. R.R. Yager and D.P. Filev: *Essentials of Fuzzy Modelling and Control.* J. Wiley & Sons, New York, 1994.
140. R.R. Yager and J. Kacprzyk: *The Ordered Weighted Averaging Operators, Theory and applications.* Kluwer Academic Publishers, Boston, Dordrecht, London, 1997.
141. R.R. Yager and A. Rybalov: Uninorm aggregation operators. *Fuzzy Sets and Systems* **80** (1996) 111–120.
142. R.R.Yager and A. Rybalov: Noncommutative self–identity aggregation. *Fuzzy Sets and Systems* **85** (1997) 73–82.
143. L.A. Zadeh: Fuzzy sets. *Inform. Control* **8** (1965) 338–353.
144. H.J Zimmermann and P. Zysno: Latent connectives in human decision making. *Fuzzy Sets and Systems* **4** (1980) 37–51.

Part 2

Theoretical Aspects of Aggregation Operators

Aggregation Based on Integrals: Recent Results and Trends

Michel Grabisch

LIP6, Université de Paris VI
8, rue du Capitaine Scott, 75015 Paris
email Michel.Grabisch@lip6.fr

Abstract. The paper presents an overview of recent results related to the use of non additive integrals (Choquet and Sugeno integrals) as aggregation operators. We essentially address two main issues, which are the treatment of negative numbers, and the case of ordinal information to aggregate. We do not explicitely presuppose any particular form of aggregation problem (e.g. multicriteria decision making), and remain at an abstract level, although decision making in general can be considered as the main motivation of the work.

1 Introduction

In the field of aggregation operators, the use of integral-based operators has recently become a whole subject, with the introduction of the Choquet integral [5] as a flexible, yet simple, way of aggregating numerical values (see e.g. [14,16]). This should not be surprising, as the idea of mean value and area under a function are closely related, and many aggregation operators could be considered as integral operators on a finite set.

Nevertheless, it seems that the main point which has caused the interest of researchers on this topic is mainly due to the very special non linear character of e.g. the Choquet integral, or more generally any fuzzy integral, that is, integral based on fuzzy measures or capacities [43]. The common feature to all these functionals is the comonotonic (pseudo-)additivity, (see results of Schmeidler [40], de Campos and Bolaños [6] , Grabisch *et al.* [28], Benvenuti and Mesiar [1] etc.) or equivalently in the finite case, the fact that the integral is (pseudo-)additive in any simplex induced by the set of permutations on the universe of discourse.

The properties of fuzzy integrals as aggregation operators have been studied at length (see e.g. [13,38,29]), and the introduction of the Möbius transform and the interaction transform have yet permitted to go deeper into the analysis of the Choquet integral [20,15,22,21]. These results will not be cited here, and we refer the reader to the above mentioned literature.

The topics we address in this survey paper are based on practical considerations arising when one deals with decision making problems, either multicriteria decision making or decision under uncertainty or risk. The first one concerns the use of negative numbers for representing values under a fixed level of indifference or neutrality, denoted zero. This is natural every time one wants to deal with concepts which can be paired, such as good/bad, attractive/repulsive, like/dislike, gain/loss, etc. Clearly, there exists a middle point, and the treatment of "negative" and "positive" numbers should be differentiated. The mathematical problem which arises here is the following: How to extend the aggregation operator defined originally for positive numbers only, to negative numbers? The question may look trivial for simple operators as the weighted mean, but several solutions may exist for more complex ones, as the Choquet integral. In this last case, the two usual extensions are called by Denneberg [7] the *symmetric* and the *asymmetric* integrals, the first one being discovered independently by Šipoš [46]. We should mention here that in decision under uncertainty or risk, the so-called Cumulative Prospect Theory [45] extends the Choquet expected utility model for negative utilities, and is based on a generalization of the Šipoš (or symmetric) integral. For a comparative use of the symmetric and asymmetric integrals in decision under multiple criteria and decision under uncertainty, see [24].

The second topic is concerned with the use of ordinal or qualitative information. In real applications in decision making, the presence of qualitative information is the rule rather than the exception, and most of practitioners perform an arbitrary transformation of the data into a numerical scale, so that the powerful numerical machinery can be used. It is an elementary fact of measurement theory [39] that such transformations could not be done without interfering with the decision process, which means that the decision process, based somewhere on some aggregation operator, has to be *ordinal*, i.e. built uniquely with comparisons and avoiding any standard arithmetical operation. The Sugeno integral [43], defined only with the use of minimum and maximum operators, seems to be the right counterpart of the Choquet integral. It is worth to be noted that the Sugeno integral has lead to a qualitative framework of decision under uncertainty and risk, with axioms similar to the ones of expected utility (see [12]).

The picture would not be complete if we did not address the problem of considering negative "numbers" on an ordinal scale, in order to define the ordinal counterparts of the symmetric and asymmetric Choquet integrals. We address this topic in the last past of the paper.

In the sequel, $\wedge, \vee$ denote min and max respectively.

2 Basic definitions and notations

Let $N = \{1, \dots, n\}$ be a finite set, representing the index set of quantities to be aggregated (states of nature or criteria, etc.). A *capacity* or *fuzzy measure*, *non-additive measure* on N is any set function $v : \mathcal{P}(N) \longrightarrow \mathbb{R}^+$ such that $v(\emptyset) = 0$ and $A \subset B \subset N$ implies $v(A) \leq v(B)$ (monotonicity). If in addition $v(N) = 1$, v is a *normalized* capacity.

The *conjugate* capacity of v, denoted $\bar{v}$, is defined by $\bar{v}(A) = v(N) - v(A^c)$.

Particular capacities of interest are the *unanimity games* u_A, with $A \subset N$. They are defined as:

$$u_A(B) := \begin{cases} 1, & \text{if } B \supset A, \\ 0, & \text{otherwise.} \end{cases}$$

We introduce some linear transformations of capacities (see [8,22] for a full presentation of this topic).

Definition 1. Let v be a capacity. For all $A \subset N$:

- the *Möbius transform* of v is given by:

$$m^v(A) := \sum_{B \subset A} (-1)^{|A|-|B|} v(B). \tag{1}$$

- the *interaction transform* is given by:

$$I^v(A) := \sum_{B \subset N \setminus A} \xi_n(|B|, |A|) \sum_{C \subset A} (-1)^{|A|-|C|} v(C \cup B), \tag{2}$$

  with $\xi_n(k,l) = \frac{(n-k-l)!k!}{(n-l+1)!}$.
- the *co-Möbius transform* is defined by:

$$\check{m}^v(A) := \sum_{B \supset N \setminus A} (-1)^{n-|B|} v(B) = \sum_{B \subset A} (-1)^{|B|} v(N \setminus B).$$

When possible, the superscripts v will be omitted.

We give some explanations about the usefulness of each transform. It is known that any capacity can be written as a linear combination of unanimity games thanks to the Möbius transform. Specifically:

$$v(A) = \sum_{B \subset N} m(B) u_B(A). \tag{3}$$

In other words, the unanimity games form a basis for the capacities, and the Möbius transform is the coordinate vector of the capacity.

The interaction transform is a generalization of the Shapley value [42], and of the interaction index I_{ij} defined for a pair of criteria i, j by Murofushi and Soneda [36]. The Shapley value of v is the vector $\phi^v := [\phi_1^v \cdots \phi_n^v]$ defined $\forall i \in N$ by:

$$\phi_i^v := \sum_{K \subset N \setminus i} \xi_n(|K|, 1)[v(K \cup \{i\}) - v(K)]. \tag{4}$$

The interaction index I_{ij} is defined by:

$$I_{ij}^v := \sum_{K \subset N \setminus \{i,j\}} \xi_n(|K|, 2)[v(K \cup \{i,j\}) - v(K \cup \{i\}) - v(K \cup \{j\}) + v(K)], \quad \forall i, j \in N. \tag{5}$$

Clearly, $\phi_i^v = I^v(\{i\})$, and $I_{ij}^v = I^v(\{i,j\})$.

Lastly, the co-Möbius transform takes its name from the fact that it is closely related to the Möbius transform of the conjugate capacity [22]:

$$\check{m}^{\bar{v}}(A) = (-1)^{|A|+1} m^v(A), \quad \forall A \subset N, A \neq \emptyset. \tag{6}$$

Its interest is more technical, although it is known in Dempster-Shafer theory under the name of "commonality function" [41]), and in possibility theory [10] under the name of "guaranted possibility measure" (see e.g. [11]). For more details on this topic, see [22,27].

We will need in the sequel the Bernoulli numbers B_k, defined by the recurrence relation

$$B_k := -\sum_{l=0}^{k-1} \frac{B_l}{k-l+1}\binom{k}{l}, k > 0, \tag{7}$$

and $B_0 = 1$. Now $\beta_k^l := \sum_{j=0}^{k} \binom{k}{j} B_{l-j}, \quad k,l = 0,1,2,\ldots$.

We introduce now integrals with respect to capacities. As we are in the finite case, we assimilate integrand functions (resp. positive integrand functions) with vectors in $\mathbb{R}^n$ (resp. $(\mathbb{R}^+)^n$).

Definition 2. Let v be a capacity, and $a = (a_1, \ldots, a_n) \in (\mathbb{R}^+)^n$. The *Choquet integral* of a with respect to v is defined by:

$$\mathcal{C}_v(a_1, \ldots, a_n) := \sum_{i=1}^{n} [a_{(i)} - a_{(i-1)}] v(A_{(i)}), \tag{8}$$

where $\cdot_{(i)}$ indicates a permutation on N so that $a_{(1)} \leq a_{(2)} \leq \cdots \leq a_{(n)}$, and $A_{(i)} := \{(i), \ldots, (n)\}$. Also $a_{(0)} := 0$.

Definition 3. Let v be a normalized capacity, and $a = (a_1, \ldots, a_n) \in [0,1]^n$. The *Sugeno integral* of a with respect to v is defined by:

$$\mathcal{S}_v(a_1, \ldots, a_n) := \bigvee_{i=1}^{n} a_{(i)} \wedge v(A_{(i)}), \tag{9}$$

with same notations as above.

3 Aggregating numerical information

In this section we suppose that the vector of quantities to aggregate belongs to $\mathbb{R}^n$, thus it may involve negative numbers. We focus on the Choquet integral and the way we can extend it to negative integrands. We discard the use of Sugeno integral, which will be studied in the section on ordinal aggregation.

3.1 Symmetric and asymmetric Choquet integral

For any $a = (a_1, \ldots, a_n) \in \mathbb{R}^n$, let us introduce a^+, a^-, the positive and the negative parts of a, that is $a^+ = (a_1^+, \ldots, a_n^+)$, $a^- = (a_1^-, \ldots, a_n^-)$, $a_i^+ = a_i \vee 0$, $a_i^- = -a_i \vee 0$. According to Denneberg [7], two integrals can be defined.

- the *symmetric integral*, denoted $\check{C}_v$, is defined by:

  $$\check{C}_v(a) = C_v(a^+) - C_v(a^-).$$

  It was originally introduced by Šipoš [46].
- the *asymmetric integral*, denoted C_v, is defined by:

  $$\check{C}_v(a) = C_v(a^+) - C_{\bar{v}}(a^-).$$

  It is the usual definition of the Choquet integral for real integrands, hence the notation.

These two integrals take their name from the following property. For any $a \in \mathbb{R}^n$,

$$C_v(-a) = -C_{\bar{v}}(a) \tag{10}$$

$$\check{C}_v(-a) = -\check{C}_v(a). \tag{11}$$

We give now the espressions of Choquet and Šipoš integrals in terms of the capacity, the Möbius transform, the co-Möbius transform, and the interaction transform, for any real-valued integrand. Proofs can be found in [25,26].

Proposition 1. *Let v be a capacity. For any $a \in \mathbb{R}^n$,*

$$C_v(a) = a_{(1)} + \sum_{i=2}^{n} [a_{(i)} - a_{(i-1)}] v(\{(i), \ldots, (n)\}) \tag{12}$$

$$\begin{aligned}\check{C}_v(a) = &\sum_{i=1}^{p-1} [a_{(i)} - a_{(i+1)}] v(\{(1), \ldots, (i)\}) + a_{(p)} v(\{(1), \ldots, (p)\}) \\ &+ a_{(p+1)} v(\{(p+1), \ldots, (n)\}) + \sum_{i=p+2}^{n} [a_{(i)} - a_{(i-1)}] v(\{(i), \ldots, (n)\})\end{aligned} \tag{13}$$

where $\cdot_{(i)}$ indicates a permutation on N so that $a_{(1)} \leq a_{(2)} \leq \cdots \leq a_{(p)} < 0 \leq a_{(p+1)} \leq \cdots \leq a_{(n)}$.

Proposition 2. *Let v be a capacity. For any $a \in \mathbb{R}^n$,*

$$C_v(a) = \sum_{A \subset N} m(A) \bigwedge_{i \in A} a_i, \tag{14}$$

$$\begin{aligned}\check{C}_v(a) &= \sum_{A \subset N} m(A) \left[\bigwedge_{i \in A} a_i^+ - \bigwedge_{i \in A} a_i^- \right] \\ &= \sum_{A \subset N^+} m(A) \bigwedge_{i \in A} a_i + \sum_{A \subset N^-} m(A) \bigvee_{i \in A} a_i,\end{aligned} \tag{15}$$

where $N^+ := \{i \in N | a_i \geq 0\}$ *and* $N^- = N \setminus N^+$.

As remarked by Calvo and Mesiar [3], equation (15) can be written as follows:

$$\check{\mathcal{C}}_v(a) = \sum_{A \subset N} m(A) \mathrm{med}\Big(\bigwedge_{i \in A} a_i, \bigvee_{i \in A} a_i, 0\Big)$$

where med indicates the median.

Proposition 3. *For any* $a \in \mathbb{R}^n$,

$$\mathcal{C}_v(a) = \sum_{A \subset N, A \neq \emptyset} (-1)^{|A|+1} \check{m}^v(A) \bigvee_{i \in A} a_i = \sum_{A \subset N} m^{\bar{v}}(A) \bigvee_{i \in A} a_i \tag{16}$$

$$\check{\mathcal{C}}_v(a) = \sum_{A \cap N^+ \neq \emptyset} (-1)^{|A|+1} \check{m}^v(A) \bigvee_{i \in A} a_i + \sum_{A \cap N^- \neq \emptyset} (-1)^{|A|+1} \check{m}^v(A) \bigwedge_{i \in A} a_i \tag{17}$$

$$\check{\mathcal{C}}_v(a) = \sum_{A \subset N^+} (-1)^{|A|+1} \check{m}^{\bar{v}}(A) \bigwedge_{i \in A} a_i + \sum_{A \subset N^-} (-1)^{|A|+1} \check{m}^{\bar{v}}(A) \bigvee_{i \in A} a_i, \tag{18}$$

where $N^+ = \{i \in N | a_i \geq 0\}$, *and* $N^- = N \setminus N^+$.

Proposition 4. *Let* $a \in \mathbb{R}^n$.

$$\begin{aligned} \mathcal{C}_v(a) = & \sum_{A \subset N} \Bigg(\sum_{B \subset N \setminus A} B_{|B|} I^+(A \cup B) \Bigg) \bigwedge_{i \in A} a_i \\ & + \sum_{A \subset N, A \neq \emptyset} (-1)^{|A|+1} \Bigg(\sum_{B \subset N \setminus A} \beta_{|B|}^{|B|} I^-(A \cup B) \Bigg) \bigvee_{i \in A} a_i \end{aligned} \tag{19}$$

$$\begin{aligned} \check{\mathcal{C}}_v(a) = & \sum_{A \subset N^+} \Bigg(\sum_{B \subset N \setminus A} B_{|B|} I^+(A \cup B) \Bigg) \bigwedge_{i \in A} a_i \\ & + \sum_{A \subset N^-} \Bigg(\sum_{B \subset N \setminus A} B_{|B|} I^+(A \cup B) \Bigg) \bigvee_{i \in A} a_i \\ & + \sum_{A \cap N^+ \neq \emptyset} (-1)^{|A|+1} \Bigg(\sum_{B \subset N \setminus A} \beta_{|B|}^{|B|} I^-(A \cup B) \Bigg) \bigvee_{i \in A} a_i \\ & + \sum_{A \cap N^- \neq \emptyset} (-1)^{|A|+1} \Bigg(\sum_{B \subset N \setminus A} \beta_{|B|}^{|B|} I^-(A \cup B) \Bigg) \bigwedge_{i \in A} a_i, \end{aligned} \tag{20}$$

where $I^+(A) = I(A)$ *if* $I(A) > 0$ *and 0 otherwise, and* $I^-(A) = I(A)$ *if* $I(A) < 0$ *and 0 otherwise.*

3.2 Main properties

We give some comparative properties of Choquet and Šipoš integrals, which are straightforward from the definitions. We will use the notation (a_A, b_{A^c}) with $a, b \in \mathbb{R}$ and $A \subset N$ for denoting a point $(x_1, \ldots, x_n)$ in $\mathbb{R}^n$ such that $x_i = a$ if $i \in A$, and $x_i = b$ otherwise.

Proposition 5. *Let $a \in \mathbb{R}^n$ and v be a capacity. The Choquet and Šipoš integrals satisfy the following properties.*

(i) extension of the capacity

$$\mathcal{C}_v(1_A, 0_{A^c}) = \check{\mathcal{C}}_v(1_A, 0_{A^c}) = v(A), \forall A \subset N. \tag{21}$$

(ii) symmetry and asymmetry. For any $a \in \mathbb{R}^n$,

$$\mathcal{C}_v(-a) = -\mathcal{C}_{\bar{v}}(a) \tag{22}$$

$$\check{\mathcal{C}}_v(-a) = -\check{\mathcal{C}}_v(a) \tag{23}$$

(iii) monotonicity. If $a_i \leq a'_i, i = 1, \ldots, n$,

$$\mathcal{C}_v(a_1, \ldots, a_n) \leq \mathcal{C}_v(a'_1, \ldots, a'_n) \tag{24}$$

$$\check{\mathcal{C}}_v(a_1, \ldots, a_n) \leq \check{\mathcal{C}}_v(a'_1, \ldots, a'_n). \tag{25}$$

(iv) invariance to positive affine transformations. For all $\alpha > 0$, for all $\beta \in \mathbb{R}$

$$\mathcal{C}_v(\alpha a_1 + \beta, \ldots, \alpha a_n + \beta) = \alpha \mathcal{C}_v(a_1, \ldots, a_n) + \beta. \tag{26}$$

(v) homogeneity. For all $\alpha \in \mathbb{R}$,

$$\check{\mathcal{C}}_v(\alpha a_1, \ldots, \alpha a_n) = \alpha \check{\mathcal{C}}_v(a_1, \ldots, a_n). \tag{27}$$

We give now characterizations of Choquet and Šipoš integrals.

Proposition 6. *[29, Theorem 6.1.1] Let us consider a functional $M_v : \mathbb{R}^n \longrightarrow \mathbb{R}$ defined with respect to a capacity v. Then M_v is the Choquet integral $\mathcal{C}_v$ for all capacities v on N, if and only if M_v satisfies the following properties:*

(i) M_v is a linear function of v, i.e. $M_v = \sum_{A \subset N} a_A\, v(A)$, where the a_A's are 2^n functions from $\mathbb{R}^n$ to $\mathbb{R}$.
(ii) M_v is non decreasing in each place.
(iii) M_v is invariant to positive affine transformations (see Proposition 5 (iv)).
(iv) M_v is an extension of v (see Proposition 5 (i)).

Similarly, we can give a characterization of Šipoš integral [25].

Proposition 7. *Let us consider a functional $M_v : \mathbb{R}^n \longrightarrow \mathbb{R}$ defined with respect to a capacity v. Then M_v is the Šipoš integral $\check{\mathcal{C}}_v$ for all capacities v on N, if and only if M_v satisfies the following properties:*

(L) M_v *is a linear function of* v, *i.e.* $M_v = \sum_{A \subset N} a_A\, v(A)$, *where the* a_A*'s are* 2^n *functions from* $\mathbb{R}^n$ *to* $\mathbb{R}$.

(In) M_v *is non decreasing in each place.*

(H) M_v *is homogeneous (see Proposition 5 (v)).*

(Sh) M_v *is invariant to positive shifts, i.e.* $\forall a \in (\mathbb{R}^+)^n$, $\forall \alpha \geq 0$, $M_v(a+\alpha) = M_v(a) + \alpha$.

(Ind) M_v *satisfies independence between positive and negative parts, i.e. for any* $a, b, c, d \in \mathbb{R}^n$, $a \geq 0, b \geq 0, c \leq 0, d \leq 0$, $\forall A \subset N$, *we have*

$$M_v(a_A, c_{A^c}) - M_v(a_A, d_{A^c}) = M_v(b_A, c_{A^c}) - M_v(b_A, d_{A^c})$$

(Ext) M_v *is an extension of* v.

The property (Ind) means that the difference $M_v(a_A, c_{A^c}) - M_v(a_A, d_{A^c})$ does not depend on a: positive scores and negative scores do not interact.

The above properties make clear the following. As remarked by Sugeno and Murofushi [44], due to properties (iv) and (v) of proposition 5, the Choquet integral maps an interval scale to an interval scale, i.e. a scale where only difference of numbers has a meaning. On the other hand, the Šipoš integral maps a ratio scale (where also the ratio of numbers is meaningful) to a ratio scale. A ratio scale has a zero with a fixed position, while its position is arbitrary for a difference scale. Hence formula (12) does not differentiate between negative and positive numbers, while the Šipoš integral makes the symmetry between positive and negative numbers. For a deeper study of this distinction in the general framework of decision making, see [24].

3.3 The case of 2-additive capacities

Capacities which can be expressed with a limited set of coefficients are of special interest in applications. k-additive capacities [18] are such that their Möbius transform vanishes for subsets of more than k elements (or equivalently: the interaction transform vanishes for subsets of more than k elements). 1-additive capacities are classical measures. The case of 2-additive capacities is of particular interest, since despite their simplicity they can model interaction between pairs of elements.

Choquet and Šipoš integrals take an interesting form when the underlying capacity is 2-additive. The following, which is a particular case of Prop. 4, can be shown [25].

Proposition 8. *Let $t \in \mathbb{R}^n$, and v a 2-additive measure. Then*

$$\mathcal{C}_v(t) = \sum_{I_{ij}>0} (t_i \wedge t_j) I_{ij} + \sum_{I_{ij}<0} (t_i \vee t_j)|I_{ij}| + \sum_{i=1}^{n} t_i(\phi_i - \frac{1}{2}\sum_{j \neq i} |I_{ij}|)$$

$$\check{\mathcal{C}}_v(t) = \sum_{i,j \in N^+, I_{ij}>0} (t_i \wedge t_j) I_{ij} + \sum_{i,j \in N^-, I_{ij}>0} (t_i \vee t_j) I_{ij}$$
$$+ \sum_{i,j \in N^+, I_{ij}<0} (t_i \vee t_j)|I_{ij}| + \sum_{i,j \in N^-, I_{ij}<0} (t_i \wedge t_j)|I_{ij}|$$
$$+ 2\sum_{i \in N^+} t_i\Big(\sum_{j \in N^-, I_{ij}<0} |I_{ij}|\Big) + 2\sum_{i \in N^-} t_i\Big(\sum_{j \in N^+, I_{ij}<0} |I_{ij}|\Big)$$
$$+ \sum_{i=1}^{n} t_i\Big(\phi_i - \frac{1}{2}\sum_{j \neq i} |I_{ij}|\Big).$$

with $N^+ = \{i \in N | t_i \geq 0\}$, and $N^- = N \setminus N^+$.

The above formulas give a clear understanding of both integrals.

- the Choquet integral is a convex sum of conjunctive, disjunctive, and linear terms. A conjunctive aggregation between two elements is performed whenever there is a positive interaction between them, and symmetrically, a disjunctive aggregation is performed when the interaction is negative. The linear part contains the Shapley index, diminished by the amount of interaction with other elements. Since any term in the summation could be zero (but not all!), the Choquet integral with respect to a 2-additive capacity provides a really flexible way of aggregation. Note that no distinction is made with respect to the sign of values.
- the Šipoš integral does basically the same, but distinguishes between positive and negative values in a very particular way. If the values are negative, the behaviour of the aggregation is exactly the opposite of the positive case, i.e. conjunctive aggregation becomes disjunctive and vice versa. Secondly, there is no interaction effect with pair of elements of different sign: they are processed separately, as one could expect from the examination of the characterization result (see Prop. 7 *(Ind)*).

4 Aggregating ordinal information

We turn now to the case of ordinal information to be aggregated. As explained in the introduction, any arithmetical operation is forbidden, and only comparisons make sense. For this reason, and due to the similarity of properties, the Sugeno integral appears to be the right counterpart of the Choquet integral in the ordinal world. Let us show how close are the two integrals by introducing the following *residuated difference operator*:

$$a \dot{-} b := \begin{cases} a, & \text{if } a > b \\ \mathbb{O}, & \text{otherwise} \end{cases} \tag{28}$$

for any $a, b \in \mathbb{R}^+$. Then, as remarked by Murofushi and Sugeno [37], the Sugeno integral can be written as follows:

$$\mathcal{S}_v(a) := \bigvee_{i=1}^{n} [(a_{(i)} \dot{-} a_{(i-1)}) \wedge v(A_{(i)})]. \tag{29}$$

The similarity with the Choquet integral is striking, replacing sum by maximum, and product by minimum.

The aim of this section is to define symmetric and asymmetric Sugeno integrals, in order to introduce the same distinction in the ordinal case as the one induced by the Šipoš integral. Clearly, the first thing to achieve is the introduction of negative "numbers" on an ordinal scale, and a suitable algebra.

Let E be a linearly ordered scale with $2k+1$ degrees defined by:

$$e_{-k} < e_{-k+1} < \cdots < e_{-1} < e_0 < e_1 < \ldots < e_{k-1} < e_k$$

centered around e_0. We speak of "negative" value for any degree with a negative index, and denote $E^+ := \{e_0, e_1, \ldots, e_k\}$ and $E^- := E \setminus E^+$. By commodity, we denote e_0 by $\mathbb{O}$ and e_k by $\mathbb{1}$. We introduce for every e_i in E its symmetrical element $-e_i$ defined $-e_i := e_{-i}$ (symmetric ordinal scale). We extend on E the usual operations $\vee, \wedge$ defined on E^+, so as to have an algebraic structure like a ring. Denoting $\otimes, \oslash$ these extended operators, called *symmetric maximum* and *symmetric minimum* respectively, it would be desirable to have properties like $-(a \otimes b) = (-a) \otimes (-b)$, $a \otimes (-b) = a \dot{-} b$, and $-(a \oslash b) = (-a) \oslash b = a \oslash (-b)$, where $\dot{-}$ is the residuated difference (28) properly extended on E. Here, $\otimes, \oslash$ play the role of addition and product respectively.

The following definitions are suitable for our purpose [23]. The *symmetric difference* is defined for any a, b on E^+ by:

$$a \dot{-} b := \begin{cases} a, & \text{if } a > b \\ \mathbb{O}, & \text{if } a = b \\ -b, & \text{otherwise.} \end{cases} \tag{30}$$

Using this definition, the symmetric maximum is defined in Table 1. Then, the symmetric minimum is defined in Table 2. A more concise definition is:

$$a \otimes b := \begin{cases} -(|a| \vee |b|) & \text{if } b \neq -a \text{ and } |a| \vee |b| = -a \text{ or } = -b \\ \mathbb{O} & \text{if } b = -a \\ |a| \vee |b| & \text{else} \end{cases} .$$

$a \ovee b$	$b < \mathbb{O}$	$b = \mathbb{O}$	$b > \mathbb{O}$
$a < \mathbb{O}$	$a \wedge b$	a	$b \veebar (-a)$
$a = \mathbb{O}$	b	$\mathbb{O}$	b
$a > \mathbb{O}$	$a \veebar (-b)$	a	$a \vee b$

Table 1. Definition of the symmetric maximum $\ovee$

$a \owedge b$	$b < \mathbb{O}$	$b = \mathbb{O}$	$b > \mathbb{O}$
$a < \mathbb{O}$	$\lvert a\rvert \wedge \lvert b\rvert$	$\mathbb{O}$	$-(\lvert a\rvert \wedge b)$
$a = \mathbb{O}$	$\mathbb{O}$	$\mathbb{O}$	$\mathbb{O}$
$a > \mathbb{O}$	$-(a \wedge \lvert b\rvert)$	$\mathbb{O}$	$a \wedge b$

Table 2. Definition of the symmetric minimum $\owedge$

Except for the case $b = -a$, $a \ovee b$ equals the absolutely larger one of the two elements a and b.

$$a \owedge b := \begin{cases} -(|a| \wedge |b|) & \text{if } \operatorname{sign} a \neq \operatorname{sign} b \\ |a| \wedge |b| & \text{else} \end{cases}$$

With these definitions, we can show the following result.

Property 1. The structure $(E, \ovee, \owedge)$ has the following properties.

(i) $\ovee$ is commutative.
(ii) $\mathbb{O}$ is the unique neutral element of $\ovee$, and the unique absorbant element of $\owedge$.
(iii) $a \ovee -a = \mathbb{O}$, for all $a \in E$.
(iv) $-(a \ovee b) = (-a) \ovee (-b)$.
(v) $\ovee$ is associative on E^+ and on E^-.
(vi) $\owedge$ is commutative.
(vii) $\mathbb{1}$ is the unique neutral element of $\owedge$, and the unique absorbant element of $\ovee$.
(viii) $\owedge$ is associative on E.
(ix) $\owedge$ is distributive w.r.t $\ovee$ in E^+ and E^-.

The associativity of $\vee$ and distributivity do not hold on E in general.

The second step is to redefine the various tranformations of capacities introduced in section 2. In this ordinal context, a capacity is defined naturally as a set function $v : \mathcal{P}(N) \longrightarrow E^+$ such that $v(\emptyset) = \mathbb{O}$, $v(N) = \mathbb{1}$, and $A \subset B$ implies $v(A) \leq v(B)$. Following Berge [2], who presents the Möbius function as a powerful tool for inversion formulas over posets, we define the *ordinal Möbius transform* of v, denoted $m^v_\vee$ as the solution of the equation:

$$v(A) = \mathop{\ovee}_{B \subset A} m^v_\vee(B). \tag{31}$$

In fact, unlike the cardinal case, there is not a single solution to this equation. Previous studies by the author [17,19] and related works of Mesiar [31] and Marichal [30] have shown that the smallest positive solution is given by:

$$m_{\vee}^{v}(A) := \begin{cases} v(A), & \text{if } v(A) > v(A \setminus i), \forall i \in A \\ \mathbb{O}, & \text{otherwise,} \end{cases}$$

and the greatest one is simply $m_{\vee}^{v} \equiv v$. Any set function comprised between these two solutions is a solution. We take as definition of the Möbius transform the lower bound, which is a non-negative function, so that we can use from now on usual $\vee$ and $\wedge$ operators. If there is no fear of ambiguity, the superscript ^{v} can be dropped.

Interestingly enough, the Möbius transform can be expressed as:

$$m_{\vee}^{v}(A) := \bigvee_{B \subset A, |A \setminus B| \text{even}} v(B) \curlywedge \bigvee_{B \subset A, |A \setminus B| \text{odd}} v(B)$$

which is an exact transcription of the cardinal case (see (1)). However, if v is not monotonic, the above formula is no more solution of (31) in general.

The ordinal Möbius transform has many interesting properties, which are parallel to the classical definition. First, any capacity can be expressed on the basis of unanimity games through the formula

$$v(A) = \bigvee_{B \subset N} m_{\vee}^{v}(B) \wedge u_B(A).$$

Let us consider now any possibility measure Π, i.e. a capacity such that $\Pi(A \cup B) = \Pi(A) \vee \Pi(B)$ for all $A, B \subset N$. Then its Möbius transform $m_{\wedge}^{\Pi}$ is:

$$m_{\wedge}^{\Pi}(\{i\}) = \Pi(\{i\}), \quad \forall i \in N, \text{ and } m_{\wedge}^{\Pi}(A) = 0, \quad \forall A \subset N, |A| > 1.$$

Thus we have the same kind of result as with probability measures in the cardinal context. Lastly, we mention that the Sugeno integral defined on E^{+} can be expressed with the ordinal Möbius transform.

$$\mathcal{S}_v(a) = \bigvee_{A \subset N} m_{\vee}^{v}(A) \wedge \left(\bigwedge_{i \in A} a_i\right).$$

This expression is very similar to (14).

Attempts have been done by the author to define an interaction transform in this ordinal context [19]. Changing in a suitable way operators in formula (2), and taking into account some properties which are counterparts of the cardinal case (see below), the following equation defines the *ordinal interaction transform* of a fuzzy measure v:

$$I_{\vee}^{v}(A) := \bigvee_{B \subset N \setminus A} [\bigvee_{\substack{C \subset A \\ |A \setminus C| \text{ even}}} v(B \cup C) \curlywedge \bigvee_{\substack{C \subset A \\ |A \setminus C| \text{ odd}}} v(B \cup C)]. \qquad (32)$$

For $A = \{i\}$, we get the *ordinal Shapley value*:

$$\phi_{\vee}^{v}(i) := \bigvee_{A \subset N \setminus i} [v(A \cup i) \dot{-} v(A)]. \tag{33}$$

The ordinal Shapley value fulfills the following properties:

(A1) $\bigvee_{i \in N} \phi_{\vee}^{v}(i) = v(N)$ (sharing of $v(N)$).
(A2) if i is such that $v(A \cup i) = v(A)$ for every $A \subset N \setminus i$, then $\phi_{\vee}^{v}(i) = 0$ (null player).
(A3) if i, j are such that $v(A \cup i) = v(A \cup j)$ for every $A \subset N \setminus \{i, j\}$, then $\phi_{\vee}^{v}(i) = \phi_{\vee}^{v}(j)$ (symmetric players).

These properties are counterparts of the properties of the (original) Shapley value [42]. However, the "maxitivity" does not hold: in general $\phi_{\vee}^{v \vee \nu} \neq \phi_{\vee}^{v} \vee \phi_{\vee}^{\nu}$, where v, ν are two capacities. Lastly, the ordinal interaction transform can be expressed through the ordinal Möbius transform by:

$$I_{\vee}^{v}(A) = \bigvee_{B \subset N \setminus A} m_{\vee}^{v}(A \cup B). \tag{34}$$

The last step is to address the definition of the Sugeno integral over E, the expression over E^{+} being given by equation (9), where $a \in (E^{+})^{n}$, and v is an ordinal capacity.

A natural approach is to define the symmetric Sugeno integral as follows:

$$\check{\mathcal{S}}_{v}(a) := \mathcal{S}_{v}(a^{+}) \circledvee (-\mathcal{S}_{v}(a^{-})) \tag{35}$$

for any $a \in E^{n}$, where $a_i^{+} := a_i \wedge \mathbb{O}$, and $a_i^{-} := (-a_i) \wedge \mathbb{O}$. It can be shown that $\check{\mathcal{S}}_{v}(a)$ can be rewritten as [23] :

$$\check{\mathcal{S}}_{v}(a) := \left[\mathop{\circledvee}_{i=1}^{p} \left(a_{(i)} \circledwedge v(\{(1), \ldots, (i)\})\right) \right] \circledvee \left[\mathop{\circledvee}_{i=p+1}^{n} \left(a_{(i)} \circledwedge v(\{(i), \ldots, (n)\})\right) \right], \tag{36}$$

for any $a \in E^{n}$, with $a_{(1)} \leq \cdots \leq a_{(p)} < 0 \leq a_{(p+1)} \leq \cdots \leq a_{(n)}$. This formula can be derived also starting from equation (13) and rewriting it as a Sugeno integral using (29). Note that the above expression is well-defined, i.e. there is no problem of associativity of $\circledvee$, since $\circledvee_{i=1}^{p}$ operates only on E^{-}, and $\circledvee_{i=p+1}^{n}$ only on E^{+}. The symmetric Sugeno integral indeed satisfies the property of symmetry:

$$\check{\mathcal{S}}_{v}(-a) = -\check{\mathcal{S}}_{v}(a) \tag{37}$$

for any capacity v and any $a \in E^{n}$.

Also, $\check{\mathcal{S}}$ can be expressed in terms of the ordinal Möbius transform, in a way which is the counterpart of (15):

$$\check{\mathcal{S}}_v(a) = \underset{A\subset N}{\circledcirc}\left(m_\vee(A) \otimes \left[\bigwedge_{i\in A} a_i^+ \curlywedge \bigwedge_{i\in A} a_i^-\right]\right) \tag{38}$$

These results show the adequacy of the whole construction.

The case of asymmetric integral can be done similarly, by putting:

$$\mathcal{S}_v(a) := \mathcal{S}_v(a^+) \circledcirc (-\mathcal{S}_{\bar{v}}(a^-)) \tag{39}$$

for any $a \in E^n$ [9]. However, a proper definition of the conjugate of v needs more structure than we have introduced so far. We introduce the *conjugation* mapping $E \longrightarrow E$, $x \mapsto \bar{x}$, such that $\bar{\bar{x}} = x$ and $x \leq y$ iff $\bar{y} \leq \bar{x}$. On a linearly ordered scale, the conjugation mapping is unique and always exists.

Then the conjugate of v is defined as $\bar{v}(A) := \overline{v(\bar{A})}$, which enables the definition of the asymmetric integral. However, it seems that this definition does not permit to derive an expression of the integral which coincides with the integral on E^+, as one would have expected. Another approach, advocated in [23], is to consider that the asymmetric integral should not make any distinction between positive and negative quantities, and thus ignores the position of the zero. Then, the zero can be put at the lower extremity, and we are working on a scale with only positive values, where the usual definition of the Sugeno integral works.

5 Concluding remarks

We have tried to give in this paper the main recent results on integral-based aggregation. These essentially concerned the use of negative quantities in aggregation, as well as the use of ordinal information, and especially *negative ordinal* quantities. Of course, we do not pretend to exhaustivity, and we believe that much more material on related topics would have been worth to present. As an example, the notion of k-additive measure (capacity), or its ordinal counterpart (k-maxitive [32,4] or k-possibility measures [17]) seems to give rise to many researches, as shown by the works of Miranda for the characterization of k -additive measures [35], and those of Mesiar on other formulations of k-additive measures, especially in the continuous case [33,34].

References

1. P. Benvenuti and R. Mesiar. Integrals with respect to a general fuzzy measure. In M. Grabisch, T. Murofushi, and M. Sugeno, editors, *Fuzzy Measures and Integrals — Theory and Applications*, pages 205–232. Physica Verlag, 2000.
2. C. Berge. *Principles of Combinatorics*. Academic Press, 1971.

3. T. Calvo, R. Mesiar, and J. Martin. Integral based aggregation of real data. In *6th Int. Conf. on Information Processing and Management of Uncertainty in Knowledge-Based Systems (IPMU)*, pages 58–62, Granada, Spain, July 1996.
4. K. Cao-Van and B. De Baets. Decomposition of Choquet and Sugeno integrals and the inverse problem. In *8th Int. Conf. on Information Processing and Management of Uncertainty in Knowledge-Based Systems (IPMU)*, pages 402–409, Madrid, Spain, July 2000.
5. G. Choquet. Theory of capacities. *Annales de l'Institut Fourier*, 5:131–295, 1953.
6. L. de Campos and M.J. Bola nos. Characterization and comparison of Sugeno and Choquet integrals. *Fuzzy Sets & Systems*, 52:61–67, 1992.
7. D. Denneberg. *Non-Additive Measure and Integral.* Kluwer Academic, 1994.
8. D. Denneberg and M. Grabisch. Interaction transform of set functions over a finite set. *Information Sciences*, 121:149–170, 1999.
9. D. Denneberg and M. Grabisch. Functions and measures with linear ordinal scales, (a-)symmetric Sugeno integral and ordinal Ky Fan distance. In *FUR X*, Torino, Italy, May 2001.
10. D. Dubois and H. Prade. *Possibility Theory.* Plenum Press, 1988.
11. D. Dubois and H. Prade. Possibility theory: qualitative and quantitative aspects. In D.M. Gabbay and Ph. Smets, editors, *Handbook of Defeasible Reasoning and Uncertainty Management Systems*, pages 169–226. Kluwer Academic Publishers, 1998.
12. D. Dubois, H. Prade, and R. Sabbadin. Qualitative decision theory with Sugeno integrals. In M. Grabisch, T. Murofushi, and M. Sugeno, editors, *Fuzzy Measures and Integrals — Theory and Applications*, pages 314–332. Physica Verlag, Heidelberg, 2000.
13. M. Grabisch. Fuzzy integral in multicriteria decision making. *Fuzzy Sets & Systems*, 69:279–298, 1995.
14. M. Grabisch. The application of fuzzy integrals in multicriteria decision making. *European J. of Operational Research*, 89:445–456, 1996.
15. M. Grabisch. Alternative representations of discrete fuzzy measures for decision making. *Int. J. of Uncertainty, Fuzziness, and Knowledge Based Systems*, 5:587–607, 1997.
16. M. Grabisch. Fuzzy measures and integrals for decision making and pattern recognition. *Tatra Mountains Mathematical Publications*, 13:7–34, 1997.
17. M. Grabisch. k-additive and k-decomposable measures. In *Proc. of the Linz Seminar*, pages 88–94, Linz, Austria, February 1997.
18. M. Grabisch. k-order additive discrete fuzzy measures and their representation. *Fuzzy Sets and Systems*, 92:167–189, 1997.
19. M. Grabisch. On the representation of k-decomposable measures. In *7th IFSA World Congress*, Prague, Czech Republic, June 1997.
20. M. Grabisch. Fuzzy integral as a flexible and interpretable tool of aggregation. In B. Bouchon-Meunier, editor, *Aggregation of evidence under fuzziness*, pages 51–72. Physica Verlag, 1998.
21. M. Grabisch. A graphical interpretation of the choquet integral. *IEEE Tr. on Fuzzy Systems*, 8:627–631, 2000.
22. M. Grabisch. The interaction and Möbius representations of fuzzy measures on finite spaces, k-additive measures: a survey. In M. Grabisch, T. Murofushi, and M. Sugeno, editors, *Fuzzy Measures and Integrals — Theory and Applications*, pages 70–93. Physica Verlag, 2000.

23. M. Grabisch. Symmetric and asymmetric fuzzy integrals: the ordinal case. In *6th Int. Conf. on Soft Computing (Iizuka'2000)*, Iizuka, Japan, October 2000.
24. M. Grabisch and Ch. Labreuche. To be symmetric or asymmetric? A dilemna in decision making. In J. Fodor, B. De Baets, and P. Perny, editors, *Preferences and Decisions under Incomplete Knowledge*, pages 179–194. Physica Verlag, 2000.
25. M. Grabisch and Ch. Labreuche. The symmetric and asymmetric Choquet integrals on finite spaces for decision making. *Statistical Letters*, submitted.
26. M. Grabisch, Ch. Labreuche, and J.C. Vansnick. On the extension of pseudo-Boolean functions for the aggregation of interacting bipolar criteria. *Mathematical Social Sciences*, submitted.
27. M. Grabisch, J.L. Marichal, and M. Roubens. Equivalent representations of a set function with applications to game theory and multicriteria decision making. *Mathematics of Operations Research*, 25(2):157–178, 2000.
28. M. Grabisch, H.T. Nguyen, and E.A. Walker. *Fundamentals of Uncertainty Calculi, with Applications to Fuzzy Inference*. Kluwer Academic, 1995.
29. J.L. Marichal. *Aggregation operators for multicriteria decision aid*. PhD thesis, University of Liège, 1998.
30. J.L. Marichal, P. Mathonet, and E. Tousset. Mesures floues définies sur une échelle ordinale. working paper, 1996.
31. R. Mesiar. k-order pan-additive discrete fuzzy measures. In *7th IFSA World Congress*, pages 488–490, Prague, Czech Republic, June 1997.
32. R. Mesiar. Generalization of k-order additive discrete fuzzy measures. *Fuzzy Sets and Systems*, 102:423–428, 1999.
33. R. Mesiar. k-order additive fuzzy measures. *Int. J. of Uncertainty, Fuzziness and Knowledge-Based Systems*, 7:561–568, 1999.
34. R. Mesiar. Three alternative definitions of k-order additive fuzzy measures. *BUSEFAL*, 83:57–62, 2000.
35. P. Miranda and M. Grabisch. Characterizing k-additive measures. In *8th Int. Conf. on Information Processing and Management of Uncertainty in Knowledge-Based Systems (IPMU'2000)*, pages 1063–1070, Madrid, Spain, July 2000.
36. T. Murofushi and S. Soneda. Techniques for reading fuzzy measures (III): interaction index. In *9th Fuzzy System Symposium*, pages 693–696, Sapporo, Japan, May 1993. In Japanese.
37. T. Murofushi and M. Sugeno. Fuzzy t-conorm integrals with respect to fuzzy measures : generalization of Sugeno integral and Choquet integral. *Fuzzy Sets & Systems*, 42:57–71, 1991.
38. T. Murofushi and M. Sugeno. Some quantities represented by the Choquet integral. *Fuzzy Sets & Systems*, 56:229–235, 1993.
39. F.S. Roberts. *Measurement Theory*. Addison-Wesley, 1979.
40. D. Schmeidler. Integral representation without additivity. *Proc. of the Amer. Math. Soc.*, 97(2):255–261, 1986.
41. G. Shafer. *A Mathematical Theory of Evidence*. Princeton Univ. Press, 1976.
42. L.S. Shapley. A value for n-person games. In H.W. Kuhn and A.W. Tucker, editors, *Contributions to the Theory of Games, Vol. II*, number 28 in Annals of Mathematics Studies, pages 307–317. Princeton University Press, 1953.
43. M. Sugeno. *Theory of fuzzy integrals and its applications*. PhD thesis, Tokyo Institute of Technology, 1974.

44. M. Sugeno and T. Murofushi. *Fuzzy measure theory*, volume 3 of *Course on fuzzy theory*. Nikkan Kōgyō, 1993. In Japanese.
45. A. Tversky and D. Kahneman. Advances in prospect theory: cumulative representation of uncertainty. *J. of Risk and Uncertainty*, 1992.
46. J. Šipoš. Integral with respect to a pre-measure. *Math. Slovaca*, 29:141–155, 1979.

Associative Aggregation Operators

W.Sander

Institute for Analysis
Technical University of Braunschweig
Pockelsstr. 14
D–38106 Braunschweig, Germany
E-mail: w.sander@tu-bs.de

1 Introduction

An aggregation process occurs in many situations like in decision making or in statistical and economic measurement by aggregating expert's opinions or by synthesizing judgements. So the typical situation is as follows :
Having n numerical values $x_1, \ldots, x_n$ lying in an interval I of real numbers, the aggregation operator M defined on I^n aggregates these numbers to a value of $\mathbb{R}$ in an appropriate way so that the properties of M represent a model of the concrete situation.
Let us mention that also aggregation operations on fuzzy sets are included in this concept. Having n fuzzy sets $A_1, \ldots, A_n$ defined on a set X with values in I the aggregation operator M, applied to the membership grades of these sets, produces an aggregated fuzzy set.

$$A(x) = M(A_1(x), ..., A_n(x)) \quad , \quad x \in X.$$

There is an extensive literature concerning these types of problems, starting with the well-known papers of Kolmogorov [16] and Nagumo [24] on generalized means. For a survey we refer to chapter 5 in the book of Fodor and Roubens [8].
The common feature of all these results is that M satisfies a system of functional equations (depending on the model).
In this paper we want to give a survey on recent results for aggregation operators M which are associative and increasing :

$$M\colon I^2 \to I, \tag{1.1}$$

$$M(x, M(x,y)) = M(M(x,y),z) \quad , \quad x,y,z \in I, \tag{1.2}$$

$$x \leq x', y \leq y' \Rightarrow M(x,y) \leq M(x',y') \quad , \quad x,x',y,y' \in I. \tag{1.3}$$

Here (and in the whole paper) I denotes the interval $[0,1]$. Moreover we have restricted M to two arguments because - due to the associativity - M can be extended to a finite number of arguments. The monotonicity of M means that the aggregation cannot decrease if one of the arguments is increasing.

It turns out that conditions (1.2) and (1.3) are too weak for a characterization of aggregation operators. Therefore we look for big subclasses of

$$\mathcal{F} = \{M: I^2 \to I \quad | \, M \quad \text{is associative and increasing}\} \tag{1.4}$$

for which a characterization is possible.
Functions $M \in \mathcal{F}$ which satisfy the boundary condition

$$M(x,1) = M(1,x) = x, x \in I \tag{1.5}$$

$$\text{or} \qquad M(x,0) = M(0,x) = x, x \in I \tag{1.6}$$

are called pseudo-triangular-norms (pseudo-t-norms for short) and pseudo-triangular-norms (pseudo-s-norms for short), respectively.
This means that 1 and 0 are neutral elements of M, respectively. Every pseudo-t-norm (pseudo-s-norm) satisfies

$$M(x,0) = 0 = M(0,x), x \in I \tag{1.7}$$

$$(M(x,1) = 1 = M(1,x), x \in I). \tag{1.8}$$

For example, (1.7) follows from the inequalities
$0 \leq M(x,0) \leq M(1,0) = 0$ and $0 \leq M(0,x) \leq M(0,1) = 0$.
Commutative pseudo-t-norms (pseudo-s-norms) are called t-norms (s-norms).
Remember that the maximal pseudo-t-norm $T_{max} = min$ and that the minimal pseudo-t-norm T_{min} (the drastic product) is given by
$T_{min}(x,y) = min(x,y) \quad if \quad max(x,y) = 1$ and $T_{min} = 0$ otherwise,
that is,

$$T_{min} \leq T \leq min.$$

(In this paper min and max denote the usual minimum and maximum of two variables). Moreover we have

$$max \leq S \leq S_{max}$$

for each pseudo-s-norm S where S_{max} is the drastic sum given by
$S_{max}(x,y) = 1 - T_{min}(1-x, 1-y)$.
Thus it is natural to introduce the class $\mathcal{M}$ of so-called averaging or internal operators :

$$\mathcal{M} := \{M: I^2 \to I: min \leq M \leq max\} \tag{1.9}$$

We get immediately the following result.

Proposition 1.1 *Let $M: I^2 \to I$ be increasing. Then we have :*
M is internal $\Leftrightarrow$ M is idempotent

Proof: If M is internal then $x = min(x,x) \leq M(x,x) \leq max(x,x) = x$.
Reversely, if M is idempotent then
$min(x,y) = M(min(x,y), min(x,y)) \leq M(x,y) \leq$
$\leq M(max(x,y), max(x,y)) = max(x,y)$.

The idempotency of M represents the unanimity property
$M(x,x) = x, x \in I$ saying that if two arguments of M are equal then the aggregation process leads to this common value.
Note that min and max are the only idempotent pseudo-t-and pseudo-s-norms, respectively.
The structure of t-norms and s-norms, respectively, but also the structure theorems for averaging operators, satisfying additional properties, are well-known and lead to many applications, for example in economics [8], probabilistic metric spaces [27], many-valued and fuzzy-logic ([11], [12], [14]), probability theory [25] and generalized measures and integrals
([11] , [13]).

Our main interest is to generalize and unify recent results on subclasses of $\mathcal{F}$ which cover simultaneously the classes of t-norms, s-norms and averaging operators.

To do this we remember that a binary operation $M: I^2 \to I$ can have at most one neutral or unit element e (that is, $M(x,e) = M(e,x) = x, x \in I$) and at most one annihilator or zero element n (that is, $M(x,n) = M(n,x) = n, x \in I$). Both e and n are idempotent elements, but there can exist in general much more idempotent elements, different from a neutral element and an annihilator.

Having this in mind we introduce two subclasses of $\mathcal{F}$ which will be defined by means of the idempotent elements

$$A_M = \{x \in I: M(x,x) = x\} \tag{1.10}$$

and a neutral element of $M: I^2 \to I$, respectively. Using the notation of sections of M, given by

$$M_x, M^y: I \to I \ , \ M_x(y) = M^y(x) = M(x,y) \ , \ x,y \in I \tag{1.11}$$

we remember first the fact that M_x and M^y are idempotent functions, if M is associative and if $x, y \in A_M$. For instance, we have

$$x \in A_M \Rightarrow M_x(M_x(z)) = M(x, M(x,z)) = M(M(x,x), z) = M_x(z), z \in I. \tag{1.12}$$

Now we weaken the idempotency of M (which means that $A_M = [0,1]$) and arrive - together with a weak regularity assumption - at

$$\mathcal{F}_1 := \{M \in \mathcal{F}: A_M \quad \text{is closed,} \quad \{0,1\} \subset A_M, \quad \forall x \in A_M:$$

$$M_x \quad \text{is continuous on} \quad I, \quad M^0, M^1 \quad \text{are continuous on} \quad I\} \qquad (1.13)$$

If for example $A_M = \{0, 1\}$ then $M(0,0) = 0$ and $M(1,1) = 1$ and the regularity assumption means that M_0, M^0 and M_1, M^1 are continuous.
The assumption that A_M is closed is satisfied if A_M is finite , but we will mainly use this assumption to apply a generalization of a structure theorem for (pseudo-) t-norms or (pseudo-) s-norms.
Concerning the second subclass of $\mathcal{F}$ we assume that $e \in (0,1)$ is a neutral element of M : These functions are called pseudo-uninorms, and they were not yet investigated, but we can underline the importance of these functions, and we present representation- and characterization-theorems for regular pseudo-uninorms. Note that pseudo-t- and pseudo-s-norms have 1 and 0 as neutral element (see (1.5) and (1.6)). In analogy to t- and s-norms we denote commutative pseudo-uninorms by uninorms (which were mainly investigated in [9].)
A natural regularity condition for pseudo-uninorms M would be :

M_x, M^y are continuous on $I \setminus \{e\}$ for all $x, y \in I$.

We weaken this condition to :

$$\begin{aligned} A_M \text{ is closed}, M^1 \text{ is continuous on } [0,e), M^0 \text{ is continuous on } (e,1] \\ M_x \text{ is continuous on } [0,e) \ \forall x \in A_M \cap [0,e), \\ M_x \text{ is continuous on } [e,1) \ \forall x \in A_M \cap (e,1]. \quad (1.14) \end{aligned}$$

Now we define

$$\begin{aligned} \mathcal{F}_2 := \{M \in \mathcal{F} \colon \exists \quad e \in I \colon \\ \text{e is a unit for M , M satisfies (1.14)}\}. \qquad (1.15) \end{aligned}$$

In both classes $\mathcal{F}_1$ and $\mathcal{F}_2$ we have assumed a weak regularity condition concerning M_x. Note that the following result is valid (see [18]).

Proposition 1.2 *If $M \colon I^2 \to I$ is increasing, then M is continuous on I^2 if and only if M_x and M^y are continuous for all $x, y \in I$.*

The interpretation of the continuity assumption is obvious :
A small change in one or two variables of M does not lead to a too big change of the aggregation operator M.
Our main aim is now to give characterizations of these two "noncommutative" classes $\mathcal{F}_1$ and $\mathcal{F}_2$ since the commutativity is not present in many applications:
If for example judgements of important persons are dominant or if criteria of nonequal importance are combined, then M is dependent upon the order of the arguments.
Thus we present in section 3 and 4 new results characterizing $\mathcal{F}_1$ and $\mathcal{F}_2$, respectively.

There is only one main idea to prove these results :
Appropriate refinements of already known arguments lead to new results, or a little bit simplified : The proofs for many known results in the "commutative" case work rather often already in the "noncommutative" case (if you overcome some technical difficulties).
In section 2 we introduce some notations and present some more or less known and unknown results for the convenience of the reader. The second reason for presenting preliminaries is to shorten and clarify the structure of the proofs in section 3 and 4.

2 Preliminaries

In this section we point out three techniques which are important for our considerations.

(1) It is well-known that the study of associative functions on an arbitrary interval [a,b], $-\infty \leq a < b \leq \infty$, can be reduced to associative functions on I = [0,1].
If $M: I^2 \to I$ and if $a, b \in \mathbb{R}$ then we introduce
$\rho_{a,b}: I \to [a,b]$ and $\rho_{a,b}^{-1}: [a,b] \to I$ by

$$\rho_{a,b}(x) := a + (b-a)x, x \in [0,1] \quad \text{and} \tag{2.1}$$

$$\rho_{a,b}^{-1}(x) := \frac{x-a}{b-a}, \quad x \in [a,b] \tag{2.2}$$

and define

$$G_M(x,y) = \rho_{a,b}\Big[M(\rho_{a,b}^{-1}(x), \rho_{a,b}^{-1}(y)\Big], x, y \in [a,b] \tag{2.3}$$

which implies

$$M(x,y) = \rho_{a,b}^{-1}\Big[G_M(\rho_{a,b}(x), \rho_{a,b}(y)\Big], x, y \in [0,1]. \tag{2.4}$$

Here $G_M: [a,b]^2 \to [a,b]$ is generated from M by $\rho_{a,b}$, and obviously, many properties of M carry over to G_M (for example, associativity, monotonicity, continuity). If M has 1 or 0 as unit then G_M has b or a as unit, respectively, that is, $M(x,b) = M(b,x) = x$ or $M(x,a) = M(a,x) = x$, for all $x \in [a,b]$. This means that (I, M) and $([a,b], G_M)$, considered as semigroups, are isomorphic with respect to the algebraic, order- and topological structure.
If, for example, M is a (pseudo-) t-norm (on I) then we say that G_M is a (pseudo-) t-norm on [a,b].
Moreover let us introduce the following notations for $c, d \in I$:

$$\varphi_c(x) = \rho_{0,c}(x) \ , \ \psi_d(x) = \rho_{d,1}(x), x \in [0,1] \tag{2.5}$$

so that
$\varphi_c^{-1}: [0, c] \to I$ and $\psi_d^{-1}: [d, 1] \to I$
are given by
$\varphi_c^{-1}(x) = \frac{x}{c}, x \in [0, c]$, $\psi_d^{-1}(x) = \frac{x-d}{1-d}, x \in [d, 1]$.
If c, d are fixed we write φ and ψ for φ_c and ψ_d, respectively.

(2) We present first a structure theorem for pseudo-t-norms and pseudo-s-norms, which seems not to be well-known, and which is probably due to Schweizer,B. and Sklar,A. (a similar version can be found in a forthcoming book of Alsina,Frank and Schweizer [1]).

Theorem 2.1 *If $M: I^2 \to I$ is a pseudo-t-norm (or pseudo-s-norm), if A_M is closed and if M_x is continuous for all $x \in A_M$, then M is representable as an ordinal sum of pseudo-t-norms (or pseudo-s-norms) none of which has interior idempotents.*

Proof: Since the set of idempotents $A_M = \{x \in I: M(x, x) = x\}$ is closed we get (if $I \setminus A_M$ is nonempty) uniquely determined open intervals
$U_n = (a_n, b_n), n \in \mathbb{N}$ (which are pairwise disjoint) such that

$$I \setminus A_M = \bigcup_{n \in \mathbb{N}} U_n. \tag{2.6}$$

Thus it is sufficient to prove that $M = min$ if (for fixed $n \in \mathbb{N}$)

$$(x, y) \in (A_M \times I) \cup ([a_n, b_n] \times \left[I \setminus [a_n, b_n]\right]). \tag{2.7}$$

First, let $M(x, x) = x$ and $y \in I$. If $x < y$ then
$x = M(x, x) \leq M(x, y) \leq M(x, 1) \leq x = min(x, y)$.
If $x \geq y$ then (using (1.6) we get
$0 = M(x, 0) \leq y \leq x = M(x, x)$. Since M_x is continuous there exists $z \in [0, x]$ satisfying $y = M(x, z)$, which implies (since $x \in A_M$)

$$min(x, y) = y = M_x(z) = M_x(M_x(z)) = M_x(y) = M(x, y). \tag{2.8}$$

Thus we have shown that

$$M(x, y) = y \quad \text{if} \quad x \in A_M \quad \text{and} \quad x \geq y. \tag{2.9}$$

Now let $x \in [a_n, b_n]$. If $y \leq a_n$ then (using $y \leq a_n \in A_M, M(a_n, a_n) = a_n$ and (2.3))

$$y = M(a_n, y) \leq M(x, y) \leq M(1, y) = y = min(x, y). \tag{2.10}$$

Finally, if $y \geq b_n$ then (again using $x \leq b_n \leq y, M(b_n, b_n) = b_n$ and (2.3))

$$x = M(b_n, x) \leq M(y, x) \leq M(1, x) = x = min(x, y). \tag{2.11}$$

Thus $M = min$ (or $M = max$) on A_M. Note that in this step of the proof no other regularity condition besides the continuity of M_x , $x \in A_M$ is needed. Now, $M_n := M|_{U_n^2}$ - the restriction of M to U_n^2 - satisfies
$M_n(x, a_n) = M_n(a_n, x) = a_n$ and $M_n(x, b_n) = M(b_n, x) = b_n, x \in U_n$ (resp. $M_n(x, b_n) = M_n(b_n, x) = b_n$ and $M_n(x, a_n) = M(a_n, x) = a_n, x \in U_n$), $M_n(x, x) \leq x$ (or $M_n(x, x) \geq x$) for all $x \in U_n$, and moreover, (U_n, M_n) is a subsemigroup of (I, M), since M_n is monotonic.
Hence we have (in the case of a pseudo-t-norm M)

$$T_n(x,y) := \alpha_n(M_n(\alpha_n^{-1}(x), \alpha_n^{-1}(y))$$
$$\text{with } \alpha_n(x) = a_n + (b_n - a_n)x \ , \ x \in I \tag{2.12}$$

is a pseudo-t-norm and the result follows :

$$M(x,y) = \begin{cases} T_n(x,y) & \text{if } (x,y) \in U_n^2 \\ \min(x,y) & \text{otherwise} \end{cases} \tag{2.13}$$

Let us remark that no summand need to be continuous or archimedean (a t- or s-norm is archimedean if for every x, y in (0,1) there is a positive integer n such that $x^n \leq y$ (resp. $x^n \geq y$). But in general the condition "M is archimedean pseudo-t-norm (pseudo-s-norm) is not equivalent to "$M(x,x) < x$ $(M(x,x) > x), x \in I$").

Remark 2.2 *Theorem 2.1 remains valid if we replace the condition "M_x is continuous for all $x \in A_M$" by "M^y is continuous for all $y \in A_M$". We have only to exchange x and y in (2.7).*

If we now assume a little bit more in Theorem 2.1 then we get the following more detailled result.

Theorem 2.3 *If $M: I^2 \to I$ is a pseudo-t-norm (pseudo-s-norm) and if M is continuous in each argument then M is representable as an ordinal sum of continuous archimedean t-norms (s-norms) so that each summand is either isomorphic to the product t-norm (s-norm) or to the Lukasiewicz t-norm(s-norm).*

Proof: Using Proposition 1.2 we know that M is continuous and thus A_M is closed and $A_M \supset \{0,1\}$. Applying Theorem 2.1 we now get that - with the notation of the proof of Theorem 2.1 - M_n is a continuous pseudo-t-norm. Using the famous result of Mostert and Shields [23] (see also Paalman-de-Miranda [26] and Ling [20]) M_n is even a continuous archimedean t-norm from which the result follows depending whether M_n is strict or non-strict (for some well-known results on t- and s-norms we refer for example to the book of Klement, Mesiar and Pap [13]).
Using the following Lemma we can immediately produce another version of Theorem 2.3.

Lemma 2.4 *Let $M: I^2 \to I$ be associative. Then the following conditions are equivalent :*
(1) M is increasing, continuous in each argument and has a neutral element 1 (or 0).
(2) M is (jointly) continuous on I^2 and satisfies

$$M(1,1) = 1 \ , \ M(x,0) = M(0,x) = 0 \ , \ x \in I \tag{2.14}$$
$$(or\ M(0,0) = 0 \ , \ M(x,1) = M(1,x) = 1 \ , \ x \in I). \tag{2.15}$$

Proof: If we assume (1) then Proposition 1.2 shows that M is continuous on I^2. Of course, $M(x,1) = M(1,x) = x, x \in I$ implies $M(1,1) = 1$. The last statement is exactly (1.7).
Reversely, (1) follows from Proposition 2.42 in [13].
The importance of Theorem 2.3 is that the functions M we are starting with, are automatically commutative. This means that many known "commutative results" are already "noncommutative results".
Combining Lemma 2.4 and Theorem 2.3 we arrive at the following result :

Theorem 2.5 *If $M: I^2 \to I$ is associative, continuous on I^2 and satisfies (2.14) (or (2.15)) then M is representable as an ordinal sum of continuous archimedean t-norms (or s-norms).*

(3) We now introduce a sometimes useful method for reducing investigations about commutative, continuous, idempotent,increasing and associative functions from I^2 into I to continuous, idempotent, increasing and associative functions from I^2 into I.
Let us now start with a continuous, idempotent, increasing and associative function $M: I^2 \to I$. We define two functions $M_1, M_2: I^2 \to I$ given by

$$M_1(x,y) := M(max(x,y), min(x,y)), x, y \in I, \tag{2.16}$$
$$M_2(x,y) := M(min(x,y), max(x,y)), x, y \in I. \tag{2.17}$$

Then we see immediately that M_1, M_2 are continuous, idempotent, increasing, but they are also commutative. Moreover

$$M_1(x,y) = M(x,y) \quad if \quad x \geq y \ , \ and \tag{2.18}$$
$$M_2(x,y) = M(x,y) \quad if \quad x \leq y \ , \tag{2.19}$$

that is, M coincide with M_1 below the diagonal and with M_2 above the diagonal in the unit square. We will show below that M_1 and M_2 are associative again.
This means that results for commutative, continuous, idempotent,increasing and associative functions can be applied separately to M_1 and to M_2, and from (2.17) and (2.18) we get results for M (the "noncommutative" result) without calculation and "drawing only a geometrical picture of M_1 and M_2"

and putting M_1 and M_2 together on the diagonal of the unit square. This will be shown by an example at the end of section 3.
We still have to show the above mentioned gap.

Theorem 2.6 *If $M: I^2 \to I$ is continuous, idempotent, increasing and associative then M_1 and M_2, given by (2.16) and (2.17), have also these properties.*

Proof: We have only to show the associativity of M_1 and M_2. Let us introduce $x \cdot y = M(x,y)$, $x \vee y = max(x,y)$, $x \wedge y = min(x,y)$.
Then the associativity of M_1 is equivalent to

$$\begin{aligned} M_1\Big(x \vee M_1(y,z), x \wedge M_1(y,z)\Big) &= M_1\Big(M_1(x,y) \vee z, M_1(x,y) \wedge z\Big) \Leftrightarrow \\ (x \vee ((y \vee z) \cdot (y \wedge z))) &\cdot (x \wedge ((y \vee z) \cdot (y \wedge z))) = \\ = (((x \vee y) \cdot (x \wedge y)) \vee z) &\cdot (((x \vee y) \cdot (x \wedge y)) \wedge z). \end{aligned} \tag{2.20}$$

We have to consider 6 cases :

$$x \le y \le z, x \le z \le y, x \le z \le y, y \le x \le z, y \le z \le x, z \le x \le y, z \le y \le x. \tag{2.21}$$

In the first case, (2.20) leads to

$$(x \vee (z \cdot y)) \cdot (x \wedge (z \cdot y)) = ((y \cdot x) \vee z) \cdot ((y \cdot x) \wedge z). \tag{2.22}$$

By using the idempotency and monotonicity we get
$x = x \cdot x \le z \cdot y$ and $y \cdot x \le z \cdot z = z$
so that (2.22) reduces to $(z \cdot y) \cdot x = z \cdot (y \cdot x)$
which is true because of the associativity of M.
The last case can be handled in the same manner (note that no continuity in these two cases is needed).
Now using also the continuity of M we get in the remaining 4 cases that (2.20) is valid. Let us show this for example in the second case of (2.21). Because of $x \le z \le y$ we have $x = x \cdot x \le y \cdot z$, and (2.20) is equivalent to

$$(x \vee (y \cdot z)) \cdot (x \wedge (y \cdot z)) = ((y \cdot x) \vee z) \cdot ((y \cdot x) \wedge z), \tag{2.23}$$

or

$$(y \cdot z) \cdot x = (y \vee z) \cdot (x \vee z) \cdot (y \wedge z) \cdot (x \wedge z). \tag{2.24}$$

This last equation is true since the left hand side is $(y \cdot z) \cdot x$ whereas the right hand side is

$$\begin{aligned} (y \vee z) \cdot (x \vee z) \cdot (y \wedge z) \cdot (x \wedge z) &= \\ y \cdot z \cdot z \cdot x &= y \cdot z \cdot x. \end{aligned} \tag{2.25}$$

Thus M_1 (and in an analogous way, M_2) is associative.

3 The structure of $\mathcal{F}_1$

Theorem 3.1 *Let $M: I^2 \to I$ and define*

$$a := M(1,0), b := M(0,1), \alpha := min(a,b) \text{ and } \beta := max(a,b). \tag{3.1}$$

Then $M \in \mathcal{F}_1$ iff M has the form

$$M(x,y) = \begin{cases} \varphi_\alpha^{-1} S_M(\varphi_\alpha(x), \varphi_\alpha(y)) & (x,y) \in [0,\alpha]^2 \\ \psi_\beta^{-1} T_M(\psi_\beta(x), \psi_\beta(y)) & (x,y) \in [\beta,1]^2 \\ max(min(x,y), min(x,a), min(y,b)) & otherwise \end{cases} \tag{3.2}$$

where S_M and T_M are ordinal sums of pseudo-s-norms and pseudo-t-norms on $[0,\alpha]^2$ and $[\beta,1]^2$, respectively, and where none of the summands has an interior idempotent.

Proof: (Necessity) Using the notations (2.5) and (3.1) we may assume $a \leq b$ (the case $a \geq b$ can be handled in the same manner).
The figure 1 presents the case $a \leq b$, whereas figure 2 presents the case $a \geq b$.

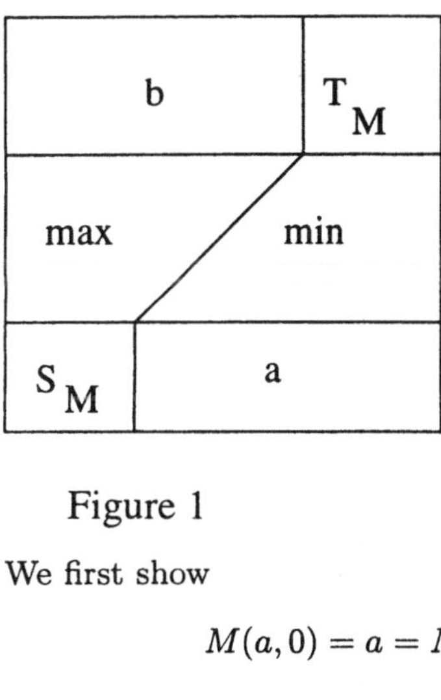

Figure 1

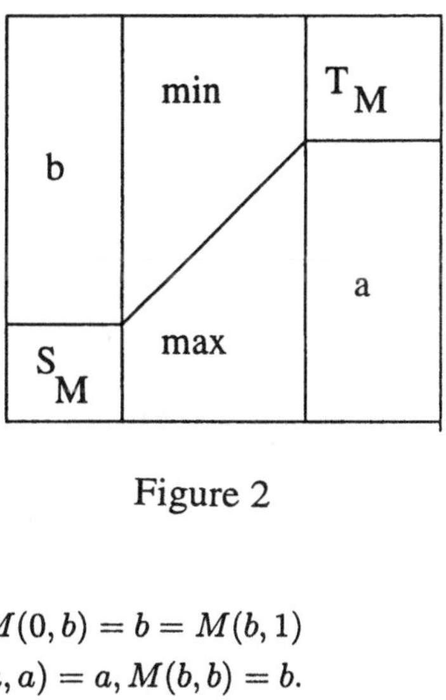

Figure 2

We first show

$$M(a,0) = a = M(1,a), M(0,b) = b = M(b,1) \tag{3.3}$$

$$\text{and } M(a,a) = a, M(b,b) = b. \tag{3.4}$$

By supposition, $0, 1 \in A_M$. Thus M_0, M_1 and M^0, M^1 are idempotent functions. Applying these functions to 1 and 0, respectively we get (3.5) (for example, $b = M_0(1) = M_0(M_0(1)) = M(0,b)$). Because of the monotonicity of M we arrive at
$a = M(a,0) \leq M(a,a) \leq M(1,a) = a$ and
$b = M(0,b) \leq M(b,b) \leq M(b,1) = b$,
that is (3.6). Thus M has the value a in $(a,0), (a,a), (1,0)$ and $(1,a)$, so that the monotonicity of M implies

$$M(x,y) = a \text{ if } (x,y) \in [a,1] \times [0,a]. \tag{3.5}$$

Similarly, we obtain

$$M(x,y) = b \text{ if } (x,y) \in [0,b] \times [b,1]. \tag{3.6}$$

Now we prove

$$M(0,y) = y, x = M(x,0), y \in [0,b], x \in [0,a], \tag{3.7}$$
$$M(1,y) = y, x = M(x,1), y \in [a,1], x \in [b,1]. \tag{3.8}$$

Indeed, we use that M_0, M_1, M^0, M^1 are continuous and that $0, 1 \in A_M$ to get for $y \in [0,b], z \in [a,1], s \in [0,a]$ and $t \in [b,1]$

$$0 = M(0,0) \leq y \leq b = M(0,b), 0 = M(1,a) \leq z \leq 1 = M(1,1),$$
$$a = M(0,0) \leq s \leq a = M(a,0), b = M(b,1) \leq t \leq 1 = M(1,1).$$

Thus there exist $\bar{y} \in [0,b], \bar{z} \in [a,1], \bar{s} \in [0,a], \bar{t} \in [b,1]$ satisfying $M_0(\bar{t}) = y, M_1(\bar{z}) = z, M^0(\bar{s}) = s, M^1(\bar{t}) = t$.
Therefore we have

$$y = M_0(\bar{y}) = M_0(M_0(\bar{y})) = M_0(y),$$
$$s = M^0(\bar{s}) = M^0(M^0(\bar{s})) = M^0(s) \tag{3.9}$$

and

$$z = M_1(\bar{z}) = M_1(M_1(\bar{z})) = M_1(z),$$
$$t = M^1(\bar{t}) = M^1(M^1(\bar{t})) = M^1(t), \tag{3.10}$$

that is, (3.7) and (3.8) are valid.
In the next step we determine M in $I \times [a,b]$.
Using (3.7) and (3.8) we arrive at (note that $a \leq b$)

$$y = M(0,y) \leq M(x,y) \leq M(1,y) = y, x \in I, y \in [a,b], \tag{3.11}$$

or equivalently,

$$M(x,y) = max(x,y) = y \ , \ 0 \leq x \leq y, y \in [a,b], \tag{3.12}$$
$$M(x,y) = min(x,y) = y \ , \ y \leq x \leq 1, y \in [a,b]. \tag{3.13}$$

An immediate calculation shows that M , given by (3.5), (3.6) and by (3.12) and (3.13) results in the third line of (3.2).
Finally let us introduce $S_M: [0,a]^2 \to [0,a]$ and $T_M: [b,1]^2 \to [b,1]$ by

$$S_M(x,y) = \varphi_a[M(\varphi_a^{-1}(x), \varphi_a^{-1}(y))], x, y \in [0,a] \tag{3.14}$$

and

$$T_M(x,y) = \psi_b[M(\psi_b^{-1}(x), \psi_b^{-1}(y))], u, v \in [b,1]. \tag{3.15}$$

We have already used that the first equality in (3.7) and (3.8) are valid on $[a, b]$. Now we use (3.7) for $x, y \in [0, a]$ and (3.8) for $x, y \in [b, 1]$ and get that S_M and T_M are pseudo-s-norms on $[0, a]$ and pseudo-t-norms on $[b, 1]$, respectively. Moreover, A_{S_M} and A_{T_M} are closed and $(S_M)_x$ and $(T_M)_x$ are continuous for all $x \in A_{S_M}$ and $x \in A_{T_M}$, respectively. Thus Theorem 2.1 gives the statement concerning S_M and T_M.
(Sufficiency) (1) If M has the form given in (3.2) then M is obviously increasing.

(2) To show the associativity let $a \leq b$ (the other case can be proved similarly) and let $x, y, z \in I$.

(a) Let $y \geq b, z \geq b$.
If $x \geq b$ then M is associative since T_M is associative.
If $x \leq b$ then $M(x, M(y, z)) = b = M(b, z) = M(M(x, y), z)$ (note that $M(y, b) \geq b$).
(b) the case $y \leq a$, $z \leq a$ can be treated in the same manner like case (a).
(c) If $a \leq z < b$ then $M(x, M(y, z)) = M(x, z) = z = M(M(x, y), z)$ since $M(t, z) = z$ for all $t \in I$ (see (3.11)).
(d) If $z < a$ and $y > a$ then $M(x, M(y, z)) = M(x, a) = a = M(M(x, y), z)$ since $M(x, y) \geq a$ and $M(x, a) = a$ (see (3.11)).
(e) If $z > b$ and $y < b$ then $M(x, M(y, z)) = M(x, b) = b = M(M(x, y), z)$ since $M(x, y) \leq b$ and $M(x, b) = b$.

(3) Now let $x \in A_M$. Then we get from (3.2):
If $x \leq a$ then
$(M_x(y) = max(x, y)$ if $y \leq b)$ and $(M_x(y) = b$ if $y \geq b.)$
If $x \leq b$ then
$(M_x(y) = a$ if $y \leq a)$ and $(M_x(y) = min(x, y)$ if $y \geq a.)$
If $a < x < b$ then
$(M_x(y) = a$ if $0 \leq y \leq a)$, $(M_x(y) = min(x, y)$ if $x \geq y > a)$,
$(M_x(y) = max(x, y)$ if $x > y \geq b)$ and $(M_x(y) = b \; if \; b < y \leq 1)$.
Thus M_x is always continuous on I if $x \in A_M$.
Similarly, M^0, M^1 are continuous on I and $\{0, 1\} \subset A_M$.

Let us now give some applications of Theorem 3.1.

(1) If $M: I^2 \to I$ is associative, increasing, continuous in each argument and if $M(0, 0) = 0, M(1, 1) = 1$ then M has the form given by (3.2) where S_M and T_M are ordinal sums of archimedean s-norms on $[0, \alpha]^2$ and archimedean t-norms on $[\beta, 1]^2$, respectively.
This is a result of Marichal in [21] which follows from Theorem 3.1 by using Theorem 2.3.

(2) Applying Theorem 3.1, Lemma 2.4 and Theorem 2.5 we get the following result:

If $M: I^2 \to I$ is associative, continuous on I^2 and satisfies (2.4) or (2.5) then M has the form given by (3.2) where S_M and T_M are ordinal sums of archimedean s-norms on $[0, \alpha]^2$ and archimedean t-norms on $[\beta, 1]^2$, respectively (note that $A_M \subset \{0, 1\}$ is both a consequence of (2.13) and (2.14)).

(3) If $A_M = [0, 1]$ in Theorem 3.1 then we get that

$$M(x, y) = max(min(x, y), min(x, a), min(y, b)) \tag{3.16}$$

for all $x, y \in I$ (since any idempotent pseudo-s-norm or pseudo-t-norm is given by max and min, respectively).

Note that (3.16) can also be written as

$$M(x, y) = min(max(x, y), max(y, a), max(x, b)). \tag{3.17}$$

Thus we get the result of Fodor [6] ((who proved the result in a more general setting). This means, that $T_M = min$ and $S_M = max$ in figures 1 and 2 on page 10. If in addition $a = M(1, 0) = 0$ or $a = M(1, 0) = 1$ then we get (because of $y \leq max(x, y)$ and $x \geq min(x, y)$)

$$M(x, y) = min(max(x, y), max(x, b), max(y, a)) = min(y, max(x, b))$$

and

$$M(x, y) = max(min(x, y), min(x, a), miny, b)) = max(x, min(y, b)),$$

that is, the weighted minimum or weighted maximum, respectively.

(4) To point out once more the idea of the proof of Theorem 3.1 and to underline the structure of $\mathcal{F}_1$ we introduce the concept of a pseudo-null-norm. From section 2 (2) we know that a pseudo-s-norm on $[0, a], a \in I$ has 0 as neutral element, and a pseudo-t-norm on $[0, b], b \in I$ has 1 as neutral element. Weakening appropriately these boundary conditions we arrive together with the associativity and monotonicity at a common generalization of pseudo-s-norms and pseudo-t-norms.

Definition 3.2 *A pseudo-null-norm is a triple (M, a, b) where $a, b \in I$, $a \leq b$ and where M is an associative, increasing operator satisfying*

$$M(x, 0) = x = M(0, x), x \in [0, a], \tag{3.18}$$
$$M(x, 1) = x = M(1, x), x \in [b, 1] \tag{3.19}$$

and either

$$M(0, y) = y = M(1, y), y \in [a, b] \quad or \tag{3.20}$$
$$M(x, 0) = x = M(x, 1), x \in [a, b]. \tag{3.21}$$

A first remark is, that indeed $a \leq b$ in Definition 3.2 : If we assume $a > b$, then we get the contradiction (using (3.18) and (3.19))

$M(0,1) \geq M(0,a) = a > b = M(b,1) \geq M(0,1)$.
Note that the conditions (3.18) - (3.20) are nothing else but (3.7) and (3.8) (which were of importance in the proof of Theorem 3.1).

If $a = b$ and if M is commutative then a pseudo-null-norm (M, a, a) is called a null-norm (see [2]). In this paper the additional condition
$M(x,a) = a, x \in I$ (a is absorbing element for M)
for a null-norm is required, but it can be omitted (see the proof of the next Theorem).

Theorem 3.3 *Let* $0 \leq a \leq b \leq 1$. *Then* (M, a, b) *is a pseudo-null-norm iff* M *is given by (3.2) where* S_M *is a pseudo-s-norm on* $[0, \alpha]^2$ *and* T_M *is a pseudo-t-norm on* $[\beta, 1]^2$.

Proof: (Necessity) We first assume that (3.18) - (3.20) are satisfied (the case that (3.18), (3.19) and (3.21) are valid, can be treated in the same manner). Defining $S_M: [0,a]^2 \to [0,a]$ and $T_M: [0,b]^2 \to [0,b]$ by (3.14) and (3.15), respectively, S_M is a pseudo-s-norm on $[0,a]^2$ and T_M is a pseudo-t-norm on $[0,b]^2$ (because of (3.19) and (3.20)).
Now, from (3.18) - (3.20) we get

$$\begin{aligned} a = M(a,0) \leq M(a,a) \leq M(1,a) = a, \\ a = M(a,0) \leq M(1,0) \leq M(1,a) = a, \\ b = M(0,b) \leq M(b,b) \leq M(b,1) = b, \\ b = M(0,b) \leq M(0,1) \leq M(b,1) = b \end{aligned} \qquad (3.22)$$

that is, M has the value a in $(a,0),(a,a),(1,0),(1,a)$ and thc valuc b in $(0,b),(0,1),(b,b),b,1)$. Thus the monotonicity of M implies again (3.5) and (3.6).
The condition (3.18) implies (3.11), or equivalently (3.12) and (3.13) so that we again arrive at the third equation in (3.2).
The sufficiency follows from the proof in Theorem 3.1.

Thus we may reformulate Theorem 3.1 as follows (using the notations of Theorem 3.1).

Corollary 3.4 $M \in \mathcal{F}_1$ *iff* M *is a pseudo-s-norm* $(\alpha = 1)$
or M *is a pseudo-t-norm* $(\beta = 0)$
or M *is a pseudo-null-norm* $(0 < \beta, \alpha < 1)$,
where in each case the occuring pseudo-s-norm and pseudo-t-norm is an ordinal sum of pseudo-s-norms and pseudo-t-norms on $[0, \alpha]^2$ *and* $[\beta, 1]^2$, *respectively and where none of the summands has an interior idempotent.*

(5) If M in Theorem 3.1 is in addition commutative then $a = b = \alpha = \beta$. Thus the third equation in (3.2) goes over into
"$M(x, y) = a$ otherwise."
Moreover, S_M and T_M are ordinal sums of s-norms and t-norms on $[0, \alpha]^2$ and $[\beta, 1]^2$, respectively, where none of the summands has an interior idempotent.
In particular, if $A_M = \{0, 1\}$ then we get a result of Mas, Mayor and Torrens [22].

(6) If M in Theorem 3.1 is in addition commutative and if $A_M = [0, 1]$ then we get the well-known result of Fung and Fu [10] :

$$M(x, y) = med(x, a, y) = \begin{cases} a & \text{if } min(x, y) \le a \le max(x, y) \\ max(x, y) & \text{if } max(x, y) \le a \\ min(x, y) & \text{if } a \le min(x, y) \end{cases} \tag{3.23}$$

where $med(x, a, y)$ is the median of $x, a, y \in I$. Note that $med(x, a, y)$ is invariant under a permutation of x, a, y and that

$$\begin{aligned} med(x, a, y) &= max(min(x, y), min(x, a), min(y, a)) \\ &= min(max(x, y), max(x, a), max(y, a)) \end{aligned} \tag{3.24}$$

Finally we remark that, starting with the result of Fung and Fu [10] we can get, using part (3) of section (2), the result of Fodor [6], as indicated by the figures 3 - 6 on page 16 .

We briefly comment these figures.

We are looking for all associative, increasing, continuous and idempotent functions $M: I^2 \to I$ and assume that we already know the structure of all these functions, if they are commutative. Since the functions M_1 and M_2 , given by (2.16) and (2.17), satisfy the assumptions of the result of Fung and Fu, we draw in figure 3 and figure 4 the functions M_1 and M_2 (see 3.23) where a,b are constants, and where w.l.o.g. $a \le b$.

In figure 5 we indicate that our noncommutative idempotent aggregation M coincides with M_1 below and with M_2 above the diagonal in the unit square. Combining now figure 3 up to figure 5 we result in figure 6, that is, we get our noncommutative aggregation "by drawing "pictures".
(Notice also the special cases a = 0 in figure 3 which results in

$$M(x, y) = min(x, y)$$

and b = 1 in figure 4 which leads to

$$M(x, y) = max(x, y).)$$

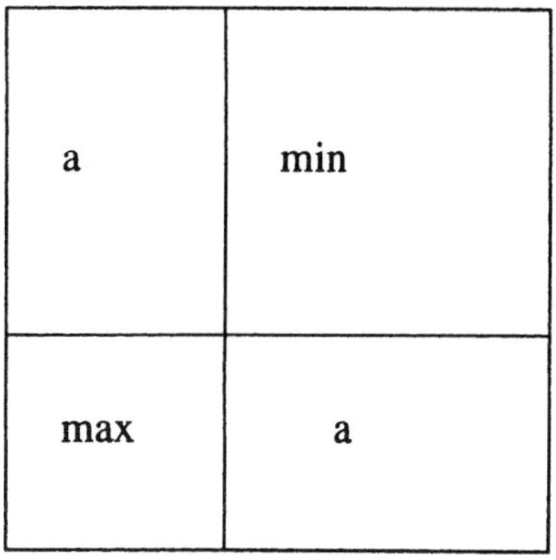

Figure 3

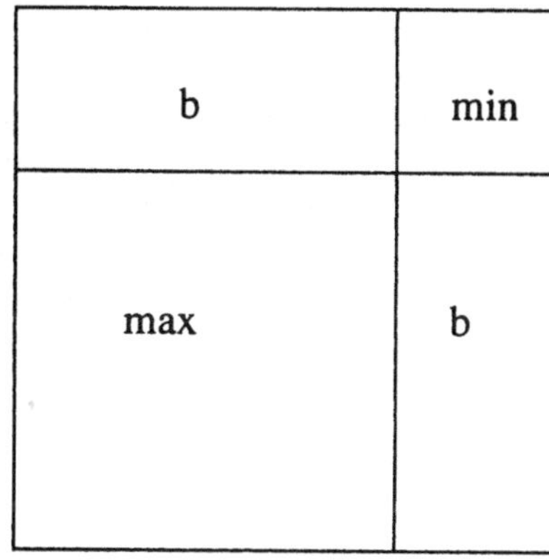

Figure 4

Figure 5

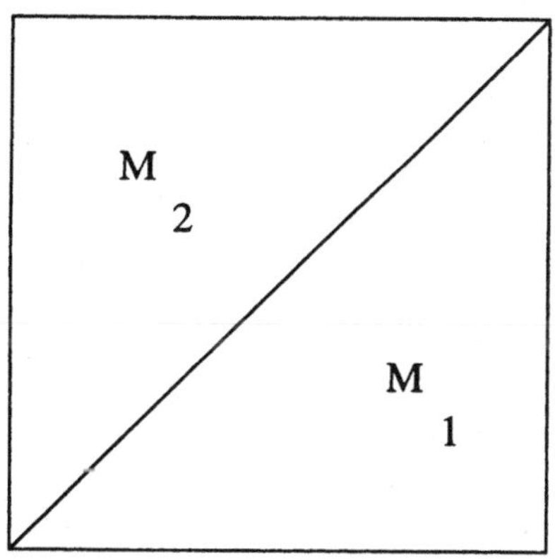

Figure 6

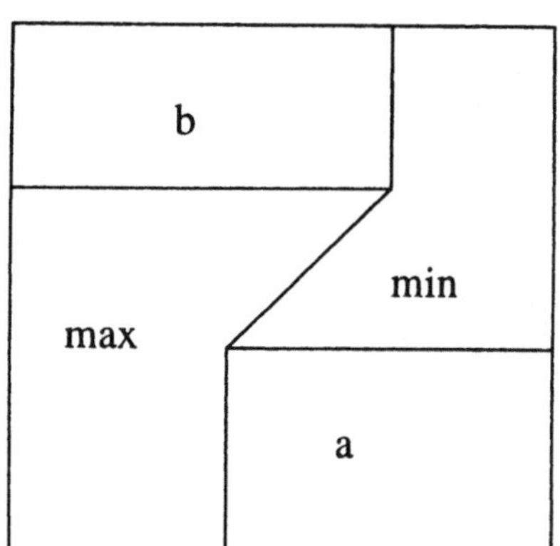

4 The structure of $\mathcal{F}_2$

In section 3 we have seen that the idempotent elements together with the values at $(0,1)$ and $(1,0)$ and a mild regularity condition lead to a characterization of associative and increasung binary operations.
Now we look for characterizations of $\mathcal{F}$ assuming that $M \in \mathcal{F}$ has an inner neutral element (which is uniquely determined). We start with a first preparatory result, which gives a first insight into the structure of a pseudo-uninorm.

Theorem 4.1 *If M is a pseudo-uninorm with inner neutral element $e \in (0,1)$ then*
(1) there exist a pseudo-t-norm $t_M: I^2 \to I$ and a pseudo-s-norm $s_M: I^2 \to I$ satisfying

$$M(x,y) = T_M(x,y) := \varphi_e(t_M(\varphi_e^{-1}(x), \varphi_e^{-1}(y)), (x,y) \in [0,e]^2, \quad (4.1)$$
$$M(x,y) = S_M(x,y) := \psi_e(s_M(\psi_e^{-1}(x), s_M(\psi_e^{-1}(y)), (x,y) \in [e,1]^2, \quad (4.2)$$
$$(2)\ min(x,y) \le M(x,y) \le max(x,y) if\ min(x,y) \le e \le max(x,y), \quad (4.3)$$
$$(3)\ x \in [0,e] \Rightarrow M(0,1) = M(x, M(0,1)), M(1,0) = M(M(1,0),x), \quad (4.4)$$
$$x \in [e,1] \Rightarrow M(0,1) = M(M(0,1),x), M(1,0) = M(x, M(1,0)), \quad (4.5)$$
$$(4)\ M(0,1) \in \{0,1\}, M(1,0) \in \{0,1\}, \{0,1\} \subset A_M, \quad (4.6)$$
$$(5)\ M(0,1) = 0, M^1 \textit{ continuous on } [0,e) \Rightarrow M(x,1) = x, x \in [0,e), \quad (4.7)$$
$$M(1,0) = 0, M_1 \textit{ continuous on } [0,e) \Rightarrow M(1,x) = x, x \in [0,e), \quad (4.8)$$
$$M(0,1) = 1, M_0 \textit{ continuous on } (e,1] \Rightarrow M(0,x) = x, x \in (e,1], \quad (4.9)$$
$$M(1,0) = 1, M^0 \textit{ continuous on } (e,1] \Rightarrow M(x,0) = x, x \in (e,1]. \quad (4.10)$$

Proof: (1) Using the notation of section 2 we define (writing $\varphi := \varphi_e$ and $\psi := \psi_e$)

$$t_M(x,y) := \varphi^{-1}M(\varphi(x), \varphi(y)), x, y \in I \quad (4.11)$$

and

$$s_M(x,y) := \psi^{-1}M(\psi(x), \psi(y)), x, y \in I. \quad (4.12)$$

An immediate calculation shows that t_M is a pseudo-t-norm on I^2 and s_M is a pseudo-s-norm on I^2 , so that $T_M: [0,e]^2 \to [0,e]$ and $S_M: [e,1]^2 \to [e,1]$ (see (4.1) and (4.2)) is a pseudo-t-norm on $[0,e]^2$ and a pseudo-s-norm on $[e,1]^2$, respectively. From the definition of t_M and s_M we get immediately (4.1) and (4.2).

(2) First observe the equivalence

$$min(x,y) \le e \le max(x,y) \Leftrightarrow (x,y) \in ([0,e] \times [e,1]) \cup ([e,1] \times [0,e]). \quad (4.13)$$

If $x \le e \le y$ then $M(x,y) \le y = max(x,y)$ and
$M(x,y) \ge M(x,e) = x = min(x,y)$.

The case $x \geq e \geq y$ can be proven in the same way.

(3) If $x \leq e$ then we obtain (using that $M(x,0) = 0 = M(0,x), x \leq e$)
$M(0,1) = M(M(x,0),1) = M(x, M(0,1))$,
$M(1,0) = M(1, M(0,x)) = M(M(1,0),x)$.
The case $x \geq e$ yields (Using that $M(x,1) = 1 = M(1,x), x \geq e$)
$M(0,1) = M(0, M(1,x)) = M(M(0,1),x)$,
$M(1,0) = M(M(x,1),0) = M(x, M(1,0))$,

(4) To prove (4) we first observe that $M(0,1) \neq e$ and $M(1,0) \neq e$ since otherwise (using (3)) we get the following contradictions :
$x = M(e,x) = M(M(0,1),x) = M(0,1) = e$ for all $x \in I$, or
$x = M(x,e) = M(x, M(1,0)) = M(1,0) = e$ for all $x \in I$.
Now, putting $x = 0$ in (4.4) we get with (1.7)
$M(0,1) = M(0, M(0,1)) = 0 \quad , \quad M(1,0) = M(M(1,0),0) = 0$,
and with $x = 1$ in (4.5) we obtain, using (1.8)
$M(0,1) = M(M(0,1),1) = 1 \quad , \quad M(1,0) = M(1, M(1,0)) = 1$.
Finally $\{0,1\} \subset A_M$ because of
$0 \leq M(0,0) \leq M(0,e) = 0 \ , \ 1 \leq M(1,e) \leq M(1,1) \leq 1$.

(5) Let us prove for example (4.9) (the other cases can be proven in the same manner using (3) and (4)) :
Because of $0 = M(0,e) < x \leq M(0,1) = 1$ and the continuity of M_0 on $(e,1]$ there exists $z \in (e,1]$ such that $x = M(0,z)$. This implies (since $0 \in A_M$) :
$M_0(x) = M_0(M_0(x) = M(0,z) = x$.

Let us discuss this result. The formulas (4.1) - (4.3) say that I^2 can be divided into 4 parts:
$I^2 = [0,e]^2 \cup [e,1]^2 \cup ([0,e) \times (e,1]) \cup ((e,1] \times [0,e))$,
where M coincides with a pseudo-t-norm T_M on $[0,e]^2$ and with a pseudo-s-norm S_M on $[e,1]^2$, but on the remaining two rectangles M behaves like an internal operator.
In the following we use for any pseudo-uninorm M with neutral element e the notation

$$I_M = [0,e) \times (e,1] \ , \ I^M = (e,1] \times [0,e). \tag{4.14}$$

We denote a pseudo-uninorm M satisfying

$$M(0,1) = a \ , \ M(1,0) = b \tag{4.15}$$

by an (a,b)-pseudo-uninorm so that (4.6) means that there are 4 types of pseudo-uninorms:

$$(0,0) - pseudo - uninorms, (0,1) - pseudo - uninorms,$$
$$(1,0) - pseudo - uninorms, (1,1) - pseudo - uninorms. \tag{4.16}$$

In order to characterize a pseudo-uninorm M the problem arises to describe more exactly the values of $M(x,y)$ for $min(x,y) \leq M(x,y) \leq max(x,y)$.

Before we prove a general representation theorem, from which we derive all following characterization theorems, we start with a result which is important for the understanding of regular pseudo-uninorms.

Theorem 4.2 *If an associative, increasing and continuous function* $M: I^2 \to I$ *has a neutral element* $e \in I$ *then* $e \in \{0,1\}$.

Proof: Let us assume that $e \in (0,1)$ so that M is a continuous pseudo-uninorm.
We apply Theorem 4.1 , (5), and consider 4 cases :
If $M(0,1) = 0$ then we get from (4.7) and (4.9) the contradictions
$1 = M(e,1) = M(lim_{x \to e-} x, 1) = lim_{x \to e-} M(x,1) = lim_{x \to e-} x = e$
and
$0 = M(0,e) = M(0, lim_{x \to e+} x) = lim_{x \to e+} M(0,x) = lim_{x \to e+} x = e.$
In an analogous way the two other cases (see (4.8) and (4.10)) lead again to contradictions.

The result in Theorem 4.2 is a result of Koch [17] which says, that the neutral element in an ordered topological semigroup on a compact interval must be one of the endpoints (we have given here another proof for this result). Note that Theorem 4.2 underlines the importance of continuous pseudo-t- and pseudo-s-norms. But it also shows that a pseudo-uninorm cannot be continuous on I^2, it must be discontinuous at least in one point of $[0,1]^2$.
We now present the announced general representation theorem for pseudo-uninorms, in which we extend some ideas of Czogala and Drewniak [4].

Theorem 4.3 *Let* $M: I^2 \to I$ *be a pseudo-uninorm with neutral element* e. *Then there exist functions* $g, h: I \to I$ *with* $g(e) = e = h(e)$, *a pseudo-t-norm* $T_M: [0,e]^2 \to [0,e]$ *and a pseudo-s-norm* $S_M: [e,1]^2 \to [e,1]$ *such that* M *has the following representation :*

$$M(x,y) = \begin{cases} T_M(x,y) & \text{if } (x,y) \in [0,e]^2 \\ S_M(x,y) & \text{if } (x,y) \in [e,1]^2 \\ min(x,y) & \text{if } 0 \le x < e, e < y < g(x) \text{ or} \\ & 0 \le y < e, e < x < h(y) \\ max(x,y) & \text{if } e < y \le 1, h(y) < x < e \text{ or} \\ & e < x \le 1, g(x) < y < e \end{cases} \tag{4.17}$$

and

$$min(x,y) \le M(x,y) \le max(x,y) \quad \text{otherwise.} \tag{4.18}$$

Moreover, we have

$g(0) - 1, h(0) - 1$ *if* M *is a (0,0)-pseudo-uninorm,*
$g(0) = 1, g(1) = 0$ *if* M *is a (0,1)-pseudo-uninorm,*
$h(1) = 0, h(0) = 1$ *if* M *is a (1,0)-pseudo-uninorm,*
$h(1) = 0, g(1) = 0$ *if* M *is a (1,1)-pseudo-uninorm.*

Proof: From Theorem 4.1 (1) we get the first two statements concerning T_M and S_M. Let us now consider the functions
$g = g_M: I \to I$ and $h = h_M: I \to I$ defined by

$$g(x) = \begin{cases} sup\{y \in [e,1]: M^y(x) = x\} & \text{if } x \le e \\ inf\{y \in [0,e]: M^y(x) = x\} & \text{if } x \ge e \end{cases} \tag{4.19}$$

and

$$h(y) = \begin{cases} sup\{x \in [e,1]: M_x(y) = y\} & \text{if } y \le e \\ inf\{x \in [0,e]: M_x(y) = y\} & \text{if } y \ge e \end{cases} \tag{4.20}$$

By definition of g and h we get $g(e) = e = h(e)$ and the second part of (4.17). Finally, if M is a (0,0)-pseudo-uninorm then $M(0,1) = 0$ and $M(1,0) = 0$ imply $g(0) = 1, h(0) = 1$ (by definition of g and h).
The other cases can be treated in the same manner :
$M(0,1) = 0, M(1,0) = 1$ imply $g(0) = 1, g(1) = 0$,
$M(0,1) = 1, M(1,0) = 0$ imply $h(1) = 0, h(0) = 1$ and
$M(0,1) = 1, M(1,0) = 1$ imply $h(1) = 0, g(1) = 0$.
Thus Theorem 4.3 is proven.

Note that in (4.19) g is a function with the independent variable x, whereas h in (4.20) is a function of the variable y.
The idea behind the definition of g and h is the following :
Obviously $M(x,e) = x = min(x,e) = M(e,x)$ if $0 \le x < e$ and
$M(x,e) = x = max(x,e) = M(e,x)$ if $e < x \le 1$. In I_M and I^M we have by Theorem 4.1 only the information that M behaves like an internal operator. Using g and h we can extend the property that $M = min$ or $M = max$ as far as possible into the domains I_M and I^M (see (4.14)).
Since we have assumed no regularity condition for M in Theorem 4.3 we cannot expect to prove regularity conditions for the functions g, h constructed by (4.19) and (4.20). We can only say that the regions

$$\begin{aligned} A_g &:= \{(x,y): 0 \le x < e < y < g(x)\} \text{ and} \\ B_h &:= \{(y,x): h(y) < x < e < y \le 1\} \end{aligned} \tag{4.21}$$

are disjoint, since otherwise there is a point $(x,y) \in [0,e) \times (e,1]$ such that $M(x,y) = max(x,y) = min(x,y)$ which is impossible.
For the same reason

$$\begin{aligned} C_h &:= \{(y,x): 0 \le y < e < x < h(y)\} \text{ and} \\ D_g &:= \{(x,y): g(x) < y < e < x \le 1\}. \end{aligned} \tag{4.22}$$

are disjoint. Note that A_g , B_h , C_h and D_g are the subgraphs of $g|_{[0,e)}, h|_{(e,1]}, h|_{[0,e)}$ and $g|_{(e,1]}$, respectively.

If we assume some type of symmetry condition, then we can show that g and h are decreasing. This is of interest since in this case the function h , which is a function of the variable y, has a "generalized" inverse. Thus the following Lemma will be important for our considerations.

Lemma 4.4 *Let $M: I^2 \to I$ be a pseudo-uninorm with neutral element e. Then the following statements are valid :*

(1) If $y > g(x)$ implies $x > h(y)$ for $x \in [0, e)$, then g is decreasing on $[0, e)$.
(2) If $y < g(x)$ implies $x < h(y)$ for $x \in (e, 1]$, then g is decreasing on $(e, 1]$.
(3) If $x < h(y)$ implies $y < g(x)$ for $y \in (e, 1]$, then h is decreasing on $(e, 1]$.
(4) If $x > h(y)$ implies $y > g(x)$ for $y \in [0, e)$, then h is decreasing on $[0, e)$.
(5) In each of the cases (1) - (4) the function M can differ (on the rectangle I_M and I^M, respectively) from min or max only on the graph of g which is equal to the graph of f.

Proof: We first prove (1) and (3). The statements (2) and (4) can be proven in an analogous way.

(1) Let $x, z \in [0, e)$ where $x \leq z$. Choose $y > g(x)$. Thus $y > g(x) \geq e > z \geq x$ and $x > h(y)$ by hypothesis. By definition of h (see (4.20)) this means $y = M_x(y)$, and we get $y = M_x(y) \leq M(z, y) \leq max(z, y) = y$, that is , $y = M(z, y)$. Since $M(z, y) = y > z$ we get $g(x) < y$ and finally $g(z) \leq inf\{y: y > g(x)\} = g(x)$.

(3) Let $v, y \in (e, 1]$ where $v \leq y$. Choose $x > h(y)$ so that we get $y < g(x)$. Because of $x < h(y) < e < v \leq y$ we arrive at $M(x, v)$ (since $M(x, y) = x \geq M(x, v) \geq min(x, v) = x$). Again using that $M(x, v) = x < v$ we obtain $h(v) \geq x$ and $h(v) \geq sup\{x: x < h(y)\} = h(y)$.

(5) We prove the statement only in case (1).
Let (x, y) be an arbitrary element from I_M. Then there are three possibilities. If $y \leq g(x)$ then $M(x, y) = x = min(x, y)$. If $y \geq x$ then (1) implies $x \geq h(y)$ so that $M(x, y) = max(x, y)$. If $y = g(x)$ we only know that $min(x, y) \leq M(x, y) \leq max(x, y)$ (see Theorem 4.3). Thus the graph of g is equal to the graph of f.
Let us now prove a result which extends the corresponding commutative version in [9].

Theorem 4.5 *Let $M: I^2 \to I$ be a pseudo-uninorm. If $M \in \mathcal{F}_2$ then there are 4 possible forms for M. In all 4 cases we have $M = T_M$ on $[0, e]^2$ and $M = S_M$ on $[e, 1]^2$, where T_M and S_M are ordinal sums of pseudo-t-norms and pseudo-s-norms on $[0, e]^2$ and $[e, 1]^2$, respectively, and where none of the summands has an interior idempotent.*
The form of M on $I_M \cup I^M$ is given by one of the following 4 cases :

$$\begin{aligned} &(a) M(x, y) = min(x, y), (x, y) \in I_M \cup I^M \\ &\quad if\ M\ is\ a\ (0,0)-pseudo-uninorm, \\ &(b) M(x, x) = max(x, y), (x, y) \in I_M \cup I^M \end{aligned} \tag{4.23}$$

$$if\ M\ is\ a\ (1,1) - pseudo - uninorm. \tag{4.24}$$

(c) If M is a (0,1)-pseudo-uninorm , then

$$M(x,y) = \begin{cases} min(x,y) & (x,y) \in I_M \\ max(x,y) & (x,y) \in I^M, \end{cases} \tag{4.25}$$

(d) if M is a (1,0)-pseudo-uninorm , then

$$M(x,y) = \begin{cases} max(x,y) & (x,y) \in I_M \\ min(x,y) & (x,y) \in I^M. \end{cases} \tag{4.26}$$

Proof: The first statement concerning T_M and S_M follows immediately from Theorem 4.1 (1) and from Theorem 2.1. Now we show
$0 = M(0,1)$ implies $M(x,1) = x, x \in [0,e)$. Indeed, if $x \in [0,e)$ then $0 = M^1(0) \leq x < M^1(e) = 1$ yields (since M^1 is continuous on $[0,e)$) that there is an element $z \in [0,e)$ satisfying $x = M^1(z)$. Thus we obtain
$M^1(x) = M^1(M^1(z)) = M^1(z) = x, x \in [0,e)$,
which means that $g(x) = 1$ for all $x \in [0,e)$ (see (4.19)).
If $0 = M_1(0)$ then we get using that M_1 is continuous on $[0,e)$ (since $1 \in A_M$)) that $M_1(y) = y$ or $h(y) = 1, y \in [0,e)$.
In the same manner we get (using $0 \in A_M$ and the continuity of M_0 and M^0 on $(e,1]$).
$M(0,1) = 1$ implies $M_0(y) = y$, that is $h(y) = 0$ if $x \in (e,1]$,
$M(1,0) = 1$ implies $M^0(x) = x$, that is $g(x) = 0$ if $x \in (e,1]$.
Combining these results we get, if M is a pseudo-uninorm, $g(x) = 1, x \in [0,e)$ (and thus $h(y) = e, y \in (e,1]$) and $h(y) = 1, y \in [0,e)$ (which implies $g(x) = e, x \in (e,1]$). This gives exactly (4.23).
In an analogous way we arrive at (4.24) - (4.26).

The following figures contain all typical elements of $\mathcal{F}_2$.

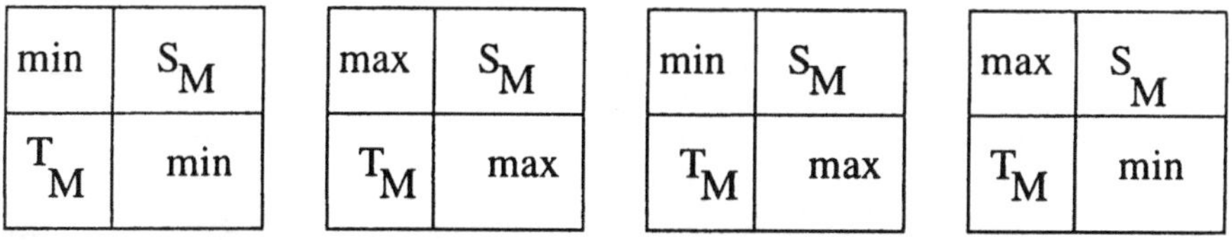

If in particular $A_M = [0,1]$ we arrive at the following figures :

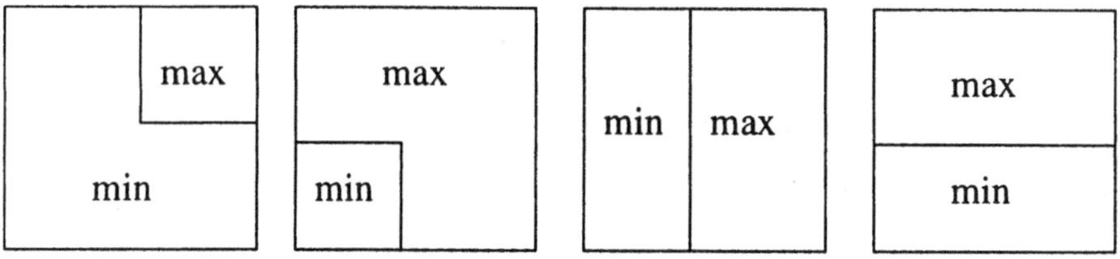

We remark that the functions $g = g_M$ and $h = h_M$, where M is a (0,0)-pseudo-uninorm from $\mathcal{F}_2$, are given by

$$g(x) = h(x) = \begin{cases} 1 & \text{if } x \in [0, e) \\ e & \text{if } x \in [e, 1] \end{cases} \tag{4.27}$$

Thus g and h are decreasing but they don't satisfy the condition given in Lemma 4.4 (1) (If $x \in [0, e)$ then $y \geq g(x)$ means that $y = 1$ so that $x > h(1) = e$ is not satisfied). But there are pseudo-uninorms which satisfy these conditions given in Lemma 4.4., for example, all idempotent pseudo-uninorms.

Now we will study this class in detail.

To formulate a first result we briefly introduce some notations for decreasing functions on I (see [27]).

Definition 4.6 *Let f be decreasing on $[a, b]$. Then $Q(f)$ is the set of functions $\bar{f}$ defined on $[f(a), f(b)]$ satisfying*

$$(1)\ \bar{f}(f(a)) = a, \bar{f}(f(b)) = b, \tag{4.28}$$

$$(2)\ if\ y \in Ranf\ then\ \bar{f}(y) \in \{x \in I: f(x) = y\}, \tag{4.29}$$

$$(3)\ if\ y \notin Ranf\ then \bar{f}(y) = sup\{x: f(x) > y\} = inf\{x: f(x) < y\}. \tag{4.30}$$

Elements $\bar{f}$ of $Q(f)$ satisfying $\{\bar{f}(x): x \in Ranf\} = \{\bar{f}(x): x \in dom\bar{f}\}$ are called quasi-inverses of f.

For properties of $Q(f)$ we refer to [13] and [27].

In the following the functions g and h denote the functions g_M and h_M, introduced in (4.19) and (4.20) (with respect to a pseudo-uninorm M).

Theorem 4.7 *Let M be an idempotent pseudo-uninorm with neutral element e. Then $g, h: I \to I$ are both decreasing , satisfy*

$$g(0) = 1 = h(0), g(1) = 0 = h(1), g(e) = e = h(e), \tag{4.31}$$

and M has the form

$$M(x, y) = \begin{cases} min(x, y) & x \in I, y \in [0, g(x)) \\ max(x, y) & x \in I, y \in (g(x), 1] \\ min(x, y)\ or\ max(x, y) & x \in I, y = g(x) \end{cases} \tag{4.32}$$

or

$$M(x, y) = \begin{cases} min(x, y) & x \in [0, h(y)), y \in I \\ max(x, y) & x \in (h(x), 1], y \in I \\ min(x, y)\ or\ max(x, y) & x = g(y), y \in I \end{cases} \tag{4.33}$$

Moreover,

$$h|_{[0,e]} \in Q(g|_{[e,1]})\ and\ g|_{[e,1]} \in Q(h|_{[0,e]}) \tag{4.34}$$

Proof: Since *min* is the only idempotent pseudo-t-norm and *max* is the only idempotent pseudo-s-norm, we arrive at (4.32) on $[0,e]^2$ and $[e,1]^2$, respectively. We now show that

$$M(x,y) \in \{min(x,y), max(x,y)\} \text{ if } x < e < y. \tag{4.35}$$

If $M(x,y) \leq e$ then $x \in A_M$ implies
$M_x(y) = M_x(M_x(y)) = min(x, M_x(y)) = x$ (since $M_x(y) \geq M_x(x) = x$).

In the same manner, $M(x,y) \geq e$ implies
(together with $y \in A_M$ and $M^y(x) \leq M^y(y) = y$))
$M^y(x) = M^y(M^y(x)) = max(y, M(x,y)) = y$.

In the same way we conclude that

$$M(x,y) \in \{min(x,y), max(x,y)\} \text{ if } y < e < x. \tag{4.36}$$

Now we consider the functions g and h (see (4.19) and (4.20)).
Since $g(e) = e = h(e)$ we see that $y = g(x) \in \{min(x,y), max(x,y)\}$ if $x \in [0,e) \cup (e,1]$ (and also $x = h(y) \in \{min(x,y), max(x,y)\}$ if $y \in [0,e) \cup (e,1]$).

Now we show that for all $x \in [0,e)$ we have :

$$y < g(x) \text{ implies } x > h(y) \text{ and } M(x,y) = max(x,y). \tag{4.37}$$

Let $y < g(x)$. Then $M(x,y) \neq x$ and thus $M(x,y) = y = max(x,y)$ (note that $y > g(x) \geq e > x$) which means that $x > h(y)$. Now Lemma 4.4 (1) shows that g is decreasing on $[0,e)$.

In the same manner we get for all $x \in (e,1]$

$$y < g(x) \text{ implies } x < h(y) \text{ and } M(x,y) = min(x,y). \tag{4.38}$$

By Lemma 4.4 (4) g is decreasing on $(e,1]$. Because of (4.35) and (4.36) we arrive at (4.32).

Using Lemma 4.4 (2) and (3) we get in the completely same way that $h: I \to I$ is decreasing with fixpoint e and that M has the representation (4.33).

The implications in (4.37) and (4.38) yield that for $y \in [h(1), e)$ we have

$$\breve{g}(y) := inf\{x \in [0,e): y > g(x)\} \geq inf\{x \in [0,e): x > h(y)\} = h(y), \tag{4.39}$$
$$\hat{g}(y) := sup\{x \in (e,1]: y < g(x)\} \leq sup\{x \in (e,1]: x < h(y)\} = h(y), \tag{4.40}$$
$$\tag{4.41}$$

or

$$\hat{g} \leq h \leq \breve{g} \text{ on } [h(1), e). \tag{4.42}$$

Using the implications in Lemma 4.4 (2) and (3), which are valid here, we get in the same manner

$$\hat{h} \leq g \leq \check{h} \ on \ (e, h(0)]. \tag{4.43}$$

where

$\hat{h} := sup\{y \in (e,1]: x < h(y)\}$ and
$\check{h} := inf\{y \in [0,e): x > h(y)\}, x \in (e, h(0)]$.

Let us now show (4.31). We already know that $g(e) = e = h(e)$.
In the following we use the last 4 statements in Theorem 4.3 and the representations given in (4.32) and (4.33) :

If M is a (0,0)-pseudo-uninorm then $g(0) = 1$ implies $h(1) = 0$ and moreover, $h(0) = 1$ implies $g(1) = 0$.
In the three other cases we arrive also at (4.31).
Let us now show (4.34). First we observe that
$h(g(e)) = h(e) = e, h(g(1)) = h(0) = 1$ and
$g(h(0)) = g(1) = 0, g(h(e)) = g(e) = e$.
Since g is decreasing we have - if $y \notin Rang$ -

$inf\{x \in [0,e): y > g(x)\} = sup\{x \in (e,1]: y < g(x)\}$ so that

$h(y) = \hat{g}(y) = \check{g}(y)$.
Finally, the third lines in (4.32) and (4.33) together with $h(1) = 0$ imply $h|_{[0,e]} \in Q(g|_{[e,1]})$.
Analogously, we obtain the second statement in (4.34) (using $h(0) = 1$).

Let us indicate that from Theorem 4.7 we can conclude many additional results. By Theorem 4.7 M is uniquely determined up to the values of the graph of g (or h). This is the important result of Czogala and Drewniak in [4]. If for example M is left-continuous on $(0,e) \times (e,1)$ and M is right-continuous on $(e,1) \times (0,e)$ the g is left-continuous on $(0,e)$ and right-continuous on $(e,1)$, respectively.
Thus M has the representation

$$M(x,y) = \begin{cases} min(x,y) & (x,y) \in ([0,e] \times [0,g(x)]) \cup ((e,1] \times [0,g(x)) \\ max(x,y) & \text{elsewhere} \end{cases} \tag{4.44}$$

Making use of Theorem 4.2.2 in [27] we get immediately the following results.

Remark 4.8 *Let $M: I^2 \to I$ be a pseudo-uninorm and let $g,h: I \to I$ be decreasing functions.*
(1) If M is left-continuous on $((0,e) \times (e,1)) \cup ((e,1) \times (0,e))$ then g is left-continuous on $(e,1)$ and h is left-continuous on $(0,e)$ and, moreover

$$g(h(y)) \geq y \ , \ y \in [h(1), e] = [0, e] \tag{4.45}$$

and

$$h(g(x)) \geq x \ , \ x \in [e, h(0)] = [e, 1]. \tag{4.46}$$

(2) If M is right-continuous on $((0,e) \times (e,1)) \cup ((e,1) \times (0,e))$ then g is right-continuous on $(e,1)$ and h is right-continuous on $(0,e)$ and, moreover

$$g(h(y)) \leq y \ , \ y \in [0, e] \tag{4.47}$$

and

$$h(g(x)) \leq x \ , \ x \in [e, 1]. \tag{4.48}$$

(3) If g is continuous on $[e,1]$ then

$h|_{[0,e]}$ is a quasiinverse of $g|_{[e,1]}$.

(4) If h is continuous on $[0,e]$ then

$g|_{[e,1]}$ is a quasiinverse of $h|_{[0,e]}$.

(5) If g is strictly decreasing on $[e,1]$, then

$Q(g|_{[e,1]}) = \{h|_{[0,e]}\}$, and $h|_{[0,e]}$ is a quasiinverse of $g|_{[e,1]}$.

(6) If h is strictly decreasing on $[0,e]$, then

$Q(h|_{[0,e]}) = \{g|_{[e,1]}\}$, and $g|_{[e,1]}$ is a quasiinverse of $h|_{[0,e]}$.

These results in Remark 4.8 allow us to characterize idempotent pseudo-uninorms if they are left-continuous or right-continuous.

Theorem 4.9 *A function $M: I^2 \to I$ is an idempotent pseudo-uninorm with neutral element e and is left-continuous on $((0,e) \times (e,1)) \cup ((e,1) \times (0,c))$ if and only if there exist decreasing functions $h: [0,e] \to [e,1]$ and $g: [e,1] \to [0,e]$ satisfying*

$$h(0) = 1, h(e) = e = g(e), g(1) = 0, \tag{4.49}$$

$$g(h(y)) \geq y, y \in [0,e], h(g(x)) \geq x, x \in [e,1], \tag{4.50}$$

such that M has the form

$$M(x,y) = \begin{cases} min(x,y) & (x,y) \in ([0,e] \times [0,h(x)]) \cup ((e,1] \times [0,g(x)]) \\ max(x,y) & (x,y) \in ([0,e] \times (h(x),1]) \cup ((e,1] \times (g(x),1]). \end{cases} \tag{4.51}$$

Proof: Necessity : This follows from Theorem 4.7 and Remark 4.8 (1) (Note that we have two representations of M in Theorem 4.7).
Sufficiency : Before we start with the proof, we illustrate the situation, given in Theorem 4.9.

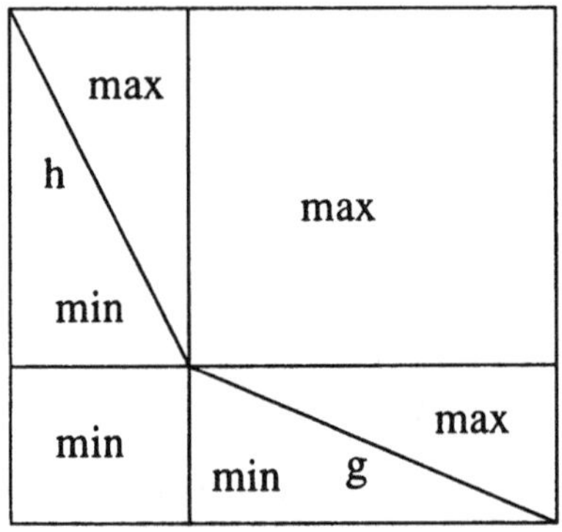

The function M, given by (4.51) and satisfying (4.49) and (4.50), is obviously left-continuous, idempotent, increasing and has the neutral element e. We have only to show that M is associative.
If $x, y, z \in I$ we distinguish 4 main cases :

$$(I)\ y \geq e, z \geq e \quad (II)\ y < e < z \tag{4.52}$$
$$(III)\ y \leq e, z \leq e \quad (IV)\ z < e < y \tag{4.53}$$

Case (I) (a) If $x \geq e$, then $M = max$ and M is associative.
(b) If $x < e$, then

$$x < e \leq y\ ,\ x < e \leq z\ ,\ M(y, z) \geq e, \tag{4.54}$$

since $y \geq e, z \geq e$ and $M(e, e) = e$).
Moreover, the associativity of M is equivalent to

$$max(M(x, y), M(x, z)) = M(M(x, z), z) \tag{4.55}$$

(indeed, since $y \geq e, z \geq e$ we have $M(y, z) = max(y, z)$ and $M(x, M(y, z)) = M(x, max(y, z)) = max(M(x, y), M(x, z)))$.
We now consider four cases in case (b) :

$$\begin{aligned}
&(b_1)\ y \geq e, z \geq e, x < e, z \leq h(x), y \leq h(x),\\
&(b_2)\ y \geq e, z \geq e, x < e, z \leq h(x), y > h(x),\\
&(b_3)\ y \geq e, z \geq e, x < e, z > h(x), y \leq h(x),\\
&(b_4)\ y \geq e, z \geq e, x < e, z > h(x), y < h(x).
\end{aligned} \tag{4.56}$$

(b_1) $max(M(x, y), M(x, z)) = max(min(x, y), min(x, z)) =$
$max(x, x) = x = min(min(x, z), max(y, z)) = min(M(x, z), M(y, z)) =$
$M(min(x, y), z) = M(M(x, y), z)$. (see (4.55))

(b_2) Because of $y > h(x) \le z$ and $M(x,y) \ge M(x,z)$ we obtain (together with (4.54))
$max(M(x,y), M(x,z)) = M(x,y) = max(x,y) = y =$
$max(y,z) = M(y,z) = M(max(x,y),z) = M(M(x,y),z).$

(b_3) Here $y \le h(x) < z$ and $M(x,y) \le M(x,z)$ imply with (4.54)
$max(M(x,y), M(x,z) = M(x,z) = M(min(x,y),z) = M(M(x,y),z).$

(b_4) $max(M(x,y), M(x,z)) = max(max(x,y), max(x,z)) = max(y,z) =$
$max(max(x,z), max(y,z)) = max(M(x,z), M(y,z)) =$
$M(max(x,y),z) = M(M(x,y),z).$

Case(II) We distinguish eight additional cases :

(a_1) $y < e < z, x \ge e, z \le h(y), y \le g(x),$
(a_2) $y < e < z, x \ge e, z \le h(y), y > g(x),$
(a_3) $y < e < z, x \ge e, z > h(y), y \le g(x),$
(a_4) $y < e < z, x \ge e, z > h(y), y > g(x),$
(A_1) $y < e < z, x < e, z \le h(y), z \le h(x),$
(A_2) $y < e < z, x < e, z \le h(y), z > h(x),$
(A_3) $y < e < z, x < e, z > h(y), z \le h(x),$
(A_4) $y < e < z, x < e, z > h(y), z > h(x),$

Note that in $(a_1) - (a_4)$ we have (note that $x \ge e, z > e$)

$$y < e \le x, M(y,z) \le M(x,z) \text{ and } M(x,z) \ge e. \tag{4.57}$$

Moreover, the associativity of M is equivalent to

$$M(x, M(y,z)) = M(y,z) \text{ in case of } (a_1) \text{ and } (a_3) \tag{4.58}$$

and to

$$M(x, M(y,z)) = M(x,z) \text{ in case of } (a_2) \text{ and } (a_4) \tag{4.59}$$

(Indeed, let us show (4.58) : Using $y \le g(x)$ we get
$M(M(x,y),z) = M(min(x,y),z) = min(M(x,z), M(y,z)) = M(y,z)$ (since $M(y,z) \le M(x,z)$). Using $y > g(x)$ we arrive at (4.59)). In an analogous way we get in $(A_1) - (A_4)$ (note that $x < e, y < e$)

$$x < e < z, M(x,y) \le M(x,z) \text{ (since } y < z\text{)}, \; M(x,y) \le e \tag{4.60}$$

and the associativity of M is equivalent to

$$min(x,y) = min(M(x,z), M(y,z)) \text{ in case of } (A_1) \text{ and } (A_2) \tag{4.61}$$

and to

$$M(x,z) = min(M(x,z), M(y,z)) \text{ in case of } (A_3) \text{ and } (A_4) \tag{4.62}$$

(Let us show for example (4.61) : We use $z \le h(y)$ and (4.60) to get
$M(x, M(y,z)) = M(x, min(y,z)) = min(M(x,y), M(x,z)) =$

$M(x, y) = min(x, y)$
and
$M(M(x, y), z) = M(min(x, y), z) = min(M(x, z), M(y, z)))$.

(a_1) Here (and also in $(a_2) - (a_4)$ and $(A_1) - (A_4)$ we have (since $x < z$)

$$M(x, y) \leq M(x, z). \tag{4.63}$$

Now, $z \leq h(y)$ implies

$$\begin{aligned} M(x, M(y, z)) = M(x, min(y, z)) = min(M(x, y), M(x, z)) = \\ M(x, y) = min(x, y) = y = min(y, z) = M(y, z), \end{aligned} \tag{4.64}$$

that is, (4.58).

(a_2) Here we make (for the first time) use of (4.50), and of $z \leq h(y)$: Thus $g(z) \geq g(h(y)) \geq y > g(x)$ implies $z < x$. Now (4.59) is valid (using (4.63) and (4.64))

$M(x, M(y, z)) = M(x, y) = max(x, y) = x = max(x, z) = M(x, z)$.

(a_3) Now we make use of (4.57), (4.63) and $z < h(y) \geq h(g(x)) \geq x$ to see that (4.53) is satisfied :

$$\begin{aligned} M(x, M(y, z)) = M(x, max(y, z)) = max(M(x, y), M(x, z)) = \\ M(x, z) = max(x, z) = z = max(y, z) = M(y, z). \end{aligned} \tag{4.65}$$

(a_4) Here (4.59) is satisfied, because (4.65) is valid here, too.

(A_1) First note that $M(x, y) = min(x, y)$ since $x < e, y < e$. Thus we get - since $x < e < z$ and $y < e < z$ - exactly (4.61) :
$min(x, y) = min(min(x, z), min(y, z)) = min(M(x, z), M(y, z))$.

(A_2) To prove (4.61) we use $h(x) < z \leq h(y)$, that is, $x > y$,to get :
$min(x, y) = y = min(z, y) =$
$min(min(x, z), min(y, z)) = min(M(x, z), M(y, z))$.

(A_3) Here we have $h(y) < z \leq h(x)$, or $y > x$. Thus $x < y < e < z$ implies $M(x, z) = min(x, z) = min(min(x, z), max(y, z)) = min(M(x, z), M(y, z))$, that is (4.62).

(A_4) Finally we get once more (4.62) :
$M(x, z) = max(x, z) = z = min(z, z) =$
$min(max(x, z), max(y, z)) = min(M(x, z), M(y, z))$.

From case (I) and (II) it is clear that case (III) and (IV) can be proven in an analogous way. Thus Theorem 4.9 is proven.

Similar to Theorem 4.9 the following result can be derived (using Remark 4.8 (2)).

Theorem 4.10 *A function $M: I^2 \to I$ is an idempotent pseudo-uninorm with neutral element e and is right-continuous on $((0,e) \times (e,1)) \cup ((e,1) \times (0,e))$ if and only if there exist decreasing functions $h: [0,e] \to [e,1]$ and $g: [e,1] \to [0,e]$ satisfying*

$$\begin{gathered} h(0) = 1, h(e) = e = g(e), g(1) = 0, \\ g(h(y)) \leq y, y \in [0,e], h(g(x)) \leq x, x \in [e,1] \end{gathered} \tag{4.66}$$

such that M has the form

$$M(x,y) = \begin{cases} min(x,y) & (x,y) \in ([0,e] \times [0,h(x))) \cup ((e,1] \times [0,g(x)) \\ max(x,y) & (x,y) \in ([0,e] \times [h(x),1]) \cup ((e,1] \times [g(x),1]). \end{cases} \tag{4.67}$$

If M in Theorem 4.9 or Theorem 4.10 is in addition commutative, then by definition

$g = h$ (see (4.19) and (4.20)),

and we get the results of DeBaets in [5] (note, that Theorem 4.9 and Theorem 4.10 always imply $g(0) = 1$ and $g(1) = 0$ if M is in addition commutative).

Let us now prove a commutative version of Theorem 4.3.

Theorem 4.11 *Let $M: I^2 \to I$ be a left- or right-continuous uninorm with neutral element e and let g or h be a decreasing function.*

Then there exist decreasing functions $G, H: I \to I, G \leq H$ with fixpoint e, a pseudo-t-norm $T_M: [0,e]^2 \to [0,e]$ and a pseudo-s-norm $S_M: [e,1]^2 \to [e,1]$ such that M has one of the following forms.

(a) If M is a (0,0)-uninorm which is left-continuous on $((0,e) \times (e,1)) \cup ((e,1) \times (0,e))$, then $G(0) = 1$, and

$$M(x,y) = \begin{cases} T_M(x,y) & (x,y)\ [0,e]^2 \\ S_M(x,y) & (x,y) \in [e,1]^2 \\ min(x,y) & (x \in [0,e), e < y \leq G(x)) \text{ or} \\ & (x \in (e,1], 0 < y \leq G(x)) \\ max(x,y) & (x \in [H(1),e), H(x) < y \leq 1) \text{ or} \\ & (x \in (e,1], H(x) < y < e) \\ \in [min(x,y), max(x,y)] & \text{otherwise.} \end{cases} \tag{4.68}$$

Moreover, G, H satisfy

$$G(H(y)) \geq y \text{ if } y \in [G(1),e] \text{ and } H(G(x)) \geq x \text{ if } x \in [e,G(0)]. \tag{4.69}$$

(b) If M is a (1,1)-uninorm, which is right-continuous on $((0,e) \times (e,1)) \cup ((e,1) \times (0,e))$, then $H(1) = 0$, and

$$M(x,y) = \begin{cases} T_M(x,y) & (x,y)\ [0,e]^2 \\ S_M(x,y) & (x,y) \in [e,1]^2 \\ min(x,y) & (x \in [0,e), e < y < G(x)) \text{ or} \\ & (x \in (e, h(0)], 0 < y < G(x)) \\ max(x,y) & (x \in [0,e), H(x) \le y \le 1) \text{ or} \\ & (x \in (e,1], H(x) \le y < e) \\ \in [min(x,y), max(x,y)] & \text{otherwise.} \end{cases} \tag{4.70}$$

Moreover, G and H satisfy

$$(G(H(y)) \le y \text{ if } y \in [G(1), e] \text{ and } H(G(x)) \le x \text{ if } x \in [e, G(0)]. \tag{4.71}$$

(c) If M is a (0,0)-uninorm, which is right-continuous on $((0,e) \times (e,1)) \cup ((e,1) \times (0,e))$, then $G(0) = 1$, and M has the representation (4.70).

(d) If M is a (1,1)-uninorm, which is left-continuous on $((0,e) \times (e,1)) \cup ((e,1) \times (0,e))$, then $H(1) = 0$, and M has the representation (4.68).

Proof: (a) Remember that $g, h: I \to I$ are given by (4.19) and (4.20). Then (4.17), (4.18) and the first and last line in the last statement of Theorem 4.3 are valid (note that a (0,1)- and a (1,0)-pseudo-uninorm cannot occur if M is commutative, since otherwise $M(0,1) = 0 = 1 = M(1,0)$ gives a contradiction). Now, using the commutativity of M, we have $g = h$, so that g and h are decreasing and left-continuous. Moreover,
h(0) = 1 if M is a (0,0)-uninorm, and
g(1) = 0 if M is a (1,1)-uninorm.
Let us illustrate the situation by the two figures 7 and 8 :

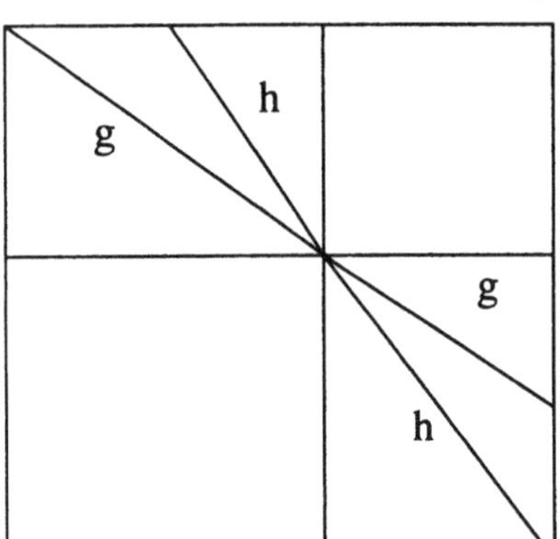

Figure 7

g(0)=1 , h(0)=1

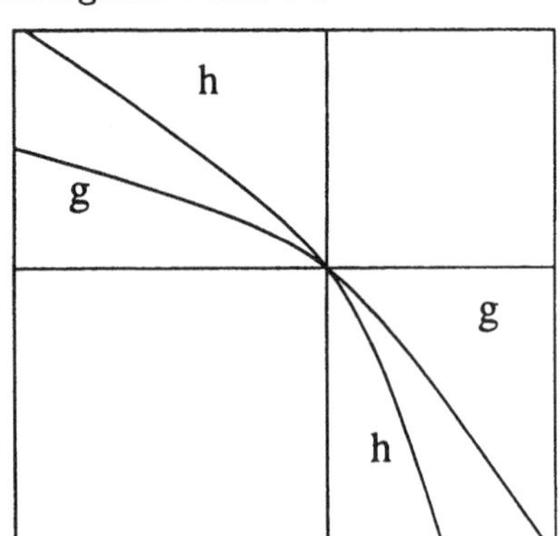

Figure 8

h(1)=0 , g(1)=0

Now let $\tilde{h}$ be the uniquely determined element in $Q(g|_{[e,1]})$ which is left-continuous on $(g(1), e)$, and let $\bar{h}$ be the uniquely determined element in $Q(g|_{[0,e]})$ which is left-continuous on $(e, g(0))$.
Note, that $\tilde{h}$ and $\bar{h}$ are also decreasing functions and left-continuous.
Now we define $G, H: I \to I$ by

$$G(x) = \begin{cases} g(x) & x \in [0, e) \\ \bar{h}(x) & x \in [e, g(0)] \\ 0 & x \in (g(0), 1] \end{cases} \tag{4.72}$$

and

$$H(x) = \begin{cases} 1 & x \in [0, g(1)) \\ \tilde{h}(x) & x \in [g(1), e] \\ g(x) & x \in (e, 1] \end{cases} \tag{4.73}$$

Then G, H are decreasing and left-continuous, $G \leq H$, and $G(e) = e = H(e)$. Since M is a (0,0)-uninorm we have $h(0) = g(0) = 1$ in (4.73) so that (4.28) leads to $G(0) = 1$ (Note that $g(1) = h(1) = 0$ in (4.74), that is $H(1) = 0$, if M is a (1,1)-uninorm).
G is left-continuous and thus $M(x, y) = min(x, y)$ if $y = G(x), x \in I$. Moreover, Theorem 4.2.2 in [27] gives again (with the above notations)

$$g(\tilde{h}(y)) \geq y, y \in [g(1), e] \text{ and } g(\bar{h}(x)) \geq x, x \in [e, g(0)] \tag{4.74}$$

which leads using the definition of G and H) to (4.69).

The cases (b) - (d) can be handled in the same manner (If M is right-continuous then we replace the uniquely determined left-continuous functions (in the above construction) by the uniquely determined right-continuous elements of $Q(g|_{[e,1]})$ and $Q(g|_{[0,e]})$, respectively).
This finishes the proof.

If in Theorem 4.11 we assume that $M \in \mathcal{F}_2$ then again the pseudo-t-norm T_M and the pseudo-s-norm S_M are ordinal sums of pseudo-t-norms and pseudo-s-norms, respectively.

We still remark, that in this paper we have investigated rather general associative aggregations, to unify most of the known results in this direction. Thus we have not given applications to De Morgan classes, to duality problems, to residuated implications or to some functional equations (see [3], [7], [15], [19]). This can be done, and results concerning these topics will appear in a forthcoming paper.
We finish this paper with two figures which explain once more the situation of Theorem 4.11 and show the connection between our new and the already known results :
If, for example, M in figure 10 is a (0,0)-uninorm, which is also idempotent, then $G = H$, and we get the results in [5] if M is left- or right-continuous.

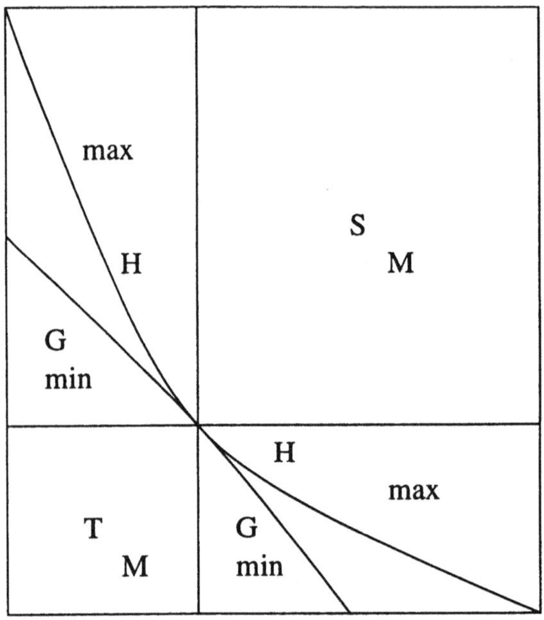

Figure 9

H(1) = 0

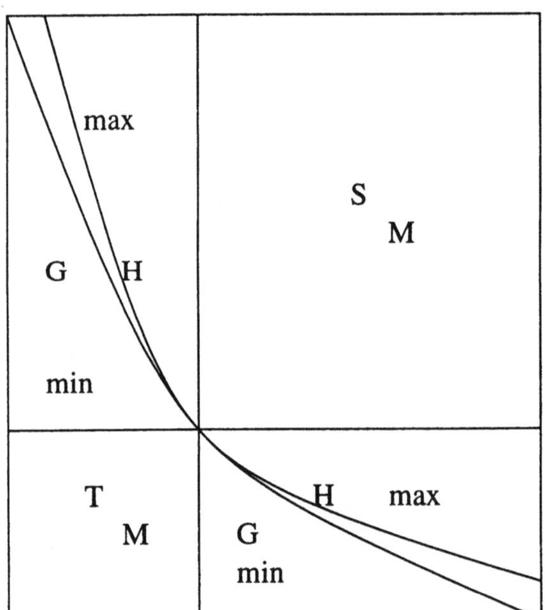

Figure 10

G(0) = 1

Finally I would like to thank A.Pavlovic from Novi Sad, Yugoslavia, for some valuable remarks.

References

[1] Alsina, C., Frank, M.J., Schweizer, B. *Associative functions on intervals.* Bookmanuscript.

[2] Calvo, T., De Baets, B. *On a generalization of the absorption equation.* Manuscript.

[3] Calvo, T., De Baets, B., Fodor, J. *The functional equation of Alsina and Frank for uninorms and nullnorms*, Fuzzy Sets and Systems 120 (2001), 15-24.

[4] Czogala, E., Drewniak, J. *Associative monotonic operations in fuzzy set theory.* Fuzzy Seta and Systems 12 (1984), 249-269.

[5] De Baets, B. *Idempotent uninorms.* Europ. J. Oper. Research 118 (1999), 631-642.

[6] Fodor, J.C. *An extension of Fung-Fu's theorem.* Internat. J. Uncertain. Fuzziness Knowledge-Based Systems 4 (1996), 51-58.

[7] Fodor, J., Calvo, T. *Aggregation functions defined by t-norms and t-conorms.* In: Studies and Fuzziness, Ed. B. Bouchon-Meunier, Physica-Verlag 1998, 36-48.

[8] Fodor, J.C., Roubens, M. *Fuzzy Preference Modelling and Multicriteria Decision Support.* Kluwer Academic Publishers, Dordrecht (1994).

[9] Fodor, J.C., Yager,R.R, Rybalov, A. *Structure of uninorms.* Internat. J. Uncertain. Fuzziness Knowledge-Based Systems 5 (1997), 411-427.

[10] Fung, L.W., Fu, K.S. *An axiomatic approach to rational decision making in a fuzzy environment.* In Zadeh, L.A., Fu, K.S., Tanaka, S., Shimura, M., editors, Fuzzy Sets and Their Applications to Cognitive and Decision Processes (1975), 227-256, Academic Press, New York.

[11] Grabisch, M., Nguyen, H., Walker, E.A. *Fundamentals of Uncertainty Calculi with Applications to Fuzzy Inference.* Kluwer Academic Publishers, Dordrecht (1995)

[12] Hajek, P. *Metamathematics of Fuzzy Logic.* Kluwer Academic Publishers, Dordrecht (1998).

[13] Klement, E.P., Mesiar, R., Pap, E. *Triangular Norms.* Kluwer Academic Publishers, Dordrecht-Boston-London (2000).

[14] Klir, G.J., Yuan, B. *Fuzzy Sets And Fuzzy Logic, Theory and Applications.* Prentice Hall PTR, Upper Saddle River-New Jersey (1995).

[15] Klement, E.P., Mesiar, R., Pap, E. *on the relationship of associative compensatory operators to triangular norms and conorms.* Internat. J. Uncertain. Fuzziness Knowledge-Based Systems 4 (1996), 129-144.

[16] Kolmogorov, A.N. *Sur la notion de la moyenne.* Accad. Naz. Lincei Mem. Cl. Sci. Fis. Mat. Natur. Sez. 12 (1930), 388-391.

[17] Koch, R.J. *Note on weak cutpoints in clans.* Duke Math.J. 24(1957), 611-615.

[18] Kruse, R.L., Deely, J.J. *Joint continuity of monotonic functions.* Amer. Math. Soc. 76 (1969), 74-76.

[19] Li, Y-M., Shi,Z-K. *Weak uninorm aggregation operators* Information Sciences 124 (2000), 317-323.

[20] Ling, C.M. *Representations of associative functions.* Publ. Math. Debrecen 12 (1965), 189-212.

[21] Marichal, J.-L. *Aggregation operators for multicriteria decision aid.*PH.D. thesis, Institute of Mathematics, University of Liege, Belgium, 1999.

[22] Mas, M., Mayor, G., Torrens, J. *t-operators.* Internat. J. Uncertain. Fuzziness Knowledge-Based Systems 7 (1999), 31-50.

[23] Mostert P.S., Shields, A.L. *On the structure of semigroups on a compact manifold with boundary.* Ann. of Math. 65 (1957), 117-143.

[24] Nagumo, M. *Über eine Klasse der Mittelwerte.*Japan. J. of Math 6 (1930), 71-79.

[25] Nelsen, R.B. *An introduction to copulas.* Volume 139 of Lecture Notes in Statistics (1999).

[26] Paalman-De Miranda, A.B. *Topological Semigroups.* Volume 11 of Mathematical Centre Tracts. Matematisch Centrum, Amsterdam (1964).

[27] Schweizer, B., Sklar, A. *Probabilistic Metric Spaces.* North-Holland, New York (1983).

Continuous WOWA Operators with Application to Defuzzification*

Vicenç Torra and Lluís Godo

Institut d'Investigació en Intel·ligència Artificial - CSIC,
Campus UAB s/n, E-08193 Bellaterra (Catalunya, Spain)
e-mail: vtorra@iiia.csic.es, http://www.iiia.csic.es/~vtorra
e-mail: godo@iiia.csic.es, http://www.iiia.csic.es/~godo

Abstract. The basic operations for combining real values are the weighted mean (WM) and the Ordered Weighted Averaging (OWA) operator. The weighted mean allows the system to compute an aggregate value from the ones coming from several sources, taking into account the reliability of each information source. Alternatively, the OWA operator allows the user to weight the values supplied in relation to their alternative ordering. The Weighted OWA operator (WOWA) allows the user to consider both aspects using two sets of weights. In this chapter we describe an extension of the WOWA operator to the continuous case after arguing its convenience. Then, we analyze the use of this operator for defuzzification.

1 Introduction

The need of techniques for aggregation of information arises in several fields of Artificial Intelligence (e.g., knowledge based systems [19], fuzzy logic controllers [31], vision [14], decision making [17]) as well as in other frameworks (e.g., management [17], scheduling [20]).

Within the framework of Artificial Intelligence these techniques are mainly used (see [22] for more details) when a system has to make a decision or when it needs a comprehensive representation of its application domain. Since combination functions are defined according to the objects to be synthesized, there exist several functions for each of the knowledge representation formalisms. For example, there are functions to combine probability distributions [12,16], fuzzy sets [3], etc.

When numerical values have to be combined, for instance in the case of fuzzy membership values, the basic operations which are considered are the weighted mean and the Ordered Weighted Averaging operator (OWA operator for short). The latter operator was introduced by Yager in [26]. Both functions are to combine the values according to a vector of weights. Their

* This is a revised and expanded version of two previous papers of the same authors: "Averaging continuous distributions with the WOWA operator" in Proc. of EFDAN'97, Dortmund (Germany) pp. 10-19.; and "On defuzzification with continuous WOWA operators" in Proc. of ESTYLF'97, Tarragona (Spain), pp. 227-232.

difference is in the meaning they assign to these weights, and therefore how these weights are combined with the input values. See [21] for a detailed analysis of their difference. In short, on the one hand, the weighted mean allows the system to compute an aggregate value from the ones coming from several sources, taking into account the reliability of each information source. Alternatively, the OWA operator allows the user to weight the values supplied in relation to their relative ordering. In this way, a system can give more importance to a subset of the input values than to another subset. For example, as it is described in [28], the OWA operator can model how are determined the ratings in certain sports in the Olympic Games: five judges give their punctuation and the average of these values is performed after dropping the extreme values). This corresponds to give zero importance to extreme values. Therefore, in the weighted mean, weights measure the importance of an information source with independence of the value that the source has captured. On the other hand, in the OWA operator weights measure the importance of a value (in relation to other values), independently from the information source that has captured it.

However, both types of weights can be needed in a single problem. As a matter of example [21], consider a robot with five sensors (whose reliability is known) to determine the distance to the nearest object in the direction of its movement. In this case, the data supplied by each source can be combined by means of the weighted mean. However, with independence of the information source that provides the measurement they are not equally important. The robot should give more importance to nearest objects than those that are further. For example, due to the orientation of the robot two of the readings can be very large. This situation is easily modeled by means of the OWA operator. The OWA operator uses a set of weights $w_1, \cdots, w_n$ where w_1 is the weight corresponding to the greatest value, and w_i corresponds to the i-th greatest value. In the particular case of the robot, a possible set of weights which gives more importance to nearest values would be $w = (0.1, 0.2, 0.3, 0.4)$. To calculate a final distance and to make a decision, the robot should consider both factors, the reliability of the sensors and the importance of small values. The Weighted OWA operator (WOWA operator for short) defined in [21] and [22] allows the user to consider both aspects using two vectors of weights.

WOWA operators are also appropriate in other settings, e.g. in constraint satisfaction problems, scheduling and decision making. We justify below only the first case, application of the operator to the other settings are similar in nature. In constraint satisfaction problems the typical case is to consider several alternatives to make a decision. Each alternative has to satisfy several criteria (constraints) at the same time. When this is analysed in the crisp setting, the selection of an alternative is achieved requiring that the solution satisfies the intersection of all constraints. When soft constraints (constraints are fuzzy valued instead of just being true or false) are considered, the intersection is substituted by a suitable conjunction operator, usually the minimum or any other t-norm (see [8] for a survey on flexible constraint satisfaction techniques). In this way, the extent an alternative satisfies all the constraints is calculated as the t-norm applied to the satisfaction degrees of each constraint. However, t-norms are very strict operators because they do not allow for compensation (as [20] point out, the use of a t-norm forbids compensation for one bad constraint satisfaction). However, as pointed out by Zimmermann and Slany in [33] and [20], humans are able to perform some kind of compensation between alternatives. Because of this, other type of operators, like aggregation operators, have also been considered in constraint satisfaction problems. These operators compensate small values with large ones using some kind of averaging. A classical example is the OWA operator cited above. However, this operator is not always appropriate because it is invariant to permutations of the inputs (they satisfy the generalized commutativity property). So, all the values (i.e., the evaluations of all constraints) are equally important, and so, all the constraints are also considered with the same importance (it has to be said that this is also the case when t-norms are considered). An aggregation function that considers the importance of each constraint is then required. The typical example of an aggregation function that allows for importances is the arithmetic mean, but this operator does not allow for compensation. In order to consider at the same time the importance of the constraints and the compensation of their validity, the WOWA operator can be used (see [8] for another approach to combine priority with fuzzy constraints).

So far, we have described why WOWA operators are important in several fields. Now we want to motivate the extension of the WOWA operators to cope with continuous scales of weights.

In management, the preference of a population of individuals over a certain product can be modeled with a random variable (for example, the price they agree to pay for a consumer good). In this case, when a decision has to be made about, a final value has to be chosen from the probability distribution. This selection can be understood as an averaging of the distribution. The easiest choice is to take the mean value of the distribution. This corresponds to consider everybody equally important from the decision point of view. However, there can be situations in which the averaging should reflect other strategies. For instance, (a) when everybody has to be able to buy the product, or (b) when the product is addressed only to a minority with high economic capabilities. In these cases, besides of considering the bare probability distribution, the averaging method has to consider the different importance of some values with respect to other values. In case (a) small prices have to be stressed, while in (b) these values can be undervaluated and only large values are relevant. The continuous WOWA operator proposed in this paper is suitable for considering both the distribution and the relative importance of the values.

A similar case can be found in robotics. A common assumption [15] is that the uncertainty in sensory data can be adequately modeled as random variables. As detailed by [1], such models can be constructed experimentally, or obtained from the characteristics of the sensor. In this case, it is needed a method to obtain a final value from a probability distribution (whether this proceeds from a single sensor or is the result of the fusion of several distributions). This final value can be seen as the combination of all the values considering their importance of high or low values. In this case, again, it is needed a combination operator for combining a probability distribution with a function describing the relative importance of the values.

Another application is defuzzification. Aggregation operators have recently (e.g. [30], [29]) been applied in the defuzzification process of fuzzy control systems. Actually, several defuzzification methods can be understood as a blending or mixing of different solutions: the defuzzification value is calculated as an aggregation of all possible solutions according to the possibility attached to each one. The use of WOWA can focus defuzzification on some of the relevant values of the output domain.

In this chapter, we describe the Continuous WOWA operator, and show its application to the defuzzification process.

The chapter begins in Section 2 with some preliminaries describing the WOWA and related aggregation operators. Then, in Section 3 the continuous WOWA operator is introduced. Section 4.2 is devoted to the use of the operator to defuzzification. The chapter finishes with some concluding remarks.

2 Averaging numerical values with WOWA operators

When numerical values have to be aggregated, for instance in the case of fuzzy membership values, the basic operations which have been considered are the *Weighted Mean* operator (WM for short) and the *Ordered Weighted Averaging* operator (OWA for short). Given a normalized n-dimensional weighting vector[1] $\boldsymbol{p} = (p_1, \ldots, p_n)$, the Weighted Mean of dimension n based on the weights $\boldsymbol{p}$ is the mapping $WM^{\boldsymbol{p}} : \mathbb{R}^n \to \mathbb{R}$ defined as

$$WM^{\boldsymbol{p}}(A) = \sum_{i=1,n} p_i \cdot a_i$$

where $A = (a_1, \ldots, a_n)$ is a set of n numerical observations. Together with their general form (the quasi-arithmetic mean), the WM operators are studied in [2].

On the other hand, OWA operators were introduced by Yager in [26,27] and are defined as follows. Let $\boldsymbol{w} = (w_1, \ldots, w_n)$ be a normalized weighting

[1] As usual, by a normalized weighting vector $\boldsymbol{p} = (p_1, \ldots, p_n)$ we mean that $p_i \in [0, 1]$ and $\Sigma_{i=1,n} p_i = 1$.

vector of dimension n and let A be a set of n numerical observations as above. In such a case, the Ordered Weighted Averaging operator of dimension n based on $\boldsymbol{w}$ is the mapping $OWA^{\boldsymbol{w}} : \mathbb{R}^n \to \mathbb{R}$ defined by

$$OWA^{\boldsymbol{w}}(A) = \sum_{i=1,n} w_i \cdot a_{\sigma(i)}$$

where $\{\sigma(1), \ldots, \sigma(n)\}$ is a permutation of $\{1, \ldots, n\}$ such that $a_{\sigma(1)} \leq a_{\sigma(2)} \leq \ldots \leq a_{\sigma(n)}$. Note that in this definition the permutation of the elements is the converse of the usual OWA definition, where the permutation is defined such that $a_{\sigma(1)} \geq a_{\sigma(2)} \geq \ldots a_{\sigma(n)}$. This change is just for technical purposes in the extension to the continuous case. OWA operators are analyzed in [11,10].

As mentioned in the Introduction, sometimes we may need to take into account the two kinds of weighting which are modelled by the WM and OWA operators, namely weights according to the reliability of the information sources (WM) and weights to the values themselves (OWA). The *Weighted OWA* operator (WOWA operator for short) defined in [21,22] allows the user to consider both aspects using two sets of weights.

Definition 1. [21] Let $\boldsymbol{p}$ and $\boldsymbol{w}$ be two n-dimensional normalized weighting vectors and let $A = (a_1, \ldots, a_n)$ be a set of n numerical observations. A mapping $WOWA^{\boldsymbol{w},\boldsymbol{p}} : \mathbb{R}^n \to \mathbb{R}$ is a Weighted Ordered Weighted Averaging operator of dimension n based on the weights $\boldsymbol{p}$ and $\boldsymbol{w}$ if it is defined by

$$WOWA^{\boldsymbol{w},\boldsymbol{p}}(A) = \sum_{i=1,n} \lambda_i \cdot a_{\sigma(i)} \tag{1}$$

where σ is a permutation as above and the weights λ_i's are defined as

$$\lambda_i = W^*(\sum_{j\leq i} p_{\sigma(j)}) - W^*(\sum_{j<i} p_{\sigma(j)})$$

with $W^* : [0,1] \to [0,1]$ being a monotonic increasing function that interpolates[2] the set of points $\{(i/n, \sum_{j\leq i} w_j)\}_{i=1,n}$ together with the point $(0,0)$.

Therefore, besides the two sets of weights $\boldsymbol{p}$ and $\boldsymbol{w}$, a WOWA operator needs an interpolating mechanism. Several interpolation methods can be used to build (non-decreasing) W^*'s, for example the method described in [5] and revisited in [24]. However, note that different interpolation functions may yield different operators.

Notice that WOWA operators generalize both WM and OWA operators. As it is shown in Figure 1, in the particular case when $\boldsymbol{w} = (1/n, \ldots, 1/n)$ we recover the WM operator and when $\boldsymbol{p} = (1/n, \ldots, 1/n)$ we recover the OWA operator. Finally if we take both $\boldsymbol{w} = \boldsymbol{p} = (1/n, \ldots, 1/n)$ what we get is just

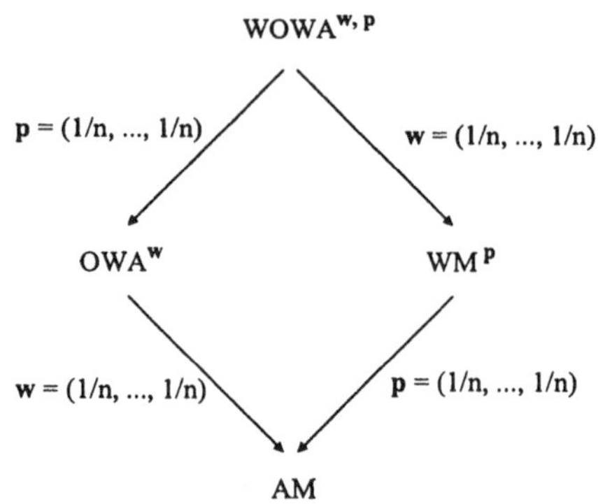

Fig. 1. Relationships among WOWA, OWA and WM

the arithmetic mean (AM). Moreover, WOWA operators have been proved to be Choquet integrals [23], like WM and OWA are so.

To avoid the non trivial task of constructing the interpolating functions W^*, there is an alternative procedure to define WOWA operators, namely to start directly with a single weighting vector $\boldsymbol{p}$ and with a non-decreasing function W^* such that $W^*(0) = 0$ and $W^*(1) = 1$. Then, the coefficients λ_i's and the final combination are computed as before. This approach is very similar to the one described by Yager in [28] to take into account importance weights for different information sources, actually there is a direct correspondence. He calls a function Q satisfying the above required properties to W^* a *regular monotonically non-decreasing quantifier*, i.e. non-decreasing, $Q(0) = 0$ and $Q(1) = 1$. For the sake of simplicity we shall use in this chapter this second procedure, and thus we redefine WOWA operators using these quantifiers.

Definition 2. Let $\boldsymbol{p}$ be a normalized weighting vector of dimension n and let Q be a regular monotonically non-decreasing quantifier. Then the WOWA operator of dimension n based on Q and p is the mapping $WOWA^{Q,\boldsymbol{p}} : \mathbb{R}^n \to \mathbb{R}$ defined as

$$WOWA^{Q,\boldsymbol{p}}(A) = \sum_{i=1,n} \lambda_i \cdot a_{\sigma(i)} \tag{2}$$

where σ is a permutation as usual and the weights λ_i's are defined now as

$$\lambda_i = Q(\sum_{j\leq i} p_{\sigma(j)}) - Q(\sum_{j<i} p_{\sigma(j)}).$$

[2] For technical reasons, the function W^* is required to be a straight line when the points can be interpolated in this way.

Obviously, both definitions are equivalent in the following sense

1. given $\boldsymbol{p}$ and $\boldsymbol{w}$, Q corresponds to the function W^* in the first definition, that is, $WOWA^{\boldsymbol{w},\boldsymbol{p}}(A) = WOWA^{W^*,\boldsymbol{p}}(A)$;
2. given $\boldsymbol{p}$ and Q, Q itself is an acceptable interpolating function for the first definition provided we define the weighting vector $\boldsymbol{w}$ by $w_i = Q(i/n) - Q((i-1)/n)$ for each $i = 1, n$.

3 Extension of the WOWA operator to a continuous domain

Since the extension of the WOWA operator to the continuous case is based in turn on the extension of the weighted mean, for the sake of clarity, we define below the weighted mean and its extension to the continuous case.

Let $\boldsymbol{p}$ be a normalized weighting vector of dimension n and consider the WM of a set of observations $A = (a_1, \ldots, a_n)$ based on $\boldsymbol{p}$,

$$WM^{\boldsymbol{p}}(A) = \sum_{i=1}^{n} p_i \cdot a_i. \tag{3}$$

Since the weights are normalized, when all the a_i's are different the function $prob$ defined as $prob(a_i) = p_i$ corresponds to a probability distribution on the set of values $\{a_i\}_{i=1,n}$. Therefore, we can assume that the pairs $(a_i, prob(a_i))$ define a random variable in a discrete domain. Notice that when there are several a_i's that are equal a similar probabilistic interpretation can be given. This case corresponds to having a new set $\{b_j\}_{j=1,m}$ with a set of probabilities $prob(b_i)$ such that $b_i \neq b_j$ for all $i \neq j$, and $prob(b_j) = \sum_{i:a_i=b_j} p_i$. With this interpretation in mind, the weighted mean (3) is nothing but the expected value of a random variable. Using the cumulative distribution F corresponding to $prob$, an equivalent expression for the weighted mean (3) is

$$WM^{\boldsymbol{p}}(A) = \sum_{j} (F(b_j) - F(b_{j-1})) \cdot b_j. \tag{4}$$

Based on this probabilistic interpretation of the discrete weighted mean, the extension to the continuous case amounts to consider a continuous random variable determined by a probability density function f on a (continuous) domain of values A in the real line $\mathbb{R}$. For simplicity reasons, in many cases we shall assume A to be just an interval $[a, b]$. In this case, the summation in (4) results in an integral and the differences $F(b_i) - F(b_{i-1})$ result in a differential dF. That is, we get

$$CWM^f(A) = \int_A x dF$$

as an expression for the continuous weighted mean, but under the usual assumptions of differentiability, $dF = f(x)dx$, which leads to the following definition.

Definition 3. Let $f : A \to [0,1]$ be a probability density function on A. The corresponding *Continuous Weighted Mean* $CWM^f(A)$ is the mean value of A with respect to f, i.e.

$$CWM^f(A) = \int_A f(x)x dx,$$

which reduces to $\int_a^b f(x)xdx$ when $A = [a,b]$.

The extension of the WOWA operator to a continuous domain will be done analogous to the extension of the weighted mean as it has been described above. The construction of the continuous WOWA is based on a cumulative probability distribution $F(x)$ (defined on a domain A) and a regular monotonically non-decreasing quantifier $Q : [0,1] \to [0,1]$. For the sake of simplicity, we assume that both $F(x)$ and $Q(y)$ are absolutely continuous, and thus there exist positive functions $f(x)$ and $q(x)$ such that

$$F(x) = \int_{-\infty}^{x} f(x')dx' \qquad (5)$$

and

$$Q(y) = \int_{-\infty}^{y} q(y')dy'. \qquad (6)$$

Here, although F and Q are actually defined in the intervals $[a,b]$ and $[0,1]$ respectively, for technical reasons we consider them to be extended to the whole real line by putting $F(x) = 0$ for $x < a$ and $F(x) = 1$ for $x > b$, and $Q(x) = 0$ for $x < 0$ and $Q(x) = 1$ for $x > 1$.

The function q will be called henceforth a *relative importance function* because it can be used to weight elements according to their cumulative probability, and thus, it allows, for instance to weight small or large values independently on the values themselves. Obviously, the only required conditions to q are to be positive and the normalization condition

$$\int_0^1 q(y)dy = 1.$$

Now, consider again the discrete WOWA operator (2)

$$WOWA^{Q,p}(A) = \sum_{i=1}^{n} \lambda_i \cdot a_{\sigma(i)}$$

where the $\lambda_i = Q(\sum_{j \leq p} \sigma(j)) - Q(\sum_{j<p} \sigma(j))$. Recall that $p_{\sigma(i)}$ corresponds to the probability of the $\sigma(i)$-th element, i.e. $p_{\sigma(i)} = prob(a_{\sigma(i)})$. Therefore, using the cumulative probability distribution F of $prob$, we can rewrite this expression as:

$$\sum_i \left[Q(F(b_i)) - Q(F(b_{i-1}))\right] \cdot b_i$$

where $b_i = a_{\sigma(i)}$. Notice that in this expression the b_i's are ordered in an increasing way. Furthermore, defining $G(b) = Q(F(b))$ we can rewrite the above expression as:

$$\sum_i \left[G(b_i) - G(b_{i-1})\right] \cdot b_i \tag{7}$$

Now, let us turn our attention to the continuous case. Since all the elements in the domain are already considered in an increasing order, we do not need in this case any previous permutation σ. For the continuous case we have assumed that both F and G are absolutely continuous. Thus $G(x) = Q(F(x))$ is also absolutely continuous and besides it satisfies $G(-\infty) = 0$ (since $F(-\infty) = 0$ and $Q(0) = 0$), $G(+\infty) = 1$ (since $F(+\infty) = 1$ and $Q(1) = 1$) and G is non-decreasing (since both F and Q are non-decreasing). Therefore, G has the same properties than F, i.e. G can be also considered as a cumulative probability distribution, and so there exists a positive function $g(x)$ such that

$$G(x) = \int_{-\infty}^{x} g(x')dx'. \tag{8}$$

Because (7) is analogous to (4), the extension to the continuous case of the expression (2) can be easily achieved, as in the case of the weighted mean, by changing the summation into an integral and replacing the differences by the differential dG. Namely, by applying the chaining rule to $G(x) = Q(F(x))$ we get

$$dG = (dQ/dF)(dF/dx)dx$$

and finally, by (5) and (6), we get $g(x) = q(F(x))f(x)$. So, it seems meaningful to define the *continuous* WOWA operator as follows.

Definition 4. Let $f : \mathbb{R} \to \mathbb{R}^+$ be a probability density function on A and let q be a relative importance function. The corresponding *Continuous WOWA* operator $CWOWA^{q,f}(A)$ is the mean value of A with respect to the probability density $g(x) = q(F(x))f(x)$, i.e.,

$$CWOWA^{q,f}(A) = \int_A q(F(x))f(x)x dx. \tag{9}$$

which reduces to $\int_a^b q(F(x))f(x)xdx$ when $A = [a, b]$.

By analogy to the discrete case, when the relative importance function gives equal importance to any value, i.e. when $q(x) = 1$ for $x \in [0,1]$ and $q(x) = 0$ otherwise, we get from (9), as a particular case, the expression given in Definition 3 for the *Continuous Weighted Mean* CWM^f with respect to the density f .

On the other hand, if in (5) we now let f correspond to the uniform probability distribution on an interval $[a, b]$, i.e. $f(x) = 1/(a-b)$ for $x \in [a, b]$

and $f(x) = 0$ otherwise, we get the following expression for the *Continuous Ordered Weighted Average* induced by q in the interval $[a, b]$:

$$COWA^q([a,b]) = \frac{1}{b-a}\int_a^b q(\frac{x-a}{b-a})xdx \tag{10}$$

Finally, if we take both f as the uniform density function on $[a, b]$ and Q as the uniform quantifier $Q(x) = x$, we come up with the mean value of the interval $[a, b]$. These relationships are depicted in Figure 2, where it can be seen that, like in the discrete case, Continuous WOWA operators generalize both Continuous Weighted Mean and Continuous OWA operators.

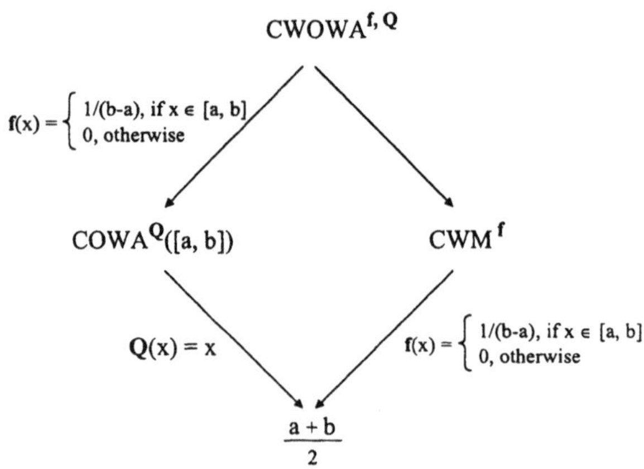

Fig. 2. Relationships among CWOWA, COWA and CWM

To conclude this section let us mention that the above approach to define continuous aggregation operators can be easily generalized to allow the average of real-valued functions instead of simple domains, in the spirit of [4]. For instance, if $\varphi : \mathbb{R} \to \mathbb{R}$ is a real function, then the average of φ over a domain A with respect to a relative importance function q and a probability density f expression (9) becomes

$$CWOWA^{q,f}(\varphi, A) = \int_{-\infty}^{+\infty} q(F(x))f(x)\varphi(x)dx. \tag{11}$$

Notice that when $\varphi(x) = x$ we recover the original expression (9). Moreover, all these continuous operators are indeed Choquet integrals (see again [4] for further details).

4 Continuous WOWA and defuzzification

4.1 Defuzzification as aggregation

By defuzzifying a fuzzy set μ in a certain domain X it is usually understood to select an element from X using the information provided by μ. Up to now, several defuzzification methods have been proposed in the literature, in a quite ad-hoc way. For instance in [7] six of the most well known methods of defuzzification, Center of Area (CoA), Center of Sums (CoS), Center of Largest Area (CLA), First of Maxima (FoM), Middle of Maxima (MoM) and Height Method (HM), are described and analyzed. Some other interesting references are for example [30], [29], [18], [25] and [6].

In [31], Yager and Filev provide a general view of the whole defuzzification process. Instead of considering a particular formulation as above, the authors describe defuzzification as a two step process in which first, (i) the membership function is converted into a probability distribution and then, (ii) an element is selected on the basis of the probability distribution. While several alternatives exist for the first step, only two main alternatives are considered in the second one: calculation of the expected value and random selection. The former is computed as usual as follows:

$$x^* = \sum_i x_i \cdot prob(x_i)$$

where $X = \{x_1, \cdots, x_n\}$ is the domain and $prob : X \to [0,1]$ is the probability distribution. As for the first step, Filev and Yager [9] introduced the so-called BAsic Defuzzification Distribution (BADD) transformations to transform fuzzy sets[3] into probability distributions. They are defined for each $\alpha \in [0, \infty]$ as:

$$prob(x_i) = \frac{(\mu(x_i))^\alpha}{\sum_{x_i \in X} (\mu(x_i))^\alpha}$$

where by convention we take $0^0 = 0$. Based on the BADD transformation and using expected value for element selection, Yager and Filev define the BADD defuzzification method:

Definition 5. [9] Let μ be a fuzzy set over $X = \{x_1, \cdots, x_n\}$, then the BADD-based defuzzification method of μ as

$$BADD_\mu^\alpha = \frac{\sum_{x_i \in X} x_i (\mu(x_i))^\alpha}{\sum_{x_i \in X} (\mu(x_i))^\alpha} \tag{12}$$

for some $\alpha \in [0, \infty]$.

[3] We assume non-empty fuzzy sets.

Another approach is to consider the whole defuzzification process as an aggregation process. This approach has been considered in, for example, [30], [29]. In this case, the domain X corresponds to the values to aggregate and the membership function corresponds to the additional information needed to aggregate the values in an appropriate way. Indeed, let A be a fuzzy set defined over a discrete domain $X = \{x_1, x_2, \ldots, x_n\}$ with $\mu_A : X \to [0,1]$ its membership function. Then, if for each x_i we define, by means of μ_A, a weight $p_i \in [0,1]$ such that $\sum_{i=1,n} p_i = 1$, the defuzzified value x^* of A can be calculated as the $WM^{(p)}(X)$, i.e.

$$x^* = \sum_{i=1,n} p_i \cdot x_i \tag{13}$$

A particular definition of the p_i's from the membership function μ_A corresponds to a particular selection of a defuzzification method. A simple, and very often used, method for defuzzification is the previously mentioned Center of Area (CoA) which corresponds to take above $p_i = \mu_A(x_i)/(\sum_{i=1,n} \mu_A(x_i))$. When, instead of discrete, a continuous domain is considered[4], the defuzzification can also be understood as an aggregation function: (13) is replaced in this case by the continuous WM according to Definition 3:

$$x^* = \int_{-\infty}^{+\infty} x f(x) dx$$

where f is in this case a probability density function. The CoA defuzzication method will obviously correspond to take

$$f(x) = k \cdot \mu_A(x), \tag{14}$$

where $k^{-1} = \int_{-\infty}^{+\infty} \mu_A(x) dx$ is a normalization factor.

In fact, the aggregation view of defuzzication is also related to Yager's approach since the expected value can be seen as an aggregation and the conversion of the membership function into a probability distribution can be seen as the conversion of the membership function into an appropriate weighting vector. Note that $BADD^1_\mu$ actually corresponds to the defuzzification of μ by the CoA method.

4.2 Using continuous WOWA for defuzzification

The approach of considering defuzzification as an aggregation procedure allows us to apply the Continuous WOWA operator to this problem. Namely, to defuzzify a fuzzy set A using a CWOWA operator, we need to specify a probability distribution f_A and a fuzzy quantifier q. The former can be defined as in (14), i.e. $f_A(x) = k \cdot \mu_A(x)$ with $k = [\int_{-\infty}^{+\infty} \mu_A(x) dx]^{-1}$, and the latter, in general, can be any arbitrary quantifier. Then, using (9), we can give a first definition of a CWOWA-based defuzzification method.

[4] From now on we consider fuzzy sets defined on the real line $\mathbb{R}$.

Definition 6. Given a fuzzy set A, with membership function μ_A, and a relative importance function q, the CWOWA-based defuzzification of μ_A with respect to q is the value

$$DF_q^{CWOWA}(A) = CWOWA^{f_{A},q}(\mathbb{R}) = k\int_{-\infty}^{+\infty} q\Big(k\int_{-\infty}^{x} \mu_A(y)dy\Big)\,\mu_A(x)\,x\,dx$$

with $k^{-1} = \int_{-\infty}^{+\infty} \mu_A(x)dx$.

Actually, this corresponds to a continuous CoA method where each value is weighted according to each accumulated probability. In this way, the defuzzification can filter the values to consider. For example, if all probabilities are to be considered, then the relative importance function is defined as $q(x) = 1$ for all $x \in [0, 1]$. On the other hand, if extreme values that have a probability (below and above respectively) less than 0.1 are not to be considered, then we should take as relative importance function $q(x) = 1.25$ for all $x \in [0.1, 0.9]$ and $q(x) = 0$ otherwise.

Note that in (14) above the normalization constant k corresponds to the cardinality $Card(A)$ of the fuzzy set A (see for instance [32,13]). Analogously, we can define the cardinality of the subset $A_{\leq x} = A \cap [-\infty, x]$ of those elements of A which are smaller or equal than a certain value x as

$$Card(A_{\leq x}) = \int_{-\infty}^{+\infty} min(\mu_A(y), \mu_{[-\infty,x]}(y))\,dy = \int_{-\infty}^{x} \mu_A(y)\,dy$$

where $\mu_{[-\infty,x]}$ stands for the crisp membership function of the interval $[-\infty, x]$. This definition permits to see $Card(A_{\leq x} \mid A) = Card(A_{\leq x})/Card(A)$ as the relative cardinality of the fuzzy subset $A_{\leq x}$ with respect to the fuzzy set A. In this way, the defuzzification method given in Definition 6, using the continuous WOWA operator and induced by a relative importance function q can be expressed as:

$$DF_q^{WOWA}(A) = [Card(A)]^{-1} \cdot \int_{-\infty}^{+\infty} q(Card(A_{\leq x} \mid A))\,\mu_A(x)\,x\,dx.$$

This expression underlines the role of q as a filtering function defined over the relative cardinality of the fuzzy set $A_{\leq x}$. In particular, if we take $q(x) = 1$, for all $x \in [0, 1]$, we are not making any filtering at all and $DF_q^{WOWA}(A)$ yields the well known expression of the Center-of-Area method of defuzzification. The usefulness of the function q is made clear in the examples of next subsection.

4.3 An example

Let us show how continuous WOWA operator can be used to defuzzify the fuzzy set A depicted in Figure 3 with the three different relative importance functions illustrated in Figure 4. We first consider the relative importance function q_1, a function that makes all the values equally important and, as already noticed in last section, coincides with the CoA defuzzification method:

$$DF1 = DF_{q_1}^{WOWA}(A) \approx 2.70$$

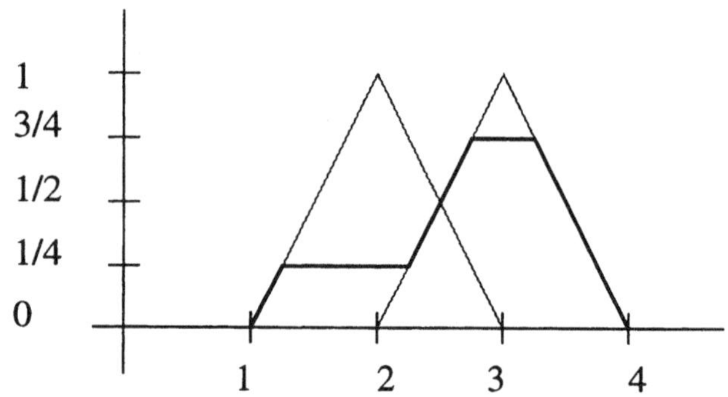

Fig. 3. Membership function of fuzzy set A

If we now defuzzify with the function q_2, we are discarding all those values outside the central interval with the 50% of the fuzzy cardinality of A. This would correspond to the case where we are not very confident in the extreme values and we prefer to concentrate on a central range by ignoring the 25% of the lowest values and the 25% of the highest values. In this case defuzzification yields the following value:

$$DF2 = DF_{q_2}^{WOWA}(A) \approx 2.82$$

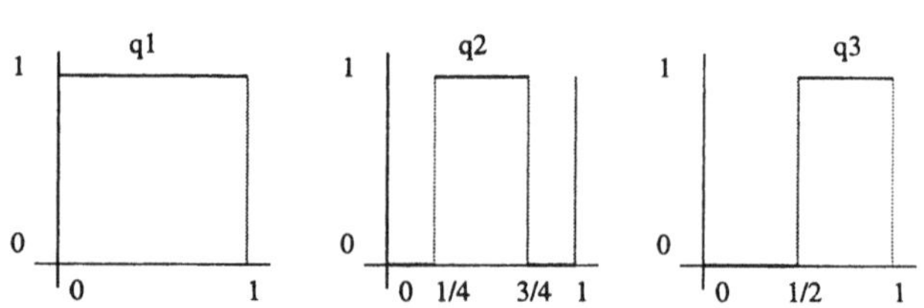

Fig. 4. Three different relative importance functions: q_1, q_2 and q_3

Finally, if we consider the function q_3 only large values are considered to be important and we discard the 50% of the lowest values. This option would correspond to a very optimistic or very pessimistic, depending on the meaning of the variable being defuzzified, criterion. The defuzzified value is then:

$$DF3 = DF_{q_3}^{WOWA}(A) \approx 3.25$$

which is, as expected, sensibly higher than the previous ones. The three defuzzified values are shown in Figure 5.

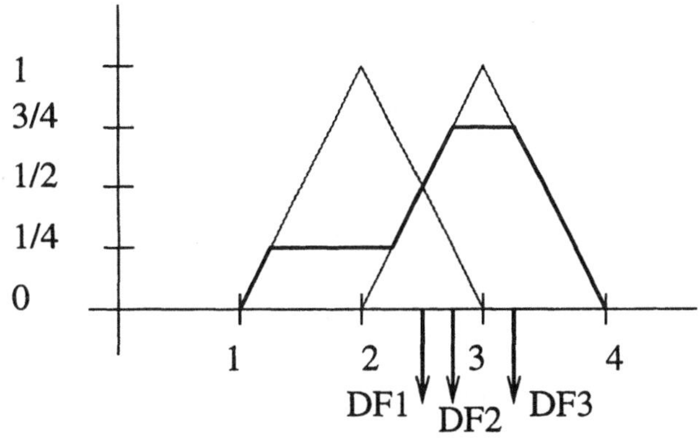

Fig. 5. Defuzzified values using q_1, q_2 and q_3

5 Discussion and final remarks

After introducing the Weighted Mean and the OWA operator and considering the meaning of the weights in both functions, we have reviewed the Weighted OWA (WOWA) operator that generalizes both.

In this paper we have proposed the extension of WOWA operators to the continuous case and we have shown its possible application for defuzzification. Although it is usual to consider only a finite set of values in relation to aggregation operators, it is also interesting to consider the possibility of aggregating a continuous set of values. Indeed, the weights that define for instance a WM are actually a probability distribution on a discrete set of values. Therefore, if instead of a discrete probability distribution we are given a continuous distribution describing the behaviour of a random variable, then it is meaningful to extend the WOWA operator to average, according to a given quantifier, these distributions into a single value.

Our approach to defuzzification with the WOWA operators is close in spirit to Yager's approach [29], where he suggests the use of heuristic knowledge to discard or focus on some part of the fuzzy set to be defuzzified in order to overcome some typical problems (e.g. ambiguity) that use to appear in fuzzy control systems. Yager's way to give more importance to some of the possible solutions than to another ones is not exactly comparable to ours. The weighting of values in [29] amounts to a kind of filtering of the membership function which yields another membership function over which a "classical" defuzzification method applies. Instead, our filtering is not directly over the membership function but on the function which at each point gives the accumulated relative cardinality.

In order to compare defuzzification methods, five criteria are considered in [7]. In our setting only the first four criteria make sense:

Continuity: a small change in the membership function should not result in a large change in the defuzzified value.

Disambiguity: the defuzzification should never provide ambiguous results (even in the case of a symmetric bimodal membership function)

Plausibility: the defuzzified value of a convex membership function should have a high degree of membership value.

Computational complexity: the less computationally complex method, the better.

The fifth criterion ("weight counting") takes into account the matching degree of the input with the membership functions of the antecedents in the fuzzy control rules. This criterion does not apply in our setting, since we are only taking into account the output membership functions and not how they have been obtained. A simple analysis shows that defuzzifying with continuous WOWA operators fits for sure the continuity and disambiguity criteria, the computational complexity may vary on the chosen function q (but in any case not very low), and the plausibility criterion is not generally satisfied like the CoA method.

As for possible extensions, it is clear that a straightforward extension of the method proposed in this paper is to incorporate the methodology proposed in [31], already mentioned in Section 4.1. They consider a defuzzification process composed of two steps: applying a BADD transformation to the membership function to generate a probability distribution and then to compute the mean value of this distribution. The BADD transformation plays a role of a weighting of values according to their plausibility. Therefore it seems rather natural to extend our method by first applying a BADD transformation to take into account the plausibility importance of values and then applying a WOWA aggregation to take into account the accumulated relative cardinalities.

Acknowledgements

The authors are grateful to Radko Mesiar for their useful comments and suggestions. The authors have been partially supported by the Spanish CICYT project SMASH: TIC-96-1038-C04-01 and TIC-96-1038-C04-04.

References

1. Abidi M.A. (1992) Fusion of Multi-dimensional Data Using Regulation, in M.A. Abidi, R.C. Gonzalez (eds.), Data Fusion In Robotics and Machine Intelligence, Academic Press, U.K., pp. 415-455.
2. Aczél J. (1984) On weighted synthesis of judgements, Aequationes Math., 27, 288-307.
3. Alsina C., Mayor G., Tomás M.S., Torrens J. (1993) A characterization of a class of aggregation functions, Fuzzy Sets and Systems, 53, pp. 33-38.
4. Calvo T., Kolesarova A., Komornikova M., Mesiar R. (2001) Aggregation operators: properties, classes and construction methods. Chapter in this monograph.
5. Chen J.E., Otto K.N. (1995) Constructing membership functions using interpolation and measurement theory, Fuzzy Sets and Systems, 73:3, pp. 313-327.
6. Detyniecki M., Yager R. R. (2000) Ranking fuzzy numbers using α-weighted valuations, Int. J. of Unc., Fuzziness and Knowledge-Based Systems, 8:5 573-591.
7. Driankov D., Hellendoorn H., Reifrank M. (1993) An introduction to Fuzzy Control. Springer Verlag.
8. Dubois D., Fargier H., Prade H. (1996) Possibility theory in constraint satisfaction problems: Handling priority, preference and uncertainty. Applied Intelligence, 6, pp. 287-309.
9. Filev D., Yager R.R. (1991) A generalized defuzzification method under BADD distributions, Intl. Journal of Intelligent Systems 6, pp. 687-697.
10. Fodor J., Marichal J.L., Roubens M. (1995) Characterization of Ordered Weighted Averaging Operators, IEEE Trans. of Fuzzy Systems, 3:2, pp. 236-240.
11. Grabisch M. (1995) Fuzzy integral in multicriteria decision making, Fuzzy Sets and Systems, 69, pp. 279-298.
12. Genest C., Zidek J.V. (1986) Combining Probability Distributions: A Critique and An Annotated Bibliography, Statistical Science, 1:1, pp. 114-148.
13. Klir G.J., Yuan B. (1995) Fuzzy Sets and Fuzzy Logic, Prentice-Hall PTR.
14. López B., Álvarez S., Millán P., Puig D., Riano D., Torra V. (1994) Multistage vision system for road lane markings and obstacle detection, Proceedings of Euriscon '94, Malaga, pp. 489-497.
15. Luo R.C., Kay M.G. (1992) Data Fusion and Sensor Integration: State-of-the-art 1990s, in M.A. Abidi, R.C. Gonzalez (eds.), Data Fusion In Robotics and Machine Intelligence, Academic Press, U.K., pp. 7-135.
16. Nelsen R.B. (1999) An Introduction to Copulas, Lecture Notes in Statistics 139, Springer, New York, 1999.
17. Nishiwaki Y., Preyssl C., Schmid S. (1994) Decision Making under uncertainty: Application of fuzzy logic to risk assessment abd nabagement of space systems, Proceedings of the Third International Conference on Fuzzy Logic, Neural Networks and Soft Computing, Iizuka (Japan), pp. 263-264.
18. Rondeau L., Ruedas R., Levrat L., Lamotte L. (1997) A defuzzification method respecting the fuzzification, Fuzzy Sets and Systems, 86, pp. 311-320.
19. Sandri S.A., Dubois D., Kalfsbeek H. (1995) Elicitation, Assesment, and Pooling of Expert Judgements Using Possibility Theory, IEEE Trans. on Fuzzy Systems, 3:3, pp. 313-335.

20. Slany W. (1996) Scheduling as a fuzzy multiple criteria optimization problem, Fuzzy Sets and Systems, 78, pp. 197-222.
21. Torra V. (1996) Weighted OWA operators for synthesis of information, Proc. of Fifth IEEE Int. Conference on Fuzzy Systems (IEEE-FUZZ'96), pp. 966-971, New Orleans, USA.
22. Torra V. (1997) The Weighted OWA operator, Int. J. of Intel. Systems, 12, pp. 153-166.
23. Torra V. (1998) On some relationships between the WOWA operator and the Choquet integral, Proceedings of the Seventh Conference on Information Processing and Management of Uncertainty in Knowledge-Based Systems (IPMU'98), 818-824, Paris, France.
24. Torra, V. (2000) The WOWA operator and the interpolation function W*: Chen and Otto's interpolation method revisited, Fuzzy Sets and Systems, 113:3, pp. 389-396.
25. Valente de Oliveira J. (1995) A set-theoretical defuzzification method. Fuzzy sets and systems, 76, pp. 63-71.
26. Yager R.R. (1988) On ordered weighted averaging aggregation operators in multi-criteria decision making, IEEE Trans. on SMC, 18, pp. 183-190.
27. Yager R.R. (1993) Families of OWA operators, Fuzzy Sets and Systems, 59, pp. 125-148.
28. Yager R.R. (1996) Quantifier Guided Aggregation Using OWA operators, Int. J. of Intel. Systems, 11, pp. 49-73.
29. Yager R.R. (1996) Knowledge-based defuzzification. Fuzzy Sets and Systems, 80, pp. 177-185
30. Yager R.R., Filev D.P. (1993) On the issue of defuzzification and selection based on a fuzzy set. Fuzzy Sets and Systems, 55, pp. 255-271.
31. Yager R.R., Filev D.P. (1994) Essentials of fuzzy modeling and control, John Wiley.
32. Zadeh L.A. (1978) PRUF - A meaning representation language for natural language, Intl. Journal of Man-Machine Studies, 10, pp. 395-460.
33. Zimmermann H.J. (1991) Fuzzy Set Theory - and Its Applications, Kluwer, Dordrecht, 2nd revised edition.

Using Importances in Group Preference Aggregation to Block Strategic Manipulation

Ronald R. Yager
Machine Intelligence Institute
Iona College
New Rochelle, NY 10801
yager@panix.com

ABSTRACT: We consider the problem of preference aggregation in group decision making. The role of the collaborative imperative used by the group in determining the form of the aggregation function is noted. We focus on one collaborative imperative, called the primal, deriving from participant autonomy, an agent doesn't have to accept a decision by the group it doesn't like. We showed that t-norms and more generally the class of uninorms having zero fixation provide appropriate aggregation operators to implement this primal collaborative imperative. We then discuss the possibility of an agent using a strategic manipulation of its preference information to get its preferred alternative. A mechanism based upon the use of an importance weighting is then suggested for modifying the construction of the group decision function to defend against strategic manipulation.

Keywords: preference aggregation, weighted uninorms, strategic manipulation, group decision making

Introduction

One application of aggregation methods is in group decision making. [1, 2]. Here each of the participants provides their preference function over the set of available alternatives. These individual preference functions are then aggregated to obtain a group preference function . The selected alternative is the one that optimizes this group preference function. Here we discuss the use of the uninorm [3, 4] as the aggregation function. An interesting issue arises in this environment as a result of the fact that each participant in the group has as their **true goal** *the maximization of their own individual preference function* and **not** necessarily the maximization of the group preference function. This situation may lead a participant to strategically manipulate the preference information they provide so as to attain their goal. In this work we consider one type of strategic preference manipulation and provide a modification of the group decision making procedure to defend against strategic manipulation. Formally this modification makes use of the ability to include importances in the aggregation process.

Group Decision Making and the Collaborative Imperative

Assume we have a set $X = \{x_1, x_2, ..., x_q\}$ of alternative one of which must be chosen by a group of *n* agents who must collaborate on the choice of this action. We shall assume each agent represents his preference in terms of a fuzzy subset over the set X. We let A_j indicate the fuzzy for agent j. We assume each agent is unaware of the preference functions of the other participants.

A group decision mechanism is a process for selecting one alternative based upon the preferences of the individuals making up the group. Any *nondiscriminatory* decision mechanism should treat all the participants in the same way, the only distinction made between the participants should be based upon information contained in their individual preference functions. A second requirement is that any alternative which is the first choice of all the participants should be a first choice of the group, this is called *Pareto Optimality*.

An approach for obtaining a group decision making is to aggregate the individual preference functions to obtain a group preference function A and then choose as the selected alternative the one with the largest value in A. One important issue here is the choice of the operator used for the aggregation of the individual preference functions. Formally this aggregation operator is a mapping $F: I^n \to I$ such that for any $x \in X$, the group preference is $A(x) = F(A_1(x), A_2(x), ..., A_n(x))$. A number of basic required properties can be associated with F:

1. Symmetry, the indexing of the participants is unimportant
2. Monotonicity, if $a_i \geq b_i$ then $F(a_1, a_2, ..., a_n) \geq F(b_1, b_2, ..., b_n)$
3. $F(1, 1, ..., 1) = 1$
4. $F(0, 0, ...0) = 0$

The first condition assures us of non-discrimination, it requires that each of the participants are treated the same. The second condition enforces the Pareto optimality as well as providing a positive association between the individual agents' preferences and the group preference. The third condition assures us that if all the participants are completely satisfied with a solution then the group is completely satisfied with the solution. The fourth condition implies that if all the participants are completely dissatisfied by a solution then the group should be completely dissatisfied.

These conditions are not very restrictive and allow many different functions for F. The actual choice of F will be a reflection of what collaborative imperative the group decides to use. That is, the group must pre-agree upon some mechanism to guide how they shall aggregated their individual preference we call this the collaborative imperative. In group decision making however, there exists a default or primal collaborative imperative that needs no pre–agreement by the participants, as it is naturally imposed by the autonomy of each of the participants. Because of agent autonomy an agent doesn't have to participate if it doesn't like the group decision. Thus the key feature of this primal collaborative imperative is that any participant doesn't have to accept a solution it doesn't like. A formal expression of this primal collaborative imperative is that a participating agent can dismiss (reject) any solution if he doesn't like it. This type of imperative can be implemented using a Min type aggregation operator, $A(x) = Min_i[A_i(x)]$. Under this imperative if there exists an agent i such that $A_i(x) = 0$ then $A(x) = 0$, any agent can singlehandedly dismiss a solution. Essentially this implies all agents must accept the solution.

More generally we can capture this primal collaborative imperative using a t-norm [5] aggregation operator T, that is $A(x) = T_i[A_i(x)]$. A t-norm is a symmetric, monotonic, associative operator with identity one. The t-norm has the property that if any $A_i(x) = 0$ then $A(x) = 0$.

While t-norms provide an aggregation with this primal characteristic they may not provide all the desired diversity we would like in a group aggregation operator. We recall that for any t–norm

$T(A_1(x), A_2(x), ..., A_n(x), A_{n+1}(x)) \leq T(A_1(x), A_2(x), ..., A_n(x))$

The implication of this is the only ability a participant has with this type of aggregation is to detract from an alternative, in no situation can the agent "support" an alternative. In particular, there exists no possibility for agents to upwardly reinforce each other, only downward reinforcement is possible [6]. Thus for example if two agents attribute low scores to an alternative $A_1(x) = 0.2$ and $A_2(x) = 0.25$ then if we use a product t-norm, $A(x) = 0.05$ and we see that these two low scores have reinforced each other to even more strongly reject the alternative. On the other hand, if two agents strongly support an alternative, $A_1(x) = 0.9$ and $A_2(x) = 0.95$ then there exists no t-norm that enables these two scores to reinforce each other and more strongly support the alternative.

As it turns out, the t-norm is not the only aggregation operation that provides the basic characteristic associated with the primal-default collaboration imperative. Another class of aggregation operators that provides this primal feature are the uninorm operators. As we shall subsequently see, these operators additionally allow for the association of other features such as reinforcement in both directions. We now turn to the uninorm aggregation operator.

Uninorm Aggregation Operators

The uninorm operators, introduced in [3, 4], provide a unification and generalization of the t–norm and t-conorm operators. In [7-9] the authors have studied in considerable detail the properties of these operators.

Definition: A uninorm is a mapping $R: I \times I \rightarrow I$ having the following properties:

1. $R(x, y) = R(y, x)$: ***Commutativity***
2. $R(x, y) \geq R(u, v)$ for $x \geq u$, $y \geq v$: ***Monotonicity***
3. $R(x, R(y, z)) = R(R(x, y), z)$: ***Associativity***
4. There exists some $g \in [0, 1]$ such that for all x, $R(x, g) = x$: ***Identity Element***

The uninorm operators have its first three properties in common with both the t-norm and t–conorm but the fourth condition is more general, it allows for an identity anywhere in the unit interval. For a t-norm T the identity is one and for a t-conorm S the identity is zero. When we want to emphasize the identity associated with a uninorm R we shall express it as R_g, g being its identity. Thus R_1 is a t-norm and

R_0 is a t-norm.

If A_j are a collection of n fuzzy subsets of X we can implement a uninorm based aggregation of these fuzzy subsets as $A = \text{Uni}(A_1, ..., A_n)$ such that for each $x \in X$

$$A(x) = R(A_1(x), A_2(x), ..., A_n(x))$$

where R is a uninorm operator.

Consider the t-norm operator T, which is also R_1. Since $T(a, 1) = a$ then monotonicity implies that $T(a, b) \leq a$. This implies that for any t-norm T, $T(a, b) \leq \text{Min}(a, b)$. This situation implies that t-norm aggregation can never increase. Specifically the addition of an argument to a t–norm aggregation never results in an increase

$$T(a_1, ..., a_n) \geq T(a_1, ..., a_n, a_{n+1}).$$

essentially $T(a, a_{n+1}) \leq a$.

Another related property of the t-norm is *zero fixation*, the appearance of zero in any argument always results in an aggregated value of zero, if there exists some $a_j = 0$ then $T(a_1, a_2, ..., a_n) = 0$. This property follows from the fact that $T(a, 1) = a$, hence $T(0, 1) = 0$ and then monotonicity implies $T(0, x) = 0$. Finally associativity leads to the complete zero fixation.

Let us now consider the t-conorm S, which is R_0. Since $S(a, 0) = a$ the monotonicity implies that $S(a, b) \geq a$, this implies that for any t-conorm S, $S(a, b) \geq \text{Max}(a, b)$. This situation implies that t-conorm aggregation can never be a decreasing type aggregation. Specifically the addition of an argument a t-conorm aggregation can never result in a decrease in aggregated value.

$$S(a_1, ..., a_n) \leq S(a_1, ..., a_n, a_{n+1}),$$

Essentially $S(a, a_{n+1}) \geq a$. Another property of the t-conorm is one fixation. The appearance of one as any of the arguments always results in an aggregated value of one, $S(a_1, ..., a_n) = 1$ if there exists an $a_j = 1$.

We now consider the issue of fixation for uninorms. We recall that the t-norm, R_1 has zero fixation, the occurrence of zero as one of its arguments assures an aggregated value of zero. The t–conorm, R_0, has one fixation, the occurrence of one as one of its arguments assures an aggregated value of one. The following result bears very strongly on this issue of fixation with respect to uninorms.

Theorem: Assume R is a uninorm with $g \in (0, 1)$ then $R(0, 1) \in \{0, 1\}$.

Proof: Let R be a uninorm with identity g. Let $R(0, 1) = a$

a) Assume $a \leq g$.

From associativity $R(0, 0, 1) = R(0, R(0, 1)) = R(R(0, 0),1)$. Since $R(0, 0) = 0$ then $R(0, 0, 1) = R(R(0, 0), 1) = R(0, 1) = a$. However it is also true that $R(0, 0, 1) = R(0, R(0, 1)) = R(0, a)$. Since $R(0, g) = 0$ and it is assumed $a \leq g$ then $R(0, 0, 1) = R(0, a) = 0$ this implies that $a = 0$.

b) Assume $a \geq g$

From associativity $R(1, 1, 0) = R(R(1, 1), 0) = R(1, R(1, 0))$. Since $R(1, 1) =1$ then $R(1, 1, 0) = R(R(1, 1), 0) = R(1, 0) = a$. However it is also true that $R(1, 1, 0) = R(1, R(1, 0)) = R(1, a)$. Since $R(1, g) = 1$ and it is assumed that $a \geq g$

then $R(1, 1, 0) = R(1, a) = 1$ this implies that $a = 1$.

The implication of this theorem is that all uninorms are such that either $R(1, 0) = 1$ or $R(0, 1) = 0$, if $R(0, 1) > g$ it must be 1, if $R(0, 1) < g$ it must be 0 and if $R(0, 1) = g$ it can be either 0 or 1. Thus we can divide the uninorms into two classes, those for which $R(1, 0) = 0$ and those for which $R(1, 0) = 1$. We should note that two possibilities are available only for $g \in (0, 1)$. For $g = 1$ then we have a t-norm and $R(1, 0) = 0$. For $g = 0$ we have a t-conorm, $R(1, 0) = 1$. Essentially we have two classes of uninorms those with $R(0, 1) = 1$ and with $R(0, 1) = 0$. Consider the class in which $R(0, 1) = 1$, in this case monotonicity implies $R(a, 1) = 1$ for all a, this implies a one fixation, $R(a_1, a_2, ..., a_n) = 1$ if $\exists\, a_j = 1$. On the other hand if $R(0, 1) = 0$ then monotonicity implies $R(0, a) = 0$ for all a and hence we have a zero fixation. We shall call these two classes zero and one fixated uninorms. We note that a zero fixated uninorm can have any identity except zero and a one fixated uninorm can have any identity except one.

An interesting relationship between the uninorm operators and mean operators can be observed. Let M be an mean operator it is well known that for these operators

$$\text{if } a_{n+1} \geq M(a_1, ..., a_n) \text{ then } M(a_1, ..., a_n, a_{n+1}) \geq M(a_1, ..., a_n)$$
$$\text{if } a_n \leq M(a_1, ..., a_n) \text{ then } M(a_1, ..., a_n, a_{n+1}) \leq M(a_1, ..., a_n)$$

We see that additional arguments have a clear effect on the aggregation process depending upon their relationship to a boundary point, the existing aggregated value, a positive effect if the additional argument's value is at least as great as the existing value and negative if its value is not greater than the existing value. We also see that in the case of the uninorm additional arguments also have a decided effect on the aggregation depending on their relationship to a boundary point, the identity, but here the boundary point doesn't vary with the prior value aggregated, it is the fixed identity. While the properties of the boundary point differs for two type of operators what is similar is their ability to both increase or decrease the current aggregated value with the inclusion of additional values. However there exists an interesting difference between these two types of operators. In the case of the uninorm the identity provides "a priori" knowledge of the effect of their scores, ie by a providing a score above the identity an agent knows 'a priori' he is helping increase support for an alternative. On the other with averaging operators an agent doesn't know whether his score will increase or decrease support for a alternative, it depends upon the scores of the other agents.

In [7] Fodor, Yager and Rybalov introduced a general class of uninorms. If T and S are t–norm and t-conorm operators then the mappings R defined by **I** and **II** below are two general classes of uninorm operators with identity $g \in [0, 1]$.

I.

(a) $R(x, y) = g\, T(\frac{x}{g}, \frac{y}{g})$ if $0 \leq x, y \leq g$

(b) $R(x, y) = g + (1 - g)\, S(\frac{x - g}{1 - g}, \frac{y - g}{1 - g})$ if $g \leq x, y \leq 1$

(c) $R(x, y) = \text{Min}(x, y)$ if $\text{Min}(x, y) \leq g \leq \text{Max}(x, y)$

II.

(a) $R(x, y) = g\,T(\frac{x}{g}, \frac{y}{g})$ if $0 \le x, y \le g$

(b) $R(x, y) = g + (1 - g)\,S(\frac{x - g}{1 - g}, \frac{y - g}{1 - g})$ if $g \le x, y \le 1$

(c) $R(x, y) = \mathrm{Max}(x, y)$ if $\mathrm{Min}(x, y) \le g \le \mathrm{Max}(x, y)$

The difference between the two classes are in item c, in the first case when one of the arguments is above the identity and the other below the identity we take the Min while in the second case we take the Max. In particular we note that for the first class $R(0, 1) = 0$ and for the second class $R(0, 1) = 1$, hence the first class are zero fixated and the second one fixated.

An example of a uninorm called the three Π operator was introduced in [4]. For this operator $R(x, y) = \frac{xy}{\bar{x}\bar{y} + xy}$ where $\bar{x} = 1 - x$. In this uninorm the identity $g = 0.5$. The case of $R(0, 1)$, which of course is the same as $R(1, 0)$, is undefined and must specified as being either one or zero Using the associativity property we get that

$$R(x_1, x_2, ..., x_n) = \frac{\prod_{j=1}^{n} x_j}{\prod_{j=1}^{n} \bar{x}_j + \prod_{j=1}^{n} x_j}$$

A generalization of this class is obtained by letting $R(x, y) = \dfrac{\frac{xy}{g}}{\frac{xy}{g} + \frac{\bar{x}\bar{y}}{1 - g}}$. This has identity g.

The three Π operator and related operators are generated uninorms as discussed in [7, 10]

Group Decision Making Using Uninorms

If $A_1, ..., A_n$ are a collection of fuzzy subsets corresponding to the preference functions of the participating agents then a possible group decision function is $A(x) = R(A_1(x), A_2(x), ..., A_n(x))$ where R is a uninorm. In order to enforce the primal collaborative imperative giving any participating agent the ability to void any alternative we must have $R(0, a) = 0$ for any a. This implies that R must be drawn from the class of zero fixated uninorm operators. This class is defined by $R(0, 1) = 0$.

One important characteristic of a uninorm is the choice of identity g. Let us look at the role of the identity when using a uninorm to aggregate preference functions. We are particularly interested in getting some idea of the semantics associated with the g. Since $R(a_1, a_2, ..., a_n, g) = R(a_1, a_2, ..., a_n)$ we see that a rating $A_j(x) = g$ can be seen as being neutral or indifferent with respect to an alternative. We see that in assigning the value g the participant is saying "let the others decide on this one, I neither support nor detract." As we have already noted scores above g supports an alternative and those below g negates an alternative. We can view g as providing a bounding, those scores above g can be seen as being supportive while those below g can be seen as being not supportive.

The uninorm operator with a given identity g provides an ideal setting for allowing the inclusion of preference information of the type which can be seen as providing strong support or weak support or strong rejection.

Let us review the overall process of group decision making. Each participants provides a preference function A_i. These preference functions are then aggregated to obtain the group preference function $A(x) = R(A_1, (x), ..., A_n(x))$. The alternative x^* that maximizes the group preference function is then selected, $A(x^*) = \text{Max}_x[A(x)]$. What must be kept in mind in this group decision making setting is that the true goal of each participant is *to maximize their individual degree of satisfaction*. That is, while the group is based upon the maximization of the group preference function, each agent's goal is to obtain as the group selection the alternative that maximizes its own preference function. In pursuit of this goal as noted in [11] an individual agent can use any available means to effect the group decision mechanism so as to optimize their individual payoff resulting from the group decision.

One avenue open to a participant to effect the group choice is to manipulate the preference function which they provide to the group decision process so as to most benefit themselves. With this in mind one desirable feature of any group decision procedure is that it should discourage potential benefits from strategic manipulation of preference information.and encourage participants to provide their true preferences. Let us see if the approach suggested is subject to strategic manipulation and if so how to remedy it.

Consider the situation in which we have *n* agents and $X = \{x_1, ..., x_q\}$. Without loss of generality let us focus on agent one, his preference function is A_1. Assume that x_1 is his most preferred alternative, $A_1(x_1) = 1$. His goal is to try to assure that no alternative other than x_1 is selected. With this goal in mind let us see how this agent can best assure himself of this. One strategy he can use is the following. Instead of providing the group with his true preference function, A_1, he can provide the group with a strategically manipulated preference function $\hat{A}_1$ in which $\hat{A}_1(x_1) = 1$ and $\hat{A}_1(x_j) = 0$ for all $x_j \neq x_1$. Using this we get a group preference function

$$A(x) = R(\hat{A}_1(x), A_2(x), ..., A_n(x)).$$

Since the uninorm performing the aggregation has a zero fixation and since the information provided by this agent has $\hat{A}_1(x_j) = 0$ for $x_j \neq x_1$ we get $A(x_j) = 0$ for all $x_j \neq x_1$. This assures agent one that no alternative will have a better score than his most preferred. Furthermore, if $R[A_2(x_1), ..., A_n(x_j)] \neq 0$, then x_j will be the clear group choice. On the other hand if $R[A_2(x_1), ..., A_n(x_1)] = 0$, then $A(x_1) = 0$ and all alternatives have $A(x) = 0$

More specifically, we see that any agent can potentially benefit from the strategic manipulation of their preference functions. This capability for strategic manipulation is a direct result of the desire to satisfy the primal characteristics, giving the agent the ability to void any alternative it doesn't want.

What is even more disturbing is that the use of this type of strategy by any two agents who don't have the same preferred alternative will lead to a situation where the group preference function is a null set, $A(x) = 0$ for x. This situation quickly renders this type of group decision making procedure useless. We must try in some way to modify the formulation of the group preference function to negate the incentive for strategic manipulation of individual preference functions.

Subsequently we shall suggest an approach to modifying the construction of the group preference function so as to accomplish this. Specifically we shall assign to each participating agent a degree of importance and use diminution of its importance to penalize an agent for strategic manipulation. In anticipation of suggesting a modification of the formulation of the group preference to address the possibility of strategic preference manipulation we turn to the issue of including importance in the uninorm aggregation.

Importances in Uninorm Aggregations

Consider the situation in which we have a collection of arguments consisting of pairs, (w_i, a_i) where a_i is a value (score) to be aggregated and $w_i \in [0, 1]$ is the associated weight. Here we want to take the uninorm aggregation of these pairs, $R((w_1, a_1), ..., (w_n, a_n)) = \bar{a}$. One approach to accomplishing this task is to transform the pairs (w_j, a_j) into effective values, $b_j = h(w_j, a_j)$ and then use the regular uninorm aggregation, $\bar{a} = R(b_1, ..., b_n)$ [12]. We note that this approach has been used in the case of weighted t-norm and t-conorm [13]. The question now becomes what is the form of function h. Let us try to identify some salient features we want to associate with h.

Here $w_i \in [0, 1]$ with $w_i = 0$ being the case of zero importance and $w_i = 1$ being the case of most importance. Since the inclusion of importance can be seen as an extension of the situation in which we have not considered importance we must first imbed the original case in the new framework. The case of not considering importance can be viewed as one in which all the arguments are of equal importance at the value at the opposite extreme from the no importance. Hence we shall denote the original case as one where $w_j = 1$ for all j. In order for this to give us the answer obtained with no consideration of importances we must have

$$R((1, a_1), (1, a_2), (1, a_n)) = R(a_1, ..., a_n),$$

this requires that

$$h(1, a) = a.$$

Consider now the case of zero importance $w_i = 0$. The basic requirement here is that any argument that has zero importance should not contribute to the aggregation. This can easily be accomplished by assuring

$$h(0, a) = g,$$

the identity of the uninorm.

Another condition required of the importance transformation function h is that it be monotonic with respect to the value a. Specifically if a and $\hat{a}$ are two arguments such that $a \geq \hat{a}$ then any modification by importance should not interchange their ordering. This requires that that for any w if $a \geq \hat{a}$ then

$$h(w, a) \geq h(w, \hat{a}).$$

Another property is that h(w, a) be bounded by its values for w = 0 and w = 1 that is $h(w, a) \in [g \wedge a, g \vee a]$, A more general condition is one that we shall call consistency. As a fourth condition we require that as we move from w = 0 to w = 1 the function h(w, a) changes in a consistent monotonic manner. Let w and $\tilde{w}$ be such that $\tilde{w} > w$ then

if $h(1, a) > h(0, a)$ $(a > g)$ then for all $\tilde{w} > w$ we have $h(\tilde{w}, a) \geq h(w, a)$

if $h(1, a) < h(0, a)$ $(a < g)$ then for all $\tilde{w} > w$ we have $h(\tilde{w}, a) \leq h(w, a)$

if $h(1, a) = h(0, a)$ $(a = g)$ then for all $\tilde{w} > w$ we have $h(\tilde{w}, a) = h(w, a)$

Assuming derivatives exist essentially this says that if a > g then $\frac{\partial h(w, a)}{\partial w} \geq 0$, if $a < g$ then $\frac{\partial h(w, a)}{\partial w} \leq 0$ and if a = g then $\frac{\partial h(w, a)}{\partial w} = 0$.

For the case of t-conorms a class of importance transformation operators satisfying these four conditions are

$$h(w, a) = T(w, a)$$

where T is a t-norm. It is clear that this satisfies the requirement we specified

1. $T(1, a) = a$
2. $T(0, a) = 0$, the identity of the t-conorm

and from the monotonicity of the t-norm

3. $T(w, a) \geq T(w, \hat{a})$ if $a \geq \hat{a}$

Since g = 0 for the t-conorm, implying a > g, the fourth condition requires that for all $\tilde{w} > w$ we have $h(\tilde{w}\, a) \leq h(w, a)$ this is satisfied by monotonicity of the t-norm.

Three examples of this importance transformation operator are:

$h(w, a) = \text{Min}[w, a]$

$h(w, a) = w \cdot a$

$h(w, a) = \text{Max}[w + a - 1, 0]$

In the case of the t-norm, uninorms where g = 1 a class of importance transformation operators is

$$h(w, a) = S(\overline{w}, a)$$

where S is any t-conorm and $\overline{w} = 1\text{-}w$, a negation. It is easy to see that satisfies the four basic requirements. Some examples of this are

$h(w, a) = \text{Max}[\overline{w}, a]$

$h(w, a) = \overline{w} + a - \overline{w}\, a = \overline{w} + a\, w$

$h(w, a) = \text{Min}[\overline{w} + a, 1]$

The preceding were examples of importance transformation operators valid for special values of g, in the case g = 0 and in the second g = 1. An example of an importance transformation operator that is valid for all g is

$$h_g(w, a) = w\, a + \overline{w}\, g$$

with $\bar{w} = 1 - w$. We see this as follows. Assume $w = 1$, $h_g(1, a) = a + 0\,g = a$. Assume $w = 0$ here $\bar{w} = 1$ and hence $h_g(0, a) = 0\,a + 1\,g = g$. Assume $a \geq \hat{a}$ then

$$h_g(w, a) = w\,a + \bar{w}\,g \geq w\,\hat{a} + \bar{w}\,g \geq h(w, \hat{a})$$

Finally we see that $\frac{\partial}{\partial w} h_g(w, a) = a - g$, thus if $a > g$ this is non-negative and if $g > a$ it's non-positive thus satisfying our consistency condition. We also note that $h_g(w, a) \in [a \wedge g, a \vee g]$. In the case in which $g = 0$ this transformation operator becomes $h_0(w, a) = wa$ which is one of forms we previously introduced. In the case when $g = 1$, he uninorm is a t-norm, $h_1(w, a) = \bar{w} + w\,a$.

Another example of an importance transformation that is valid for all g is

$$h_g(w, a) = (w \wedge a) \vee (\bar{w} \wedge g) \vee (a \wedge g)$$

where $\wedge$ and $\vee$ are the Min and Max operators respectively. We see this as follows. If $w = 0$ then

$$h_g(0, a) = 0 \vee g \vee (a \wedge g) = g$$

If $w = 1$ then

$$h_g(w, a) = a \vee 0 \vee (a \wedge g) = a$$

If $a_1 > a_2$ then

$$h_g(w, a_1) \geq hg(w, a_2).$$

Consider now the fourth requirement.

i.) if $a > g$ then $h_g(w, a) = (w \wedge a) \vee (\bar{w} \wedge g) \vee g = (w \wedge a) \vee g$ from this we see $h(w_1, a) \geq h(w_2, a)$ if $w_1 > w_2$

ii) if $g > a$ then $h(w, a) = a \vee (\bar{w} \wedge g)$ from this we see $h(w_1, a) \leq h(w_2, b)$ if $w_1 > w_2$.

Using Importance Weights to Deter Strategic Manipulation

In this section we shall consider a modification of the procedure used to aggregate the individual preference functions to obtain . This modification will still use the individual preference functions, but it will include a mechanism to combat the kind of strategic manipulation described above. In the preceding we calculated the group preference function A as

$$A(x) = R(A_1(x), A_2(x), ..., A_n(x))$$

where R is a uninorm. We now propose to associate with each A_j an importance weight w_j and calculate

$$A(x) = R((w_1, A_1(x)), (w_2, A_2(x)), ..., (w_n, A_n(x)))$$

Thus A is obtained by a weighted uninorm aggregation.

We now turn to the issue of associating an importance weight w_i with each agent. Let $v_i = \sum_{x \in X} A_i(x)$, the total of the membership grades associated with the agent, it can be seen as the total satisfaction allocated by a participating agent. Since each fuzzy subset is assumed normal $1 \leq v_i \leq q$ where q is the number of alternatives.

We shall let $u_i = v_i - 1$, it is the case that $u_i \geq 0$. We now associate with each agent a degree of importance,

$$w_i = \frac{u_i}{Max_j[u_j]}$$

An agents importance is directly related to the total degree of satisfaction it distributes. The more easily satisfied an agent the more important, "happy" agents get more importance. On the other hand an agent who completely discounts all the alternatives except their favorite has an importance weight of zero. As we shall see in using this type of importance weighting a penalization has been introduced based upon an agent's total score. It is further noted that the determination of the importance weights requires no additional information other then the already provided individual preference functions.

Let us now see how this importance weighted aggregation works to penalize strategic manipulation.. Using importance weights our group preference function becomes $A(x) = R_g((w_1, A_1(x)), (w_2, A_2(x)), \ldots\ldots., (w_n, A_n(x)))$. Here R_g is a uninorm with identity g, w_j is the importance of agent j and $A_j(x)$ is the satisfaction agent attributes to alternative x. Letting h_g indicate our importance transformation function we have $A(x) = R_g(h_g(w_1, A_1(x)), h_g(w_2, A_2(x)), \ldots\ldots., h_g(w_n, A_n(x)))$.

Without loss of generality assume that agent 1 tries to manipulate the situation by supplying a preference function $\hat{A}_1$ in which $\hat{A}_1(x_1) = 1$ and $\hat{A}(x_j) = 0$ for all $x_j \neq x_1$. In this case $v_1 = 1$ and $u_1 = 0$ and his importance becomes zero. Thus for any x $h_g(w_i, \hat{A}_1(x)) = h_g(0, A_1(x)) = g$ and therefore agent one plays no role in the aggregation since g is the uninorm identity. Thus we see that this approach penalizes an agent that tries to strategically manipulate its preference functions.

Let us look at some special cases when g = 1, R_g is a t-norm. In the case of the Min t-norm and the importance transformation $h(w_j, A_j(x)) = \overline{w}_j \vee A_j(x)$ we get $A(x) = Min_j[(\overline{w}_j \vee A_j(x))]$.where

$$w_j = \frac{\sum_x A_j(x) - 1}{Max_i\left[\sum_x A_i(x) - 1\right]}.$$

We note in the case when we didn't include penalization $A(x) = Min_j[A_j(x)]$ it is as if $w_j = 1$ for all agents. In the case in which we use the product t-norm and the importance transformation $h(w_j, A_j(x)) = \overline{w}_j + w_j A_j(x) = 1 - w_j \overline{A_j(x)}$ we get

$$A(x) = \prod_{j=1}^{n} [(1 - w_j\overline{A}_j(x))]$$

The effect of the inclusion of any importance weight based upon an agents total support for the alternatives can be seen as essentially modifying the underlying collaborative imperative used by the group. Using our original aggregation an agent to reject any alternative it doesn't want, as a matter of fact an agent can eliminate as many alternatives as it likes. With the introduction of an importance weight determined by the agent's total score the aggregation imperative becomes somewhat more complex Under this new imperative we keep the agents ability to eliminate any (or even a few) alternatives it doesn't want but we reduce its ability to eliminate many alternatives. This allows an agent to very effectively enforce a preference function such as

"**I don't want** alternative x_j"

but not enforce a preference function of the type

"**All I want** is alternative x_j"

Let us look at this. Consider the uninorm R with identity g and let us use as our importance transformation $h(w, a) = wa + \bar{w}g$. In this case we get

$$A(x_i) = R_g[(w_1 A_1(x_i) + \bar{w}_1 g), \ldots\ldots, (w_n A_n(x_i) + \bar{w}_n g)]$$

Letting E_j be the fuzzy subset such $E_j(x_i) = w_j A_j(x_i) + \bar{w}_j g$, we see then

$$A(x_i) = R_g[E_1(x_i), E_2(x_i), \quad , E_n(x_i)]$$

We shall denote E_j as "effective preference function of agent j."

Consider first the case in which a participating agent has a preference function, "I don't want x_1". In this case, $A_j(x_1) = 0$ and $A_j(x_k) = 1$ for all $k \neq 1$ and v_j, the total of the agents score will be very large, $v_j = q - 1$. We can assume that $v_j \approx Max_i[v_i]$ and therefore $w_j \approx 1$ and hence $E_j(x_i) \approx A_j(x_i)$. Thus agent j's effective preference function will be very close to its provided preference function. It is essentially implementing the specified preference function.

Consider the case in which an agent has a preference function "All I want is x_1". In this $A_j(x_1) = 1$ and $A_j(x_k) = 0$ for $k \neq 1$. Here $v_j = 1$ and $u_j = 0$ and therefore $w_j = 0$ and hence $E_j(x_i) = w_j A_j(x_i) + \bar{w}_j g = g$. In this situation our agent's effective preference function becomes effectively "No effect."

In the beginning we described our primal or default collaboration imperative as any "any agent can reject any alternative." With the inclusion of this importance weight we are still allowing the agent this capability, rejection of any alternative, but we are adding an additional restraint on his ability to use this. Our new collaborative imperative can be seen as

"An agent can reject any alternative but it can't reject too many alternatives."

With this slight modification we are providing a defense against strategic manipulation.

An Alternative Method Determining Importances

In the preceding we have described the use of an importance weight to curtail an agent's ability to unilaterally determine the group's decision by rejecting many alternatives. We have done this by penalizing an agent for rejecting too many options. We have suggested determining this importance weight as $w_j = \frac{u_j}{Max_k[u_k]}$ where $u_j = v_j - 1$ and where $v_j = \sum_i A_j(x_i)$.

Here we shall suggest another way of calculating agent importance which allows for a penalization based upon rejecting to many alternatives. As we shall see this method allows for a more linguistic interpretation and is more in the spirit of Zadeh's idea of computing with words. In this method we specify the allowable number of alternatives an agent can reject without substantial penalization. This information can be expressed using terms such as *at most k* or *a few*." Formally we express this information in terms of a fuzzy subset. These fuzzy subsets will be over the space of the number of alternatives or the unit interval. Consider the first case. Let $N = \{0, 1, ..., q\}$ where q is the number of alternatives. We express this information as a fuzzy subset $Q: N \to [0, 1]$ such that for any $y \in N$ we let $Q(y)$ indicate the acceptability of rejecting y alternatives. For example if $n = 7$ and we have "at most 3 alternatives can be rejected by an agent." then we Q such that

$$Q(y) = 1 \quad \text{for } y \leq 3$$
$$Q(y) = 0 \quad \text{for } y > 3$$

Fuzzy subsets used to represent this type of information will have the properties:

1. $Q(0) = 1$, **2.** $Q(N) = 0$ and **3.** $Q(k_1) \geq Q(k_2)$ if $k_1 < k_2$.

We also can represent this kind of information in terms of fuzzy subsets $\hat{Q}$ in which $\hat{Q}:[0, 1] \to [0, 1]$. In this case for any $y \in [0,1]$, $\hat{Q}(y)$ indicates the degree to which an agent is allowed to reject y proportional of the alternatives. The difference between this and the preceding is that in this case we describe things in terms of proportion while in the preceding is expressing things in terms of actual numbers of elements. In this case $\hat{Q}$ must satisfy similar conditions as Q:

1. $\hat{Q}(0) = 1$, **2.** $\hat{Q}(1) = 0$ and **3.** $\hat{Q}(y_1) \geq \hat{Q}(y_2)$ if $y_1 < y_2$

Once having this description of the allowable rejections by an agent we use the OWA operator [14] to calculate the importance weight of an agent. Consider some agent A_m and let $A_m(x_i)$ be the satisfaction of this agent with alternative x_i. We shall let b_k be the kth largest of these values. We shall denote B the vector whose components are the b_k. We calculate the importance weight of the mth participant using an OWA aggregation their preference values

$$w_m = OWA_Q[A_m(x_1), ..., A_m(x_n)] = \sum_{j=1}^{n} d_k b_k$$

We recall the d_k are called the OWA weights and are such that $d_k \in [0, 1]$ and $\Sigma_k d_k = 1$. These d_k are obtained from the information specifying the allowable number of rejections in the following manner. If Q is defined as a fuzzy subset of N then we define

$$d_k = Q(n - k) - Q(n-(k-1))$$

and if $\hat{Q}$ is defined as a proportion then

$$d_k = \hat{Q}(1 - \frac{k}{n}) - \hat{Q}(1 - \frac{k-1}{n}).$$

Example:. Assume we have seven alternatives and we allow an agent to reject "at most 3." In this case $Q(y) = 1$ for $y = 0, 1, 2, 3$ and $Q(y) = 0$ for $y = 4, 5$, and $6, 7$. Using this we have $d_k = Q(7 - k) - Q(7 - k + 1)$ hence:

$d_1 = Q(7 - 1) - Q(7) = 0$, $d_2 = Q(7 - 2) - Q(7 - 1) = 0$, $d_3 = Q(7 - 3) - Q(7 - 2) = 0$

$d_4 = Q(7 - 4) - Q(7 - 3) = 1$
$d_5 = d_6 = d_7 = 0$

In this case our OWA aggregation lead to an importance weight $w_m = b_4$, the fourth largest of the membership grades in A_m. Let us see that using b_4 is appropriate. If the fourth largest score allocated by this agent is not a rejection, b_4 has a high value. Since $b_k \geq b_4$ for $k < 4$ all the scores above are not rejected and we satisfy our condition of not rejecting more then three alternatives. If b_4 is rejected, it is low, then since for $k > 4$ $b_k \leq b_4$ we have rejected more than 3 alternatives and our agent has violated the constraint.

More generally if our specification is a crisp constraint of the form

an agent can reject at most q alternatives out of the n available

we will get an OWA vector such that $d_{n-q} = 1$ and all other components equal zero. In this case the importance weight of an agent is simply the value of the n - qth largest preference value supplied by that agent. In the case in which our specification is not so crisp the weights will be more distributed.

By using the type of formulation for including importance weights we can model a group aggregation process with a collaboration imperative of the form

Any agent can reject any alternative it doesn't want but it can't reject too many.

Here we can model "not too many" as a fuzzy subset of the unit interval.

It is interesting to note that this approach affords us a mechanism to provide some distinction between the participants by allowing them to reject different numbers of alternative. This is can be implemented by assigning them different Q's.

Conclusion

The problem of preference aggregation in group decision making was considered. We indicated the importance of the collaborative imperative used by the group in determining the form of the aggregation function. We focused on one collaborative imperative, called the primal, deriving from participant autonomy, an agent doesn't have to accept a decision by the group it doesn't like. We showed that t-norms and more generally the class of uninorms having zero fixation provide appropriate aggregation operators to implement this primal collaborative imperative. The central feature of this primal collaborative imperative is the ability of an agent to unilaterally dismiss any alternative by giving it a low score in its preference function. It was then noted that this feature carried with it the possibility for strategic manipulation by the participants. Specifically a agent can strongly under value alternatives it doesn't like and have them dismissed. A mechanism was suggested for modifying the aggregation of the participant preference to defend against this type of strategic manipulation. This mechanism involved the association of an importance weight with each participant calculated in a way that diminishes an agents importance when engaged in strategic manipulation.

References

[1]. Kacprzyk, J. and Fedrizzi, M., Multiperson Decision Making Using Fuzzy Sets And Possibility Theory, Kluwer: Hingham, MA, 1990.

[2]. Kacprzyk, J., Nurmi, H. and Fedrizzi, M., Consensus Under Fuzziness, Kluwer Academic Publishers: Boston, MA, 1997.

[3]. Yager, R. R., "Misrepresentations and challenges: A response to Elkan," IEEE Expert August, 41-42, 1994.

[4]. Yager, R. R. and Rybalov, A., "Uninorm aggregation operators," Fuzzy Sets and Systems 80, 111-120, 1996.

[5]. Klement, E. P., Mesiar, R. and Pap, E., Triangular Norms, Kluwer Academic Publishers: Dordrecht, 2000.

[6]. Yager, R. R. and Rybalov, A., "Full reinforcement operators in aggregation techniques," IEEE Transactions on Systems, Man and Cybernetics 28, 757-769, 1998.

[7]. Fodor, J. C., Yager, R. R. and Rybalov, A., "Structure of uni-norms," International Journal of Uncertainty, Fuzziness and Knowledge-Based Systems 5, 411-427, 1997.

[8]. De Baets, B. and Fodor, J., "On the structure of uninorms and their residual implicators," Proc. Eighteenth Linz Seminar on Fuzzy Set Theory, Linz, Austria, 81-87, 1997.

[9]. De Baets, B., "Uninorms: The known classes," in Fuzzy Logic and Intelligent Technologies for Nuclear Science and Industry, edited by Ruan, D., Abderrahim, H. A., D'hondt, P. and Kerre, E. E., World Scientific: Singapore, 21-28, 1998.

[10]. Klement, E. P., Mesiar, R. and Pap, E., "On the relationship of associative compensatory operators to triangular norms and conorms," International Journal of Uncertainty, Fuzziness and Knowledge-Based Systems 4, 129-144, 1996.

[11]. Yager, R. R., "Penalizing strategic preference manipulation in multi–agent decision making," Technical Report# MII-2018 Machine Intelligence Institute, Iona College, New Rochelle, NY, 2000.

[12]. Yager, R. R., "Uninorms in fuzzy systems modeling," Fuzzy Sets and Systems, (To Appear).

[13]. Yager, R. R., "Structures generated from weighted fuzzy intersection and union," Journal of the Chinese Fuzzy Systems Association 2, 37-58, 1996.

[14]. Yager, R. R., "On ordered weighted averaging aggregation operators in multi-criteria decision making," IEEE Transactions on Systems, Man and Cybernetics 18, 183-190, 1988.

Part 3

Applications of Aggregation Operators

Aggregation Operators in Engineering Design

Endre Pap

Institute of Mathematics
University of Novi Sad
21000 Novi Sad, Yugoslavia
E-mail: pap@unsim.ns.ac.yu , pape@eunet.yu

Abstract. Engineering design is conducted with incomplete and imperfect information, and in this paper there are presented some of the tools for decision making under risk and uncertainty and the application of these tools to engineering design. First it is presented an axiomatization of engineering design based on von Neumannn and Morgenstern axiomatization. Then it is given a general definition of decision making problem which enables to apply also fuzzy systems and non-additive measures. Special attention is taken on different aggregation operators which can model the decision making in engineering design. A procedure for finding the global maximum as well as some procedures for identification of non-additive measure are presented.

1. Introduction

Engineering design is recently recognized as a decision-making process at the core [Tribus 1969, Sage 1977, de Neufville 1990, Hazelrigg 1996, 1997]. This view is also consistent with the Accreditation Board Engineering and Technology's (ABET) definition of engineering design. Engineering design is conducted with incomplete and imperfect information, yet most traditional design approaches treat the design problem are deterministic. We present here tools for decision making under risk and uncertainty and apply the tools to engineering design. Decision making under uncertainty and risk goes back to Bernoulli, 1738. It has three main elements ([Hazelrigg 1997a]): identification of the options, determination of expectations on each option and the expression of values. The main decision rule is : the preferred decision is the option whose expectation has the highest value. Classical decision theory [Luce and Raiffa 1957] separates expectations and values - a common mistake is to make them equal. The decision making involve options, expectations and values. An engineering decision cannot be done in the absence of human values, whereas problems in the science are solved in the absence of options and values. The purpose of values in decision making is to rank order alternatives. This ranking is managed by a preference relation, which is connected by the usual real order relation through the utility function (see subsections 2.1 and 2.2). The person who has a transitive preference relation usually is called rational, in the opposite case he is irrational. The famous

Arrow's Impossibility Theorem [Arrow 1963] states that a group consisting only of rational individuals need not exhibit transitive preferences.

John von Neumann and Oscar Morgenstern [1953] determined a value measure to rank order alternatives under uncertainties and risk. Their utility is based on a lottery, where there are more desired outcomes and less desired outcomes. They proposed that if the probability of the more desired outcome occurring tends unity, the utility of lottery should approach the utility of the more desired outcome, and conversely. Hence the utility of the lottery should lie between the utilities of the less and more desired outcomes. John von Neumann and Oscar Morgenstern developed from this notion and six axioms [von Neumann, Morgenstern 1953, Hazelrigg 1997b] a utility theory. The main result is that the utility of a lottery is the sum of the utilities of all possible outcomes of the lottery weighted by their probabilities of occurrence. Based on von Neumann and Morgenstern axioms Hazelrigg [1997b] has introduced an axiomatic approach to engineering design (subsection 2.1).

A decision is made under the risk if the only available knowledge related the outcome states is the probability distribution. This can be used in the optimization of the utility function. If the knowledge about the probabilities of the outcome is unknown, then the decision have to be made under uncertainty. In engineering design one of the most critical problems is the preliminary design decision when the design is imprecise and most costly [Whithey 1988, Antonsson, Otto 1995, Otto, Antonsson 1991]. In such situations, fuzzy decision making can be used to handle this vagueness. This fuzziness can be modeled in different ways: fuzzy sets (membership function) [Otto, Antonsson 1991 (The Method of Imprecision), Zimmermann, Sebastian 1993,1994,1995] (subsection 2.3) , fuzzy measures (Choquet and Sugeno integrals) [Grabisch et al. 1995, Pap 1995, Wang, Klir 1992] (subsection 2.2). There are other design methodologies as optimization tools (linear, nonlinear, integer programming , multi-objective optimization - e.g., with weighted sum technique [Osycska 1984]), probability methods [Taguchi 1986]).

Aggregation operators play important role in many different approaches to decision making [Otto, Antonsson 1991, Grabisch et al. 1995, Fodor et al 1994, Klement et al. 2000a]. Therefore we will investigate them in details and manage many useful relations between them (section 3). In many systems (specially intelligent) the aggregation of incoming data plays the main role. The aggregation operators form a fundamental part of multi-criteria decision making, engineering design, expert systems, pattern recognition, neural networks, fuzzy controllers, genetic algorithms, etc., see [Grabisch et al. 1995, Fodor, Roubens 1994,Yager, Filev 1994, Otto, Antonsson 1991, Klement et al. 2000a].

Engineering decision making needs more general mathematical models, which involve also non-additive measures. Previously used additive probability measures could not model some situations as e.g. the Ellsberg Paradox, see [Grabisch et al. 1995]. For the non-additive set function (measure) m

defined on a σ-algebra Σ of subsets of a set X (for finite X it is usually taken $\Sigma = \mathcal{P}(X)$, the family of all subsets), the difference $m(A \cup B) - m(B)$ depends on B and can be interpreted as the effect of A joining B, [Grabisch et al. 1995, Pap 1995, Wang, Klir 1992]. A monotone set function m with $m(\emptyset) = 0$ is usually called fuzzy measure, [Sugeno 1974].

More than the contribution of the extension principle of the fuzzy sets [Terano et al. 1992], fuzzy connectives ([Dyckhoff et al. 1984, Grabisch et al. 1995, Klement et al. 1996, Zimmermann et al. 1980]) and fuzzy measures are important in the problem of the modeling of the behavior of decision makers. Utility theory [Quiggin 1993, Grabisch et al. 1995, Fodor et al. 1994] deals with preference relations describing the decision behavior, and as the basis of decision theory, is well axiomatically based on the fuzzy measures and Choquet integral [Fodor et al. 1994, Grabisch et al. 1995].

For a non-negative measurable function f the Choquet integral on $A \in \Sigma$ with respect to a fuzzy measure m is defined by

$$(C) \int_A f\,dm = \int_0^\infty m(A \cap F_\alpha)\,d\alpha,$$

where $F_\alpha = \{x | f(x) \geq \alpha\}$ is the α-cut of f. The Sugeno integral of f on A with respect to m is defined by

$$(S) \int_A f\,dm = \sup_{\alpha \in [0,+\infty]} (\min(\alpha, m(A \cap F_\alpha))).$$

Both integrals are usually called fuzzy integrals. The properties of the aggregation operators in engineering design represented by the Choquet and Sugeno integrals are of special interest (section 3.3). The advantage (using the interrelation between factors) of using fuzzy measures and corresponding integrals is illustrated by the following simple example (see [Wang, Klir 1992]). Suppose we evaluate three TV sets using only two quality factors: "picture", x_1 and "sound", x_2 and the corresponding weights $w_1 = 0.7$ and $w_2 = 0.3$. TV 1 has (x_1, x_2) as $(1, 0)$, TV 2 $(0, 1)$ and TV 3 has $(0.45, 0.45)$. Using the method of weighted mean, the wrong synthetic evaluation, gives for TVs: $0.7, 0.3, 0.45$, i.e., the first TV is the best, although it is without sound. Taking an importance measure m as $m(\{x_1\}) = 0.3, m(\{x_2\}) = 0.1, m(\{x_1, x_2\}) = 1, m(\emptyset) = 0$ the Sugeno integral gives for TVs: $0.3, 0.1, 0.45$. Hence the third TV is the best choice. Using the method of weighted mean, no choice of the weights would lead to the expected result, since the factors are taken independently.

We shall use an unified mathematical tool (we call it pseudo-analysis) for treating problems with uncertainties, nonlinearity and optimization [Baccelli et al. 1992, Maslov et al. 1992, Pap 1995, Wang, Klir 1992].

The advances of new powerful computers, wireless communication, next-generation Internet, will enable concurrent engineering teams to obtain the main design criteria early in engineering design process. This will enable a

rapid accommodation of new engineering design paradigms and improvement in the design process.
To identify the model for fuzzy integrals require the parameter identification of the suitable fuzzy measure. A consistent theory for the semantically interpretation of fuzzy measure is still an open research area as is the development of practical identification algorithms (4.2). Namely, the Bremmermann's limit (see subsection 4.2) and the exponential nature of the number of coefficients of a fuzzy measure forces us that even in the case of powerful computers we have to find a good (not always the best) and approximate solution.
There are some procedures for finding the approximation of the global maximum and minimum of a real function. Pap and Surla [1997, 2000] presented a procedure which works even for non-continuous functions (subsection 4.1).

The main contribution of this paper is to give proposed tasks for future research for each considered subject.

2. Engineering design

2.1. Axiomatic approach based on von Neumann-Morgenstern theory

First we start with an axiomatic approach to engineering design which guarantees a rational treatment of all information that the designer uses for the design and enables a rational decision making [von Neumann and Morgenstern 1953, Hazelrigg 1997a,b]. We shall use the symbol $\succ$ for the relation "is preferred to", the symbol $\sim$ for "is indifferent to" and the symbol $\succeq$ for "is preferred or indifferent to".

Axiom 1: *The axiom of deterministic decision making.* Given a defined set of options from which to choose, each with a known and deterministic outcome, the decision maker's preferred choice is that option whose outcome is most desired.

Axiom 2: *Ordering of alternatives.* Preference and indifference orderings hold between any two outcomes, and they are transitive.

Axiom 3: *Reduction of compound lotteries.* Any compound lottery is indifferent to a simple lottery with the same outcomes and associated probabilities (for the notion of the lottery see [von Neumann, Morgenstern 1953]).

Axiom 4: *Continuity.* Given outcomes of a lottery ordered by preference from A_1 through A_r, there exists a number u_i such that each outcome A_i is indifferent to a lottery containing only A_1 and A_r.

With mathematical symbols

$$A_i \sim [u_i A_1, (1 - u_i) A_r] = \hat{A}_i,$$

where $\hat{A}_i$ is the lottery.

Axiom 5: *Substitutability.* In any lottery L, $\hat{A}_i$ is substitutable for A_i.

Axiom 6: *Transitivity.* Preference and indifference among lotteries are transitive relations.

Axiom 7: *Monotonicity.* A lottery $[pA_1, (1-p)A_r]$ is preferred or indifferent to a lottery $[p'A_1, (1-p')A_r]$ if and only if $p \geq p'$.

Axiom 8: *Reality of engineering design.* All engineering designs are selected among the set of potential designs that are explicitly considered.

As consequence of these axioms there were deduced the following three important theorems [Hazelrigg 1997b].

Theorem 1: The expected utility theorem. *Given a pair of options, each with a range of possible outcomes and associated probabilities of occurrence, that is, two lotteries, the preferred choice is the option (or lottery) that has the highest expected utility.*

Theorem 2: The substitution theorem. *A decision maker is indifferent between a lottery L and a certainty outcome whose utility is equal to the expected utility of the lottery.*

Theorem 3: Arrow's Impossibility Theorem. *Groups consisting of rational people are not necessarily rational.*

The given eight axioms of engineering design provide a rigid framework for engineering design to be able to achieve maximum effectiveness, taking uncertainty and risk explicitly. There is need to examine Axioms 1 and 8 in more details, since they are specific for engineering design (Axioms 2 through 7 are the von Neumann-Morgenstern axioms and they were widely debated, e.g., [Luce and Raiffa 1957]), although they seems obvious (see [Hazelrigg 1997b]). It is important to examine further consequences of this axiomatic approach , especially the consequences of three fundamental theorems. It has to adopt the utility function defined on the set of consequences of acts, with probabilities attached to the different states for the purposes of engineering design. We shall investigate the relation with the case of imprecise design description (2.3).

Possibilistic mixtures strikingly differ from probabilistic measures, because they account for a non-convex structure. However, changing sum into maximum, and the product into a triangular norm, possibilistic mixtures have properties that parallel those of probabilistic mixtures. In particular, possibilistic mixtures form the underpinnings of possibilistic utility theory, as proposed by Dubois and Prade in [1995] and more recently systematized by Dubois, Godo et al. [1998]. There is need to examine the corresponding axioms of engineering design based on possibility theory.

Beyond possibilistic and probabilistic mixtures, only a form of hybridization is possible such that the mixture is possibilistic under a certain threshold, and probabilistic above. It have to examine the corresponding axioms of engineering design based on hybrid probabilistic-possibilistic theory (see [Dubois et al. 2000]).

2.2. Decision problem, multicriteria and subjective evaluation

We formulate now the decision problem in a wider sense than in the previous section [Grabisch et al. 1995].

A **decision problem** is a 5-uple $(A, \Theta, \phi, X, \succeq)$ with:

- A: set of *alternatives or acts*, among which the decision maker must choose;
- X: set of *consequences, or results.* These consequences come from the choice of an alternative;
- Θ: set of the *states of the world.* According to the state of the world $\theta \in \Theta$ (usually unknown), the consequences of the choice of an alternative $a \in A$ may differ;
- ϕ: $A \times \Theta \to X$ specifies for each state of the world θ and each alternative a the resulting consequence $x = \phi(a, \theta)$;
- $\succeq$: weak order relation on X, i.e. a binary relation verifying:
  (i) $x \succeq y$ or $y \succeq x$, $\forall x, y \in X$
  (ii) $\succeq$ is transitive, i.e. $(x \succeq y,\ y \succeq z)$ implies $x \succeq z$;

$\succeq$ is the preference relation which characterizes the decision maker, and may be strict $(x \succ y)$, and indifference $(x \sim y)$.

The basic idea behind utility theory is to transform the weak order $\succeq$ on X into a usual order $\geq$ on real numbers, by mean of so called *utility function* $u : X \to \mathbb{R}$ whose fundamental property is: $x \succ y \Leftrightarrow u(x) > u(y)$; when the property is verified, u is said to represent $\succeq$.

There are two subproblems:

- **multicriteria decision** - the state of the world is supposed as known and X is multidimensional; a consequence is $x = (x_1, \ldots, x_n)$, where $x_i \in X_i$ is an element of the set representing a criterium;

- **decision under uncertainty** - monodimensional set of the consequences is considered when the true state of the world is unknown.

The uncertainty measure on Θ might be known, and as the set of consequences X is monodimensional, elements of A are acts $f : \Theta \to X$ and the preference relation is defined on the set of acts.

Probability measures can be identified from real experiments by the frequency interpretation, but not when subjective probability measures are concerned. Subjective belief on an event to occur or which has occurred can be modeled as subjective probability measure on the states of the world, so that the set of acts can be compared.

Different authors have examined, see [Grabisch et al. 1995], the conditions for the existence of the utility functions $u : X \to \mathbb{R}$ and the unique probability measure P^* on Θ representing the preference relation $\succ$ on A through the expected utility E of the action with respect to P^* (Savage)

$$f \succ g \Leftrightarrow E[u(f(\theta_i)), P^*] > E[u(g(\theta_i)), P^*].$$

Anscombe and Aumann (1963) restrict to the case of money lotteries (the set of consequences), i.e., the case of the simple probability measures on $\mathbb{R}$ as the elements of X ($P = (p_1, y_1; \ldots; p_n, y_n)$, y_i may be negative), where p_i

is the probability of getting y_i dollars. Then a non trivial preference relation $\succeq$ is a weak order, continuous, wekly monotonic and satisfies independence if and only if there exists a unique probability P^* on Θ and a utility function $u : \mathbb{R} \to \mathbb{R}$ such that

$$f \succeq g \Leftrightarrow \int Eu(f(\theta_i))\, dP^*(\theta_i) \geq \int Eu(g(\theta_i))\, dP^*(\theta_i).$$

The previous result puts specific requirements on the preference relation and if the requirements are not so strict (independence is replaced with co-monotonic independence - for this and other definitions [Grabisch et al. 1995] may be consulted), decisions may be compared by the Choquet integrals with respect to the unique fuzzy measure m^* on Θ

$$f \succeq g \Leftrightarrow (C) \int Eu(f(\theta_i))\, dm^* \geq (C) \int Eu(g(\theta_i))\, dm^*.$$

(Each of the utility functions is unique up to the positive linear transformation).

Clear interpretation of the Choquet integral as the generalization of the expectation in decision making is an important result. The Choquet integral comparison for the decisions also exists when the independence is replaced with max-min independence existing for the relations of which either the best lottery (money lotteries are the elements of X) is assigned or the worst lottery is assigned for each pair of decisions considered for the independence. If there is no information about uncertainty measure on Θ, there are different classical criteria to make the decision, that can all be represented by the Choquet integral with respect to the suitable fuzzy measure.

Multicriteria evaluation is present in a very large field of applications, as resource allocation, design of new goods, environmental planning, quality control, evaluation of creditworthiness, etc, [Grabisch et al. 1995, Wang and Klir 1992].

There are two approaches to multiattribute utility theory:

the ordinal approach compare alternatives two by two, but the formed Hasse diagram might be with objects not all comparable to each other as the transitivity property of the preference relation is lost in the aggregation step;

the cardinal approach, followed in the sequel, gives an absolute evaluation to every alternative for a given criterion and that real values are aggregated to get a global evaluation; all objects are comparable, whatever their characteristics.

The problem considers a list of *similar objects* $S = \{s_1, \ldots, s_n\}$ described by the *set attributes* $Z = \{z_1, \ldots, z_p\}$. The *information*, precise or not, about the values of the attributes for each object exists. Each object is then evaluated according to a set of *criteria* $X = \{x_1, \ldots, x_p\}$, which are defined by the decision maker himself. The general form of expressing a criterion is supposed to be $x_j \ : \ z_j$ is A_j, where A_j is a fuzzy set on the universe of z_j. To solve the problem, *marginal* (partial) and *global* evaluation are to be performed.

Each object s_i is evaluated with respect to one particular criterion x_j defining a mapping $h_j : S \to [0,1]$. It tells the degree of an object from S to satisfy the criterion x_j. $h_j(s_i)$ is the degree of compatibility between the fuzzy set A_j and the value of x_j for object s_i.

The mapping may be defined by $h_j(s_i) = A_j(z_j^i)$ when the precise value z_j^i of attribute z_j of s_i is known. When $\tilde{z}_j^i$ as a fuzzy number (see [Dubois, Prade, 1980]) better represents the value of the attribute, the definition of Zadeh based on the extension principle

$$\widetilde{h_j(s_i)}\,(y) = \sup_{z_j | y = A(z_j)} \tilde{z}_j^i(z_j).$$

results with the fuzzy set considering the compatibility between the two fuzzy sets.

For the object with all the marginal evaluations $h_1(s_i), \ldots, h_p(s_i)$, the problem is to determine a single evaluation with respect to all the criteria. The global value is subjective as the decision makers differ in their decisions when the criteria are the same:

weight on criteria express the ideas about what is important and what could be neglected;

behavior can be tolerant or disjunctive-oriented when "it is sufficient that some criteria are met", or intolerant or conjunctive-oriented if it is declared that "all criteria must be equally well met".

t-norms (triangular norm is a binary associative and monotone operation on the interval $[0,1]$ with the neutral element 1 - [Klement et al. 2000a], first introduced in the theory of probabilistic metric spaces), quasi-Sugeno or fuzzy t-conorm integrals can be the operators used over the marginal evaluation values for they are normalized and the utility function should be determined up to the positive linear transformation.

Further investigation of non probabilistic multidimensional problems [Grabisch et al. 1995, Fodor and Roubens 1994, Klementet al. 2000a] using different types of aggregation operators (see section 3) for the utility function will be managed. There is need for further characterization the class of aggregation operators with respect to different types of fuzzy measures (decomposable measures, possibility measure, belief measure, etc.) related to Choquet, Sugeno integrals and integrals based on decomposable measures (Sugeno-Murofushi, Wang-Klir, Weber, Pap, Maslov) and the adaptation of them for the purposes of engineering design. Although has been recently a proliferation in the development of parts of the theory of non-additive measures with analogous theorems to the classical measure theory [Denneberg 1994, Pap 1995, Grabisch et al. 1995, Wang, Klir 1992] there are still missing some important parts as Radon-Nikodým type theorem and Ergodic type theorem. There are only initial results available that are related to Radon-Nikodým theorem, see [Candeloro 1987, Mesiar, Šipoš 1993, Sugeno-Murofushi 1987,

Pap 1995] and conditional fuzzy measure [Grabisch et al. 1995], [Benvenuti, Mesiar (to appear)]. With respect to Ergodic theorem (see [Blume (to appear)], at this moment solution for practical applications appears to be in using special fuzzy measures as belief and plausibility measures which can be characterized by families of probabilities or we can use measures which are distorted probabilities.
We expect that it will be investigated the further extension of eight axioms from section 2.1 (based on von Neumann-Morgenstern axioms) using special fuzzy measures as was done for possibility measure in [Dubois et al. 1997].

2.3. Fuzzy systems

An overall evaluation of design alternatives have two parts: their partial evaluation and the importance of the criteria taken into account. The first step consists in the determination of fuzzy sets representing partial evaluation of the alternatives. Since there are many different judgments with respect to the expression of the suitability of variety of alternatives there is need for some methods for this purpose. The Analytical Hierarchy Process [Saaty 1978] and some other matrix methods as [Pugh 1990] are very convenient tools for that purpose. At the second step all partial information is aggregated into a final rating.
In engineering design the preliminary design decisions are very important although the design description is still imprecise. Fuzzy design methods are convenient for representing and manipulating design imprecision [Otto, Antonsson 1991, Zimmermann, Sebastian 1993, 1994, 1995]. By the Method of Imprecision [Otto, Antonsson 1991, Scott, Antonsson 1995] contraints can be imprecise permitting to choose preferences over a range of values. This method was specially developed for engineering design and implies that the trade-off combination functions (aggregation operators) have to satisfy the boundary conditions, monotonicity, continuity, annihilation and idempotency, section 3. Then it follows by [Otto, Antonsson 1991, Scott, Antonsson 1997] that any weigthed quasi-linear mean that satisfies the annihilation property is design-appropriate. A weighted aggregation operator which is continuous, strictly monotonic, idempotent and bisymmetrical ($\mathbf{A}_2(\mathbf{A}_2(x,y),\mathbf{A}_2(u,v)) = \mathbf{A}_2(\mathbf{A}_2(x,u),\mathbf{A}_2(y,v)))$ has the representation [Aczél 1966, Fodor, Roubens 1994]

$$\mathbf{M}^f_{\omega_1,\ldots,\omega_n}(x_1,\ldots,x_n) = f^{-1}\left(\frac{\omega_1 f(x_1)+\cdots+\omega_n f(x_n)}{\omega_1+\cdots+\omega_n}\right),\ \omega_i > 0(i=1,\ldots,n),$$

where f is a strictly monotone continuous function (section 3). By this representation it is possible to construct special convenient families of aggregation operators [Otto, Antonsson 1991, Fodor, Roubens 1994, Grabisch et al. 1995].

There is need for collecting and re-analyzing existing methods of multi-objective decision-making [Scott and Antonsson 1995, Otto and Antonsson

1991, Thurston 1991, Hsu et al. 1995, Knosala et al. 1992, Muller et al. 1994, Chen et al. 1995, Zimmermann and Sebastian 1993, 1994, 1995]. Special attention has to be taken on the properties of the aggregation operators and the relation with Choquet and Sugeno integrals [Fodor, Roubens 1994, Grabisch et al. 1995, Klement et al. 2000a]- section 3.3. There is a need for a user-friendly computer system which will enable engineering designer to manage the entire complex process of choice and ranking within the set of alternatives with an ability involving different types of uncertainties as fuzzy (through membership functions and fuzzy measures), probabilistic and deterministic. It has to be compared the uncertainty measures : probability, fuzzy sets and fuzzy measure in the spirit of [Drakopoulos 1995], where he proved that probabilities have a higher representational power than fuzzy sets (with respect to max-min) and possibilities (special fuzzy measure) for finite domain, but the cost of this is high computational complexity and reduced computational efficiency (they have equal representational power when their domains are infinite).

2.4. Pseudo-analysis as a base for solving nonlinear equations in optimization, control theory, soft computing

There is a mathematical background, which we shall call pseudo–analysis, for treating problems with uncertainty, nonlinearity and optimization in mathematics and soft computing. Namely, instead of the usual plus and/or product structure of real numbers, other operations (pseudo-operations) on extended reals are considered. For example, max-min, max-plus, max- product or operations generated by some additive generator g are included, and specially triangular conorm- triangular norm [Baccelli et al. 1992, Cuninghame–Green 1979, Golan 1992, Klement et al. 2000a, Kuich 1986, Pap 1995]. Certain parts of such mathematical analysis have been developed in analogy with the classical mathematical analysis as for example measure theory, integration, integral operators, convolution, Laplace transform, see books [Klement, Mesiar, Pap 2000, Maslov et al. 1992, Pap 1995] and papers [A. Marková-Stupňanová 1999, Pap 2000, Pap, Štajner 1999, Pap, Ralević 1998]. Many problems in fuzzy logic, fuzzy sets , neural nets, fuzzy-neural nets, multicriteria decision making, etc. can be treated by these mathematical tools [Grabisch et al. 1995, Pap 1995, Pap et al. 1998, Wang et al. 1992] .

One of the recent results is related to the natural extension of von Neumann-Morgenstern utility theory using pseudo-additive measures [Dubois et al. 2000]. It turns out that the only reasonable pseudo-additive measures m admitting an independence-like concept (i.e., $m(A_1 \cap \ldots \cap A_n) = T_{i=1}^n m(A_i)$) are based on conditionally distributive pairs (S, T) (see [Klement et al. 2000a]) of t-conorms and t-norms, namely:

(a) probability measures (and T = product);
(b) possibility measures (and T is any continuous t-norm);

(c) suitably normalized hybrid set-functions m such that there is $a \in]0,1[$ which gives for A and B disjoint

$$m(A \cup B) = \begin{cases} m(A) + m(B) - a \text{ if } m(A) > a, m(B) > a, \\ \max(m(A), m(B)) \text{ otherwise.} \end{cases}$$

Let (S, T) be a pair of continuous t-conorm and t-norm, respectively, of the form $(< S_M, S_L >, < T_1, T_P >)_a$. Let u_1, u_2 be two utilities taking values in the unit interval $[0,1]$ and let μ_1, μ_2 be two degrees of plausibility from $\Phi_{S,a}$. Then we define *the optimistic hybrid utility function* by means of the hybrid mixture as

$$U(u_1, u_2; \mu_1, \mu_2) = S(T(u_1, \mu_1), T(u_2, \mu_2)).$$

In the paper [Dubois et al. 2000] it is examined in details this utility function.

There are also many applications of pseudo-analysis in different fields as optimization, nonlinear differential and difference equations, economics, game theory [Baccelli et al. 1992, Bellman 1962, Maslov et al. 1992, Pap 1995, Pap et al. 1998]. In this way there are covered with a single theory using unifying methods different problems (usually nonlinear) from many important fields (system theory, optimization, differential equations, difference equations, etc.). Important fact is that this approach gives also solutions in the form which are not achieved by other theories (for example, Bellman difference equation, Hamilton–Jacobi equation with non–smooth Hamiltonian important in the control theory) and avoids the use of the so called "viscosity solution" method [Lions 1982], which gives only upper and lower solutions. In some cases it enables for mostly nonlinear equations to obtain exact solutions in the similar form as for the linear equations. Some obtained principles as for example the pseudo-linear superposition principle allows us to transfer the methods of linear equations to the nonlinear equations.

Pseudo-analysis uses many mathematical tools from different fields as functional equations, variational calculus, measure theory, probability, functional analysis, optimization theory, semiring theory, etc. It is still in its developing form.

2.5. Example of engineering design in aero-thermo-elastic problems

The innovation of the design process is currently one of the main topics by major aerospace companies such as Boeing, McDonell Douglas, and Lockhead Martin [Noor et al. 1997]. There is need for better designs for complex systems with reduced design cycle and development times. Intelligent simulation-based design gives unprecedent flexibility for developing and improving aerospace products and processes. Using the computational intelligence by associated tools of soft computing (neurocomputing, fuzzy logics, genetic algorithms) it is exploited the tolerance for imprecision and uncertainty in real aerospace vehicles obtaining tractability, robustness, and low solution cost.

3. Aggregation operators

The aggregation operators are applied in multi-criteria decision making, engineering design, expert systems, pattern recognition, neural networks, fuzzy controllers, genetic algorithms, etc., see [Grabisch et al. 1995, Fodor, Roubens 1994, Yager, Filev 1994, Otto, Antonsson 1991, Mesiar, Komorniková 1997, Klement et al. 2000a].

We restrict to the inputs and outputs from the unit interval $[0,1]$. The case of any other closed interval can be straightforwardly derived from the $[0,1]$–case.

3.1. General properties

An *aggregation operator* $\mathbf{A}$ is a non–decreasing mapping

$$\mathbf{A} : \bigcup_{n \in \mathbb{N}} [0,1]^n \to [0,1]$$

fulfilling the following conditions:

(i) $0 \leq x_i \leq y_i \leq 1, i = 1, \ldots, n$ imply $\mathbf{A}(x_1, \ldots, x_n) \leq \mathbf{A}(y_1, \ldots, y_n)$; (ii) $\mathbf{A}(x) = x$ for all $x \in [0,1]$; (iii) $\mathbf{A}(0, \ldots, 0) = 0$ and $\mathbf{A}(1, \ldots, 1) = 1$.

Property (i) is *the monotonicity* and properties (ii) and (iii) are *the boundary conditions.* Each aggregation operator $\mathbf{A}$ can be represented as a system $(\mathbf{A}_n)_{n \in \mathbb{N}}$ of n–ary operators $\mathbf{A}_n, n \in \mathbb{N}$, on the unit interval, where $\mathbf{A}_1$ is the identity operator on $[0,1]$ and each $\mathbf{A}_n, n \geq 2$, is non–decreasing and $\mathbf{A}_n(0, \cdots, 0) = 0, \mathbf{A}_n(1, \cdots, 1) = 1$.

Depending on the field of application, several additional properties can be required and/or examined, such as commutativity, associativity, continuity, idempotency, compensation, cancellativity, etc., see [Klement et al. 2000a, Mesiar, Komorniková 1997]. Note, e.g., that the associativity of an aggregation operator $\mathbf{A}$ means that the binary operator $\mathbf{A}_2$ is associative and its corresponding n–ary extensions (for $n > 2$) are just the relevant n–ary operators $\mathbf{A}_n$. Therefore, an associative aggregation operator $\mathbf{A}$ is fully determined by $\mathbf{A}_2$.

If $\mathbf{A}$ is an aggregation operator, then the operator $\mathbf{DA} : \bigcup_{n \in \mathbb{N}} [0,1]^n \to [0,1]$ defined by

$$\mathbf{DA}(x_1, \ldots, x_n) = 1 - \mathbf{A}(1 - x_1, \cdots, 1 - x_n)$$

is called the *dual operator* of $\mathbf{A}$.

$\mathbf{DA}$ is also an aggregation operator.

Example 1

a) Let $\mathbf{A}_2$ be a t–norm (or a t–conorm) and let $\mathbf{A}$ be derived from $\mathbf{A}_2$ by the usual extension procedure. Then $\mathbf{A}$ is a commutative associative operator with a neutral element 1 (or 0). The most important continuous t-norms are $T_{\mathbf{M}}(x,y) = \min(x,y)$, $T_{\mathbf{P}} = xy$, $T_{\mathbf{L}} = \max(x+y-1, 0)$ and the corresponding dual operators (t-conorms): $S_{\mathbf{M}}(x,y) = \max(x,y)$; $S_{\mathbf{P}} = x + y - xy$, $S_{\mathbf{L}}(x,y) = min(1, x+y)$.

b) One of the most applied aggregation operator is the arithmetic mean $\mathbf{M}$,

$$\mathbf{M}(x_1, \ldots, x_n) = \frac{\sum_{i=1}^{n} x_i}{n} \quad (= \frac{SUM}{CARD}).$$

$\mathbf{M}$ is a commutative aggregation operator which is quasi-associative, i.e., a function of associative operators, namely of SUM and $CARD$, idempotent, i. e., $\mathbf{M}(x, \cdots, x) = x$ for all $x \in [0,1], n \in \mathbb{N}$, continuous, cancellative. Moreover, for any value $t \in]0,1[$, any given input $(x_1, \cdots, x_n)$ can be compensate by an additional input $(y_1, \ldots, y_m)$ so that $\mathbf{M}(x_1, \ldots, x_n, y_1, \cdots, y_m) = t$.

For any *idempotent* aggregation operator $\mathbf{A}$, i.e., $\mathbf{A}(x_1, \cdots, x_n) = x$ for all $x \in [0,1]$, we have $T_{\mathbf{M}} \leq \mathbf{A} \leq S_{\mathbf{M}}$.
Fuzzy design methods are convenient for representing and manipulating design imprecision [Otto, Antonsson 1991, Zimmermann, Sebastian 1993, 1994, 1995]. The Method of Imprecision [Otto, Antonsson 1991, Scott, Antonsson 1995] was specially developed for engineering design and implies that the trade-off combination functions (aggregation operators) have to satisfy the boundary conditions, monotonicity, continuity, annihilation and idempotency, where annihilation means that if one argument (the preference for any one attribute of the design sinks to zero) of the aggregation operator is zero then the value of the aggregation operator (the overall preference of the design) is zero. For weighted aggregation operator $\mathbf{M}^f_{\omega_1,\ldots,\omega_n}(x_1, \ldots, x_n)$ if the weights $\{\omega_i\}$ are given with respect to a ratio scale, then ω_i are not uniquelly determined, since any other system of weights $\{\omega_i'\}$ with $\omega_i' = C\omega_i$ for a positive rational number is convenient, e.g., $\omega_i' = \frac{\omega_i}{\sum_i \omega_i}$. Specially important cases are $f(x) = x, f(x) = \log x$ and $f(x) = x^s$. The last case is interesting since it generates a parametrized family of aggregation functions. For $s > 0$ the annihilation property fails, but it can be handled in the engineering design in the way that it can be assumed that preferences less than some small ε are not relevant for the designer [Scott, Antonsson 1997]. If the weighted quasi-arithmetical mean is commutative then it has the following representation

$$\mathbf{M}_f(x_1, \ldots, x_n) = f^{-1}(\frac{1}{n}\sum_{i=1}^{n} f(x_i))$$

where $f : [0,1] \to \mathbb{R}$ is a continuous strictly monotone function. Recall that quasi-arithmetical mean include the arithmetical mean, the geometric mean, the quadratic mean, the harmonic mean, etc. If the range of f is bounded, then $\mathbf{M}_f$ can be represented also by an appropriate Choquet-like integral, see 3.3.
There are design-appropriate aggregation operators which are not weighted quasi-arithmetical means, namely, they are not strictly monotone.

It is important to investigate aggregation operators which are convenient for engineering design, i.e., with the properties: idempotency, continuity, annihilation and bisymmetry (very useful property in the absence of commutativity and associativity). Special attention have to be taken on non-strict case, since strict monotonicity is not required in engineering design and it is incompatible with the annihilation property. Crucial fact in this classification will be the value $\mathbf{A}(0,1)$ which is either 0 or 1 or a number between 0 and 1. In the first case, with the commutativity property, we divide such aggregation operators on three classes: min, transforms with an automorphism of the geometric mean and the ordinal sums of such transforms.

3.2. Aggregation operators based on triangular norms, compensatory operators

Starting from a given t-norm and/or t-conorm, several useful operations on $[0,1]$ can be introduced. The conditions (i) - (iii) in 3.1 required for an aggregation operator $\mathbf{A}$ are the genuine properties of triangular norms and conorms. Therefore a large class of aggregation operators can be derived as a combination of t–norms and conorms. As far as the commutativity is another common property of both t-norms and conorms, also the corresponding aggregation operators will keep the commutativity. However, the associativity is preserved only in some special cases.

From the application point of view, there exist suggestions to use the special aggregation operators, so-called *compensatory operators* in order to model intersection and union in many-valued logic. A motivation for the use of such operators can be found in [Zimmermann and Zysno 1980, 1983]:

> "The interpretation of a decision as the intersection of fuzzy sets, computed by applying any t-norm based operator implies that there is no compensation between low and high degrees of membership. If, on the other hand, a decision is defined to be the union of fuzzy sets, represented by some t-conorm, full compensation is assumed. Managerial decisions hardly ever represent any of these extremes."

The main goal of compensatory operators is to model an aggregation of incoming values. If two values are aggregated by a t-norm then there is no compensation between low and high values. On the other hand, a t-conorm based aggregation provides the full compensation. None of the above cases covers the real decision making. To avoid such inaccuracies, [Zimmermann and Zysno 1980] suggested two kinds of so-called compensatory operators. The first of them was γ-operator, $\Gamma_\gamma : \cup_{n\in\mathbb{N}}[0,1]^n \to [0,1]$, $\gamma \in [0,1]$, $n \geq 2$

$$\Gamma_\gamma(x_1,\ldots,x_n) = (\prod_{i=1}^{n} x_i)^{1-\gamma}(1 - \prod_{i=1}^{n}(1-x_i))^\gamma.$$

Here parameter γ indicates the degree of compensation. Note that γ-operators are a special class of exponential compensatory operators [Klement et al.

2000a]. For a given t-norm T, t-conorm S (not necessarily dual to T) and parameter γ indicating the degree of compensation, the exponential compensatory operator $E_{T,S,\gamma} : [0,1]^n \to [0,1]$, $n \geq 2$, is defined by

$$E_{T,S,\gamma}(x_1,\ldots,x_n) = (T(x_1,\ldots,x_n))^{1-\gamma}(S(x_1,\ldots,x_n))^{\gamma}.$$

It is obvious that γ-operator is based an $T_{\mathbf{P}}$ $S_{\mathbf{P}}$, $\Gamma_\gamma = E_{T_{\mathbf{P}},S_{\mathbf{P}},\gamma}$. Further note that $E_{T,S,\gamma}$ is a logarithmic convex combination of T and S and up to the case when $\gamma \in \{0,1\}$ it is non-associative.Another class of compensatory operators proposed in [Zimmermann-Zysno 1980] are so-called convex-linear compensatory operators.

We have proposed an *associative class of compensatory operators* in [Klement et al. 1996]. The degree of compensation is ruled by two parameters, namely by the neutral element e and the compensation factor k. Let T be a given strict t-norm with additive generator $f, f(\frac{1}{2}) = 1$, and let S be a given strict t-conorm with an additive generator $g, g(\frac{1}{2}) = 1$. For a given $e \in]0,1[$, $k \in]0,+\infty[$, we define an associative compensatory operator

$$C(T,S,e,k) = C : [0,1]^2 \setminus \{(0,1),(1,0)\} \to [0,1]$$

by

$$C(x,y) = h^{-1}(h(x) + h(y)),$$

where $h : [0,1] \to [-\infty,+\infty]$ is a strictly increasing bijection such that

$$h(x) = \begin{cases} kf(\frac{x}{e}) & \text{if } x \in [0,e] \\ g(\frac{x-e}{1-e}) & \text{if } x \in]e,1]. \end{cases}$$

Note that on the square $[0,e]^2$, C coincides with the t-norm $T_e = (< 0,e,T >)$ (the ordinal sum, see [Klement et al. 2000a]). On the square $[e,1]^2$, C coincide with the t-conorm $S_e = (< e,1,S >)$. On the remainder of its domain, it is $T_{\mathbf{M}} < C < S_{\mathbf{M}}$, and note that small values of parameter k increase the values of C (limitedly to $S_{\mathbf{M}}$) while the large values of k decrease the values of C (limitedly to $T_{\mathbf{M}}$). On whole its domain, C is bounded, $T_e \leq C \leq S_e$. The conjunctive and disjunctive extensions of associative compensatory operators to the whole unit square (defining $C(0,1) = C(1,0) = 0$ in the conjunctive case, and $C(0,1) = C(1,0) = 1$ in the disjunctive case) are another examples of Yager and Rybalov's uni-norms, see [Klement et al. 2000a]. Note that the above mentioned ordinal sum-like construction of compensatory operators may be applied to arbitrary pair of continuous Archimedean t-norm and t-conorm (with possible loss of associativity).

There is a special interest for further investigation of uni-norms (specially the problem of general characterization) and compensatory operators, specially associative compensatory operators, which are independent of grouping the inputs in engineering design. Special attention will be taken on aggregation operators with additive generator f (associative and non-associative cases), see [Kolesárová, Komorniková 1999], and the analogous representations as were obtained for the special case of the weighted quasi-arithmetical mean.

3.3. Integral based aggregation operators

The basic idea of any integral is to aggregate the values of some function on a given universe (inputs) into a single value (output). The correspondence between the special aggregation operators and the special types of integrals was studied, e.g., in [Grabisch et al. 1995, Fodor and Roubens 1994, Benvenuti, Mesiar 2000].

Let $(m_n)_{n\in\mathbb{N}}$ be a system of fuzzy measures $m_n : \mathcal{P}(X_n) \to [0,1], X_n = \{1,\cdots,n\}, m_n(X_n) = 1$. Then the operator $\mathbf{A} : \cup_{n\in\mathbb{N}}[0,1]^n \to [0,1]$, defined by

$$\mathbf{A}(x_1,\ldots,x_n) = (C)\int_{X_n} f dm_n,$$

where the right-hand side is a Choquet integral of the function $f : X_n \to [0,1], f(i) = x_i, i = 1,\cdots,n$, with respect to the fuzzy measure m_n, is an aggregation operator.

The class of the Choquet integral based aggregation operators corresponds to the idempotent operators stable under increasing linear transformations on the unit interval, which are commonotone additive, i.e., $\mathbf{A}(\varphi(x_1),\ldots,\varphi(x_n)) = \varphi(\mathbf{A}(x_1,\ldots,x_n))$ whenever $\varphi : [0,1] \to [0,1], \varphi(x) = ax + b, a > 0$, and

$$\mathbf{A}(x_1 + y_1,\ldots,x_n + y_n) = \mathbf{A}(x_1,\ldots,x_n) + \mathbf{A}(y_1,\ldots,y_n)$$

whenever $(x_i - x_j)(y_i - y_j) \geq 0$ for all $i,j \in X_n$. Each Choquet integral based aggregation operator $\mathbf{A}$ is continuous. The commutativity of $\mathbf{A}$ depends on the properties of the underlying fuzzy measures and it is equivalent with the property that $m_n(B)$ depends only on n and the cardinality of B. In that latest case we get just the OWA operators (order weighted aggregation operators) of Yager, see [Yager and Filev 1994], including $\mathbf{M}, T_\mathbf{M}, S_\mathbf{M}$, MED.

Example 2.

a) Let $m_n(B) = \frac{\text{card } B}{n}, n \in \mathbb{N}, B \subset X_n$. The corresponding aggregation operator is the arithmetic mean $\mathbf{M}$.
b) Let m_n be a probability measure on $X_n, m_n(i) = w_{i,n}$. The corresponding aggregation operator is the weighted arithmetic mean $W, W(x_1,\ldots,x_n) = \sum_{i=1}^{n} w_{i,n}x_i$. W is commutative only if $w_{i,n} = \frac{1}{n}, i \in X_n$, i.e., when $W = \mathbf{M}$.
c) Let

$$m_n(B) = \frac{(\text{card } B + 1)\text{card } B}{(n+1)n}, n \in \mathbb{N}, B \subset X_n.$$

The corresponding aggregation operator $\mathbf{A}$ is an OWA-operator,

$$\mathbf{A}(x_1, \ldots, x_n) = \sum_{i=1}^{n} \frac{2i}{(n+1)n} y_i,$$

where $(y_1, \ldots, y_n)$ is a non-decreasing permutation of $(x_1, \ldots, x_n)$.

We can generalize the Choquet integral approach by means of so called Choquet-like integrals [Klement et al. 2000a, Mesiar 1995, Pap 1995]. Recall that we distinguish two types of Choquet-like integral. In the first case, the integral is based on a generator $g : [0, \infty] \to [0, \infty]$ and for our purposes we should require $g(1) = 1$. However, in this case, the corresponding aggregation operator can be derived by means of the classical Choquet integral and the bijection $g : [0,1] \to [0,1]$. So, e.g., the quadratic mean corresponds to the Choquet-like integral based on the generator $g(x) = x^2$, where the corresponding fuzzy measures m_n on X_n are given by

$$m_n(B) = \sqrt{\frac{\text{card } B}{n}}, \quad n \in \mathbb{N}, B \subset X_n.$$

If the Choquet-like integral is based on the supremum $\vee = \oplus$ and some pseudo-multiplication $\odot$ with a neutral element $e = 1$ (e.g., $\odot = T$ is some t–norm), then

$$\mathbf{A}(x_1, \ldots, x_n) = \bigvee_{i=1}^{n} y_i \odot m_n(B_i),$$

where $(y_1, \ldots, y_n)$ is a non-decreasing permutation of $(x_1, \cdots, x_n)$ and $B_i = \{j \in X_n; x_j \geq y_i\}$. The weighted quasi-arithmetical mean for $f(0) = 0$ can be represented by the Choquet-like (t-conorm) integral [Grabisch et al. 1995, Mesiar 1995] with respect to additive fuzzy measure.

There is need for the investigation of the relation of the aggregation operators with different types of non-additive (fuzzy) integrals. Further investigation on application of the non-additive integrals (Sugeno, Choquet and their generalizations) in the decision (subjective, multi–criteria) theory as an aggregation operator. These integrals have the advantage with respect to other aggregation operators that the non–additivity of the considered measures takes into account the interaction between criteria. For wider applications the identification of the non-additive measure is crucial. At this moment there are many different approaches by statistics, neural networks, genetics algorithms, combinatorial optimization, etc. (see 4.2).

4. Some tools for decision making in engineering design

4.1. Global maximum and minimum of a real function

There are some procedures for finding the approximation of the global maximum or minimum of a real function. We present a procedure [Pap and Surla

1997, 2000, Chichinadze 1968, 1969] which works even for non-continuous functions. It is applied on functions with values in $[0,1]$, but it works also for more general cases. We define a function $H_f : [0,1] \to [0,\infty]$ by

$$H_f(\alpha) = m(A_\alpha),$$

where $m(A_\alpha)$ is the Lebesgue measure ([Halmos 1950]) of the α-level set $A_\alpha = \{x|\ f(x) \geq \alpha\}$ of the function f. We recall that a subset K of $\mathbb{R}^n$ is convex if for every $t \in [0,1]$ also $tx + (1-t)y$ is in K whenever x and y are in K. Specially for $\mathbb{R}$ the only convex sets are the intervals.

The following theorem [Pap and Surla 2000] gives the important properties of the function H_f.

Theorem.

Let f be a real function defined on $X \subset \mathbb{R}^n$ such that $f \in L_p(X)$ and the set X consists of a family $\{B_t\}_{t\in F}$ of non-empty and non-singleton connected bounded pairwise disjoint subsets B_t of X such that on each B_t the α-cuts of the function f are convex.

Then the function $H_f(\alpha) = m(A_\alpha)$ has the following properties

(i) The function $H_f(\alpha)$ is non-increasing;

(ii) If additionally the function f has the property (NZ): that does not exist $\alpha, \beta \in [0,1], \alpha < \beta$, such that

$$m(\{x|\ x \in \mathbb{R}^n, \alpha \leq f(x) < \beta\}) = 0,$$

then the function H_f is strictly decreasing;

(iii) If additionally to the suppositions in (ii) the function f has the property (CZ): that for each $\alpha \in [0,1]$

$$m(\{x|\ f(x) = \alpha\}) = 0,$$

then the function H_f is continuous.

We remark that there are discontinuous functions which satisfy the condition (NZ) (see Example 1.).

The main property of the function H_f which corresponds to the function f is its non-increasigness. The second crucial property of the function H_f is given in the next property.

Let f be a function with the same properties (first sentence) as in the previous theorem. If $m(\{x|\ f(x) = \alpha_{\max}\}) = 0$, then the zero of the function H_f is the global maximum of f.

We can now conclude in order to find the global maximum of the function f it is sufficient to find the zero of the function H_f. Unfortunately the analytical form of the function H_f, which can serve as a direct tool for finding the height of the function f, is available only for very limited cases, usually for simpler f and lower dimension n. For this reason we have to proceed with approximation and extrapolation of the function H_f.

There are several statistical methods for finding the approximate solution of mathematical problems (Monte-Carlo and related methods, genetic algorithm). In this spirit is also the following procedure [Pap, Surla 1997, 2000].

We suppose that $f \in L_p(X)$ and that the set $X \subset \mathbb{R}^n$ consists of a family $\{B_t\}_{t\in F}$ of non-empty and non-singleton connected bounded and pairwise disjoint subsets B_t of X, where F is finite, and such that on each B_t the α-cuts of the function f are convex, i.e., f has a finite number of local maximums. The value of the function H_f at α can be calculated by the integral

$$H_f(\alpha) = \int_X \mathbf{1}_{A_\alpha}(x_1, x_2, ..., x_n)\, dx_1 dx_2 \ldots dx_n,$$

where $\mathbf{1}_{A_\alpha}$ is the characteristic function of the set A_α. We shall calculate this integral by statistical methods in the following way. Suppose, for the simplicity, that $m(X) = 1$. Consider the probability

$$p_i = \frac{m(A_{\alpha_i})}{m(X)}.$$

Hence $p_i = H_f(\alpha)$. We can find the probability p_i in the usual way. Namely, we can take k statistical experiments in which r_i times occurs the case $f(x) \geq \alpha_i$. We define the statistical probability p_i^{st} in the usual way $p_i^{st} = \frac{r_i}{k}$. This gives us the procedure for finding $m(A_{\alpha_i})$. Namely, taking the values for the variables $x_1, x_2, ..., x_n$ randomly with the same probability for each coordinate we calculate the value $f(x_1, x_2, ..., x_n)$ and compare it with α_i. We repeat this procedure k-times. Then we extract the points $x = (x_1, x_2, ..., x_n)$ for which $f(x) \geq \alpha_i$ and count them by the number r_i. In this way by k statistical experiments we obtain that the number $\frac{r_i}{k}$ gives us the approximate value of $m(A_{\alpha_i})$. In this way we shall consider the statistical function $H^{st}(\alpha) = \frac{r}{k}$ instead of the function $H_f(\alpha)$ (we omit in H^{st} the index f).

This procedure can be used not only for finding the global maximum $\alpha_{\max}$ but also for finding the coordinates $x_1^{\max}, x_2^{\max}, ..., x_n^{\max}$ of this global maximum.

The first example will illustrate that the algorithm works also for a non-continuous function. The second example presents the algorithm for a function defined on $\mathbb{R}^2$. The third example shows that the algorithm works even for functions which are close to fractals.

Example 1. We define the function f on the real interval $[0, 2]$ in the following way $f(x) = -x^2/6 + x/3$ for $0 \leq x \leq 0.5, f(x) = x/6$ for $0.5 < x \leq 1, f(x) = x^2/6 + x/3$ for $1 < x \leq 1.5, f(x) = x/6$ for $1.5 < x \leq 2$. We have obtained the following results in two iterations.

The first iteration. Inputs: the interval for random variables $[0, 2]$; $k = 500$; $l =$

$10; p = 2$. Outputs: the global maximum of f is 0.8330198 for $x = 1.449106$.
The second iteration. Inputs: the interval for random variables $[1.4, 1.6]; k = 600; l = 10; p = 2$. Outputs: the height of f is 0.869598 for $x = 1.49351$.

Example 2. The function f is defined on $[3,7] \times [-1.5, 2.5]$ in the following way

$$f(x,y) = 45^{-1}(-x^2 + 12x - 11 - 10\cos\frac{\pi}{2}x - 8\sin\pi x + \frac{1}{\sqrt{2\pi}}\exp(-\frac{(y-0.5)^2}{2}).$$

Inputs: random variables $x \in [3,7]$ and $y \in [-1.5, 2.5]; k = 1800; l = 10; p = 3$. Outputs: the maximum of f is 0.910305 for $x = 5.62650$ and $y = 0.49596$. The exact value for the maximum is 0.910274. So the error in our calculation is 0.034 .

Example 3. We find the global maximum of the function defined by the Weierstrass non-differentiable function, which is a fractal. Let $\lambda > 1$ and $1 < s < 2$ [Falconer 1990]. We define the function f on the real interval $[0,1]$ in the following way

$$f(x) = \frac{1}{\lambda^{s+1}} \sum_{k=0}^{\infty} \lambda^{(s-2)k} \sin(\lambda^k x)$$

for $x \in [0,1]$. We remark that the Hausdorff dimension of the Weierstrass function is still unknown (it is supposed that it is equal s).
Since we are able to generate with computers only finite sums we restrict ourselves on function of the form

$$f(x) = \frac{1}{\lambda^{s+1}} \sum_{k=0}^{m} \lambda^{(s-2)k} \sin(\lambda^k x),$$

where $m = j$ for which $(\lambda^{(s-2)j} \sin(\lambda^j x))^2 \leq 10^{-12}$, for all $x \in [0,1]$. We take specially $s = 1.3$ and $\lambda = 1.5$. We have obtained the following results in two iterations.
The first iteration. Inputs: the interval for random variables $[0,1]; k = 900; l = 10; p = 2$. Outputs: the height of f is 0.80634 for $x = 0.69036$.
The second iteration. Inputs: the interval for random variables $[0.6, 1]; k = 1000; l = 10; p = 2$. Outputs: the height of f is 0.84476 for $x = 0.80604$.

We have presented a universal method for finding the global maximum of a function defined on $\mathbb{R}^n$. The advantages of the method is applicability on multi-dimensional cases and even on non-continuous cases, when classical gradient methods are not applicable.
The presented procedure (with other existing methods for finding global maximum) can be applied also in other optimization problems. There are possibilities for further modifications of the preceding procedure first of all in the approximation part using better approximation and algorithms on more powerful computers.

4.2. Identification of the fuzzy measure

As the mapping, the fuzzy integral is defined by a set of 2^n (for n elements basic set X) parameters and a t-conorm system. The word "identification" has the origin in the system theory and is preferred to the word "learning", though the algorithms for finding the appropriate fuzzy measure could be the learning samples minimizing certain criterion. Usually, the model identified has the form of a suitable integral as the global evaluation of the criteria in the problems of the decision theory. Unknown measure to be identified can be regarded as the part of the parameter identification.

The contribution of every element of an index set $I = \{1, \ldots, n\}$ is characterized and the known result stated as the introduction to the problem of connecting some kind of interaction among attributes with some kind of non-additivity of the fuzzy measure.

As the intuitive facts, it is accepted that super-additivity between attributes i and j entails a strengthening interaction, support between i and j and that sub-additivity between i and j entails a weakening interaction, destructive effect between i and j. The recent definition of an interaction index by Murofushi, Soneda (see [Grabisch et al. 1995]) can help the formalization of the former intuitive facts.

The environmental evaluation in Japan [Nakamori et al. 1995] is based on the identification according to Murofushi's suggestion when the norm to be minimized gives the form of the convex quadratic programming problem to the problem of the identification. The fuzzy measure coefficients are identified for the subsets of a set with the fixed number of elements and can be represented with the *power set lattice* as the fuzzy measure is monotone. A new algorithm [Grabisch 1995a] is the improvement of the algorithms based on the characteristic of the integral to be affected only by the measure coefficients of the path in the power set lattice (joining sets $\emptyset$ and X) by the modification of the nodes not affected before to get the most *equilibrated lattice* in which the distance between the neighbors should be as equal as possible. By the changed criterion the algorithm is improved as well.

In the '90-ties has begun the communication between the scientists following different approaches in the simulated evolution [Fogel 1994]. For some problems the Sugeno λ-measure (see [Wang, Klir 1992]) is suitable to be identified as the parameters of the model [Knoll et al.1995] and the genetic algorithms [Lee et al. 1995] may be used to identify the measure known by the partial values.

With the construction of the recurrent neural networks by the (generalized acquisition of recurrent links) algorithm [Angeline 1994], changed to model the monotonicity of the fuzzy measure by the introduction of the positive independent variables, called connection variables, the fuzzy measure is identified [Klir et al. 1995] when the model has the form of the Sugeno integral.

When the model is the Sugeno integral, it is in the form of the S-T fuzzy relational equation, where T and S are suitable t-norm and t-conorm [Pedrycz 1993] and may be expressed by Sugeno integral suitable to form the upper level of the neural network, for the nodes representing the subsets of the partial evaluations. Given algorithm need to be changed for the consideration of the model in the general form of *the fuzzy t-conorm integral* ([Murofushi, Sugeno 1991], see [Grabisch et al. 1995]).

The need to encounter observations ordered in time is frequent. *Time series data* [Sugeno and Kwon 1995] are used to determine the structure (input variables and their relationships should be found) and parameters embedded in the structure (which is part of the identification that is the optimization problem).

Fuzzy measures may approximate process with interactions among their inputs, so they model subjective evaluation process. Genetic algorithms are used [Sugeno and Kwon 1995] to select a subset of regression variables. The algorithm, due to Yoneda et al., see [Grabisch et al. 1995], combines the semanticall analysis of the fuzzy measure and the optimization approach. The constraints are the monotonicity constraints, constraints from the training data, and the constraints coming from semanticall considerations: importance of criteria expresses the decision maker's preference or his knowledge and the dependency and support between criteria (interaction) are modeled with suitable parameter values equivalent to the linguistic representation.

A recent result in reducing the number of parameters for a non-additive measure is achieved by introducing a special kind of non-additive measure, so called k-order additive measure, but which is still enough general to cover many applications [Grabisch 1996, 1997, 2000, Kolesárová 1999, Marichal 1998, Mesiar 1999a,b]. Let the basic set X be finite. A set function $m : \mathcal{P}(X) \to [0, \infty]$ for $k \in \mathbb{N}$ is *k-order additive measure* whenever

$$M_m(A) = \sum_{B \subset A} (-1)^{\text{card } (A \setminus B)} m(B) = 0$$

for all $A \subset X$ such that card $(A) > k$. We remark that $M_m : \mathcal{P}(X) \to [-\infty, \infty]$ is the Möbius transform of m, see [Pap 1995].

Computational complexity related to complex systems has one limit of information processing , so called Bremermann's limit [Bremermann 1962, Klir and Folger 1992] which is about 10^{93} bits (counting the estimated age and mass of the Earth). This causes that many problems dealing with decision-making are transcomputational problems. Exponential nature of the number of coefficients of a fuzzy measure (2^n for the set of n elements) even in the case of powerful computers forces us to find solutions in some modified forms, i.e., instead of requiring the best solution we can take a good solution and an approximate solution instead of precise solution. We need a softening of the requirements of the problem of the identification of the fuzzy measure using heuristic methods and/or approximate methods. There is need to develop for powerful computers some procedures which will include fuzzy-neural networks, genetic algorithms [Wang et al. 1997], fuzzy relational equations, optimizations based on the procedure presented in 4.1, the Shapley value [Aumann and Shapley 1974, Grabisch et al. 1995, Pap 1995], k-decomposable measures [Grabisch 1997], hybrid measures [Dubois, Pap, Prade 2000].

Conclusions

We summarize the proposed tasks for investigation of methods for engineering decision making.

1. Develop theoretical background of axiomatic approach to engineering design and develop the applications of the methods of fuzzy systems, specially fuzzy measures and integrals
2. Develop further the decision-making theory using the existing results, procedures, tools
3. Develop engineering convenient aggregation operators and pseudo-analysis in optimization, soft computing, control theory
4. Develop algorithms for finding the approximate of global maximum and minimum
5. Develop algorithms for the identification of fuzzy measures (statistical, genetic algorithms, neural nets, nonlinear programing, etc.)
6. Develop global software package for powerful computers for decision making in the engineering design independent of the area of applications.

References

1. J. Aczél [1966] *Lectures on Functional Equations and their Applications.* New York, Academic Press.
2. P. J. Angeline, G. M. Saunders, J. B. Pollack [1994] An Evolutionary Algorithm that Constructs Recurrent Neural Networks, *IEEE Transactions on Neural Networks*, Vol. 5, No. 1, 54-65.
3. F.J. Anscombe, R.J. Aumann [1963] A Definition of Subjective Probability, *Ann. Math. Statist.* **34**, 199–205.
4. E. Antonsson, K. Otto [1995] Imprecision in Engineering Design, *ASME Journal of Mechanical Design*, jun., 25-32.
5. K.J. Arrow [1963] *Social Choice and Individual Values*, 2. ed., Yale University Press, New Haven, CT.
6. R.J. Aumann, L.S. Shapley [1974] *Values of Non-Atomic Games*, Princeton University Press, Princeton, New Jersey.
7. F. Baccelli, G. Cohen, G.J. Olsder, J.P. Quadrat [1992] *Synchronization and Linearity: an Algebra for Discrete Event Systems.* New York, John-Wiley and Sons.
8. R. E. Bellman, S.E. Dreyfus [1962] *Applied Dynamic Programming*, Princeton University Press, Princeton.
9. P. Benvenuti, R. Mesiar [2000] Integrals with Respect to a General Fuzzy Measure, Chapter in Monograph (Eds. M. Grabisch, T. Murofushi, M. Sugeno) *Fuzzy Measures and Integrals-Theory and Applications*, volume 40 of *Studies in Fuzziness*, Physica-Verlag, Heidelberg, 205-232.

10. P. Benvenuti, R. Mesiar (to appear) Pseudo-additive measures and triangular norm-based conditioning, *Annals of Math. and Artificial Intelligence.*
11. F. Blume (to appear) Ergodic theory, (Ed. E.Pap) Handbook of Measure Theory, North-Holland.
12. H.J. Bremermann [1962] Optimization through evolution and recombination, In "Self Organizing Systems" (Eds. M.C.Yovits et al.), Spartan Books, Washington, 93-106.
13. L.M. de Campos, M. J. Bolanos [1992] Characterization and comparison of Sugeno and Choquet integrals, *Fuzzy Sets and Systems* **52**, 61–67.
14. D. Candeloro, [1987] Radon-Nikodym derivatives and conditioning in fuzzy measure theory, *Stochastica* **11**, 107-120.
15. A. Chateauneuf [1991] On the use of capacities in modeling uncertainty aversion and risk aversion, *J. Math. Econom.* **20**, 343–369.
16. J. Chen, K. Otto [1995] A Tool for Imprecise Calculations in Engineering Design, Proc. of the Fourth IEEE, Vol. I, International Conference on Fuzzy Systems,389-394.
17. V.K. Chichinadze [1969] The Ψ-transform for solving linear and non-linear programming problems, *Automatica* **5**, 347-355.
18. V.K. Chichinadze [1968] Some new approach for solving the problems of mathematical programming, In: Proceedings of IFAC Symposium on Optimal Systems Planing, The IEEE, Inc, New York.
19. G. Choquet [1953] Theory of capacities, *Ann. Inst. Fourier (Grenoble)* **5**, 131–295.
20. R. A. Cuninghame-Green [1979] *Minimax algebra.* Lecture Notes in Economics and Math. Systems 166, Berlin, Heidelberg, New–York, Springer–Verlag.
21. D. Denneberg [1994] *Non–Additive Measure and Integral*, Kluwer Academic.
22. G. Devedžić, E. Pap [1997] Linguistic decision making model for machine tools evaluation and ranking by Choquet integral, Proc. of VII IFSA World Congress, Vol. I , Prague, 503-508.
23. E.Pap, G. Devedžić [1999] Multicriteria-multistages linguistic evaluation and ranking of machine tools, *Fuzzy Sets and Systems* **102**, 451-461.
24. A. DiNola, S.Sessa, S.,Pedricz, W., Sanchez, E. [1989] *Fuzzy Relation Equations and Their Applications to Knowledge Engineering.* Dordrecht, Kluwer Academic Publishers.
25. J.A. Drakopoulos [1995] Probabilities, possibilities, and fuzzy sets, *Fuzzy Sets and Systems* **75**, 1-15.
26. D.Dubois, J.C. Fodor, H. Prade, M. Roubens [1996] Aggregation of decomposable measures with applications to utility theory. *Theory and Decision* **41**, 59-95.
27. D. Dubois, L. Godo, H. Prade, A. Zapico [1998] Making decision in a qualitative setting: from decision under uncertainty to case-based decision.(Eds. A.G. Cohn, L. Schubert, S.C. Shapiro) Proceedings of the Sixth International Conference "Principles of Knowledge Representation and Reasoning" (KR ' 98), Morgan Kaufman Publishers, Inc, San Francisco, 594-605.
28. D. Dubois , E. Pap, H. Prade [2000] Hybrid probabilistic-possibilistic mixtures and utility functions, (Eds. B. De Baets, J. Fodor, P. Perny) Preferences and Decisions under Incomplete Knowledge , Springer-Verlag, 51-73.
29. D. Dubois, H. Prade [1995] Possibility theory as a basis for qualitative decision theory. In Proc. of the 14th Int. Joint Conf. on Artificial Intelligence (IJCAI '95), Montreal, 1924-1930.

30. D. Dubois, H. Prade [1998] Possibility theory: qualitative and quantitative aspects. The Handbook of Defeasible Reasoning and Uncertainty Management Systems Vol I (Ed. Ph. Smets), Kluwer, Dordrecht, 169-226.
31. D. Dubois, H. Prade (to appear) Qualitative possibility functions and integrals, (Ed. E.Pap) Handbook of Measure Theory, North-Holland.
32. D. Dubois, H. Prade [1982] A class of fuzzy measures based on triangular norms, *Internat. J. Gen. System* **8**, 43–61.
33. D. Dubois, H. Prade H. [1980] *Fuzzy sets and systems: Theory and applications*. New–York, Academic Press.
34. H. Dyckhoff, W. Pedrycz [1984] Generalized means as model for compensative connectives, *Fuzzy Sets and Systems* **14**, 143– 154.
35. P. Fishburn [1970] *Utility Theory for Decision Making*, J. Wiley and Sons.
36. J. Fodor, M. Roubens [1994] *Fuzzy Preference Modeling and Multi–Criteria Decision Aid*, Norwell, MA: Kluwer Academic Publisher.
37. K. Falconer [1990] *Fractal Geometry*, John Wiley and Sons, New York.
38. D. B. Fogel [1994] An Introduction to Simulated Evolutionary Optimization, *IEEE Transactions on Neural Networks*, Vol. 5, No. 1, 3-14.
39. J.S. Golan [1992] *The Theory of Semirings with Applications in Mathematics and Theoretical Computer Sciences*. New York, Pitman Monog. and Surveys in Pure and Appl. Maths. 54, Longman.
40. S. Gottwald [1986] Characterizations of the solvability of fuzzy equations. *Elektron. Informationsverarb, Kybernetik* EIK **22**, 67–91.
41. M. Grabisch [1995a] A New Algorithm for Identifying Fuzzy Measures and its Application to Pattern Recognition, *Proc. IEEE*, 145-150.
42. M. Grabisch [1995b] On Equivalence Classes of Fuzzy Connectives - The Case of Fuzzy Integrals, *IEEE Transactions on Fuzzy Systems*, Vol. 3, No. 1, 96-109.
43. M. Grabisch [1997a] On the Representation of k–Decomposable Measures, Proc. of VII IFSA World Congress, Vol. I , Prague, 478-483.
44. M. Grabisch [1997b] k-order additive discrete fuzzy measures and their representation, *Fuzzy Sets and systems* **92**, 167-189.
45. M. Grabisch [2000] The Interaction and Möbius Representation of Fuzzy measures on Finite Spaces, k-Additive Measures, Chapter in Monograph (Eds. M. Grabisch, T. Murofushi, M. Sugeno) *Fuzzy Measures and Integrals-Theory and Applications*, volume 40 of *Studies in Fuzziness*, Physica-Verlag, Heidelberg, 70-93.
46. M. Grabisch, H. T. Nguyen, E. A. Walker [1995] *Fundamentals of Uncertainity Calculi with Application to Fuzzy Inference*, Kluwer Academic Publishers, Dordrecht/Boston/London.
47. M. Grabisch, M. Schmitt [1995] Mathematical Morphology, Order Filters and Fuzzy Logic, Proc. IEEE, 2103-2108.
48. H. Imaoka [1995] A Proposal of Opposite-Sugeno Integral and a Uniform Expression of Fuzzy Integrals, Proc. IEEE, 583-590.
49. K. Ishii, M. Sugeno [1985] A Model of Human Evaluation Process Using Fuzzy Measure, *Int. J. Man–Machine Studies* **22**, 19–38.
50. G. A. Hazelrigg [1996] The Implications of Arrow's Impossibility Theorem and Approaches to Optimal Engineering Design, *J. of Mechanical Design*, June, Vol. 118, 161-164.
51. G. A. Hazelrigg [1997a] A Framework dor Decision-Based Engineering Design, Technical report.

52. G. A. Hazelrigg [1997b] An Axiomatic framework for Engineering Design, Technical report.
53. Y.-L. Hsu, Y.-F. Lin, T.-L. Sun [1995] Engineering Design Optimization as a Fuzzy Control Process, Proc. of the Fourth IEEE International Conference on Fuzzy Systems, Vol. IV, 2001-2008.
54. R.L. Keeny [1974] Multiplicative Utility Functions, *Operation Research* 1, 23-25.
55. E. P. Klement, R. Mesiar, E. Pap [1996] On the Relationship of Associative Compensatory Operators to Triangular Norms and Conorms, *Internat. J. Uncertain., Fuzziness and Knowledge Based Systems* **4**, 129-144.
56. E.P. Klement, R. Mesiar, E. Pap [2000a] *Triangular Norms*, Kluwer Academic Publishers, Dordrecht.
57. E.P. Klement, R. Mesiar, E. Pap [2000b] Integration with respect to decomposable measures, based on a conditionally distributive semiring on the unit interval. *Internat. J. Uncertain. Fuzziness Knowledge-Based Systems* 8, 701-717.
58. E.P. Klement, S. Weber [1991] Generalized measures. *Fuzzy Sets and Systems* **40** , 375-394.
59. G.J. Klir, T.A. Folger [1992] *Fuzzy Sets, Uncertainty, and Information*, Prentice-Hall International, Inc.
60. G. J. Klir, B. Yuan, J. F. Swan-Stone [1995] Constructing Fuzzy Measures from Given Data, VI IFSA World Congress, Sao Paulo, Brazil, Vol. I, 61-64.
61. A. Knoll, A. Wolfram [1995] Searching Correspondences in Colour Stereo Images - An Approach Using the Fuzzy Integral, Proc. IEEE, 2033-2037.
62. R. Knosala, W. Pedrycz [1992] Evaluation of design alternatives in mechanical engineering, *Fuzzy Sets and Systems* 47, 269-280.
63. A. Kolesárová [1999] Aggregation of k-oder maxitive fuzzy measures, *Internat. J. Uncertain. Fuzziness Knowledge-Based Systems* **6**, 569-576.
64. A. Kolesárová, M. Komorniková [1999] Triangular norm-based iterative compensatory operators, *Fuzzy Sets and Systems* 104, 109-120.
65. W. Kuich [1986] *Semirings, Automata, Languages.* Berlin, Springer-Verlag.
66. K.-M. Lee, H. Leekwang [1995] Identification of λ-fuzzy Measure by Genetic Algorithms, *Fuzzy Sets and Systems* **75**, 301-309.
67. P.L. Lions [1982] *Generalized solutions of Hamilton-Jacobi equations.* London, Pitman.
68. R.D. Luce, H. Raiffa [1957] *Games and Decisions*, John Wiley & Sons, Inc, New York.
69. M.K. Luhandjula [1982] Compensatory operators in fuzzy linear programming with multiple objectives, *Fuzzy Sets and Systems* **32**, 45–79.
70. J.L. Marichal, [1998] *Aggregations operators for multicriteria decision aid*, PhD. thesis, University of Liege.
71. A. Marková-Stupňanová [1999] A note to the idempotent functions with respect to pseudo-convolution, *Fuzzy Sets and Systems* **102**, 417-421.
72. V.P. Maslov, S.N. Samborskij (eds.) [1992] *Idempotent Analysis.* Advances in Soviet Mathematics 13, Providence, Rhode Island, Amer. Math. Soc..
73. R. Mesiar [1995] Choquet–like integrals. *J. Math. Anal. Appl.* **194** , 477-488.
74. R. Mesiar [1999a] Generalizations of k-oder additive discrete fuzzy measures, *Fuzzy Sets and Systems* **102**, 423-428.
75. R. Mesiar [1999b] k-order additive measures, *Internat. J. Uncertain. Fuzziness Knowledge-Based Systems* **6**, 561-568.

76. R. Mesiar, B.DeBaets [2000] New construction methods for aggregation operators, Proc. IPMU'2000, Madrid, 701-707
77. R. Mesiar, M. Komorniková [1997] Aggregation operators. XI Conference on Applied mathematics (Eds. D. Herceg, K. Surla), Institute of Mathematics, Novi Sad, 193-211.
78. R. Mesiar, J. Šipoš [1993] Radon-Nikodym-like theorem for fuzzy measures, *J.Fuzzy Math.* **1**, No. 4, 193-199.
79. K. Muller, M. Thargen [1994] Applications of Fuzzy Hierarchies and Fuzzy MADM Methods to Innovative System Design, Proc. of the Third IEEE International Conference on Fuzzy Systems, Vol. I, 364-367.
80. T. Murofushi, M. Sugeno [1991] Fuzzy t–conorm integrals with respect to fuzzy measures: generalization of Sugeno integral and Choquet integral, *Fuzzy Sets and Systems* **42**, 57–71.
81. T. Murofushi, M. Sugeno [1993] Some quantities represented by the Choquet integral, *Fuzzy Sets and Systems* **56**, 229–235.
82. T. Murofushi, M. Sugeno, M. Machida [1994] Nonmonotonic fuzzy measures and the Choquet integral, *Fuzzy Sets and Systems* **64**, 73–86.
83. Y. Nakamori, N. Iwamoto [1995] Analysis of Environmental Evaluation Structure by the Choquet Integral Model, Proc. IEEE, 695–700.
84. D. Nauck, F. Klawonn, R. Kruse [1994] *Neuronale Netze und Fuzzy- Systeme.* Braunschweig/Wiesbaden, Vieweg.
85. R. de Neufville [1990] *Applied System Analysis*, McGraw-Hill, Inc., Singapore.
86. A.K. Noor, S.L. Venneri, J.M. Housner, J.C. Peterson [1997] A virtual environment for intelligent design, *Aerospace America*, april, 29-35.
87. T. Onisawa, M. Sugeno, Y. Nishiwaki, H. Kawai, Y. Harima [1986] Fuzzy Measure Analysis of Public Attitude Towards the Use of Nuclear Energy, *Fuzzy Sets and Systems* 20, 259–289.
88. A. Osycska [1984] *Multi-Criterion Optimization in Engineering with Fortran Examples*, Halstad Press, New York.
89. K. Otto, E. Antonsson [1991] Trade-Off Strategies in Engineering Design, *Research in Engineering Design* **3** (**2**), 87-104.
90. E. Pap [1991] On non–additive set functions, *Atti. Sem. Mat. Fis. Univ. Modena* **39**, 345–360.
91. E. Pap [1993] Solution of nonlinear differential and difference equations. Proc. of EUFIT '93, Vol. 1 (1993), 498-504.
92. E. Pap [1994a] The range of null–additive fuzzy and non–fuzzy measures, *Fuzzy Sets and Systems* **65**, 105–115.
93. E. Pap [1994b] Nonlinear difference equations and neural nets. *BUSEFAL* **60**, 7-14.
94. E. Pap [1995] *Null–Additive Set Functions*, Kluwer Academic Publishers, Dordrecht/Boston/London.
95. E. Pap [1997] Pseudo-analysis as a mathematical base for soft computing, *Soft Computing* **1**, 61-68.
96. E. Pap [2000] Pseudo-convolution and its applications, Chapter in Monograph (Eds. M. Grabisch, T. Murofushi, M. Sugeno) *Fuzzy Measures and Integrals-Theory and Applications*, volume 40 of *Studies in Fuzziness*, Physica-Verlag, Heidelberg, 171-204.
97. E. Pap, K. Jegdić [2000] Pseudo-Analysis and Its Applications in Railway Routing, *Fuzzy Sets and Systems* **116**, 103-118.

98. E. Pap, N. Ralević [1998] Pseudo-Laplace Transform, *Nonlinear Analysis* **33**, 533-550.
99. E. Pap, I. Štajner [1997] Pseudo-convolution in the theory of optimization, probabilistic metric spaces, fuzzy numbers, system theory, Proc. of IFSA World Congress, Vol. I , Prague, 491-496.
100. E. Pap, I. Štajner [1999] Generalized pseudo-convolution in the theory of probabilistic metric spaces, information, fuzzy numbers, system theory, *Fuzzy Sets and Systems* **102**, 393-415.
101. E. Pap, D. Surla [1997] A general algorithm for finding the height of the membership function based on the Lebesgue integral, Proc. of IFSA World Congress, Vol. I , Prague, 511-514.
102. E. Pap, D. Surla [2000] Lebesgue measure of α-cuts approach for finding the height of the membership function, *Fuzzy Sets and Systems* **111**, 341-350.
103. E. Pap, D. Vivona [2000] Non-commutative and non-associative pseudo-analysis and its applications on nonlinear partial differential equations,*J. Math. Anal. Appl.* **246**, 390-408.
104. W. Pedrycz [1989] *Fuzzy Control and Fuzzy Systems.* New York, Research Sud. Press, Taunton Wiley.
105. W. Pedrycz [1993] Fuzzy Neural Networks and Neurocomputations, *Fuzzy Sets and Systems* **56**, 1–28.
106. W. Pedrycz [1993] s–t Fuzzy Relational Equations, *Fuzzy Sets and Systems* **59**, 189–195.
107. M. Pincus [1968] A closed form solution of certain programming problems, *Operation Research* 690.
108. H. Prade, C. Testemale [1987] Application of Possibility and Necessity Measures to Documentary Information Retrieval, in: Z. Wang, G.J. Klir [1992] Fuzzy Measure Theory.
109. S. Pugh [1990] *Total Design*, Addison-Wesley, New York.
110. J. Quiggin [1993] *Generalized Expected Utility Theory*, Kluwer Academic.
111. T.L. Saaty [1978] Exploring the interfaces between hierarchies, multiple objectives and fuzzy sets, *Fuzzy Sets and Systems* **1**, 57-68.
112. A.P. Sage [1977] *Methodology for Large-Scale Systems*, McGraw-Hill, Inc., New York.
113. E. Sanchez [1984] Solution of fuzzy equations with extended operations. *Fuzzy Sets and Systems* **12**, 237–248.
114. M.J. Scott, E.K. Antonsson [1995] Aggregation Functions for Engineering Design Trade-offs, 9th International ASME Conference on Design Theory and Methodology, Boston, MA, 389-396.
115. M.J. Scott, E.K. Antonsson [1997] Aggregation functions for engineering design trade-offs, *Fuzzy Sets and Systems* (in print).
116. G.A. Shafer [1976] *A Mathematical Theory of Evidence*, Princenton Univ. Press, Princenton, N.J.
117. T. M. Strat [1990] Decision analysis using belief functions, in: Z. Wang and G.J. Klir [1992] Fuzzy Measure Theory.
118. M. Sugeno [1974] Theory of fuzzy integrals and its applications. Ph. D. Thesis, Tokyo Institute of Technology.
119. M. Sugeno, T. Murofushi [1987] Pseudo-additive measures and integrals. *J. Math. Anal. Appl.* **122**, 197–222.
120. M. Sugeno, S.–H. Kwon [1995] A New Approach to Time Series Modeling With Fuzzy Measures and the Choquet Integral, Proc IEEE.

121. G. Taguchi [1986] *Introduction to Quality Engineering*, Asian Productivity Organization, Unipub, White Plains, NY.
122. H. Tahani, J. M. Keller, Information fusion in computer vision using the fuzzy integral, in: Z. Wang, and G.J. Klir [1992] Fuzzy Measure Theory.
123. T. Terano, K. Asai, M. Sugeno [1992] *Fuzzy Systems Theory and Its Applications*, Academic Press, Inc.
124. D. L. Thurston [1991] A Formal Method for Subjective Design Evaluation with Multiple Attributes, *Research in Engineering Design* **3 (2)**, 105-122.
125. M. Tribus [1969] *Rational Descriptions, Decisions and Designs*, Pergamon Press, Elmsford, New York.
126. J. von Neumann, O. Morgenstern [1953] *The Theory of Games and Economic Behavior*, Third Edition, Princeton University Press, Princeton, NJ.
127. J. Wang [1995] Determining Fuzzy Measures by Using Statistics and Neural Networks, VI IFSA World Congress, Sao Paulo, Brazil, Vol. II, 519–521.
128. Z. Wang, G. J. Klir [1992] *Fuzzy Measure Theory*, Plenum Press, New York and London.
129. Z. Wang, K. Leung, J. Wang [1997] Genetic algorithms used for determining nonadditive set functions in information fusion, Proc. of IFSA World Congress, Vol. I , Prague, 518-521.
130. S. Weber [1984] $\perp$-decomposable measures and integrals for Archimedean t-conorm. *J. Math. Anal. Appl.* **101** , 114–138.
131. D.E. Whithey [1988] Manufacturing by Design, *Harvard Business Review* **66**, **4**, 83-91.
132. R.R. Yager [1991] Connectives and quantifiers in fuzzy sets, *Fuzzy Sets and Systems* 40, 39–75.
133. R.R. Yager, D. P. Filev [1994] *Essentials of fuzzy modelling and control*, John Willey, New York.
134. L.A. Zadeh [1978] Fuzzy sets as a basis for a theory of possibility, *Fuzzy Sets and Systems* **1**, 3–28.
135. H.-J. Zimmermann [1991] *Fuzzy Set Theory and Its Applications*, Kluwer, Boston.
136. H.-J. Zimmermann, H.-J. Sebastian [1993] Optimization and Fuzziness in Problems of Design and Configuration, Proc. of the Second IEEE International Conference on Fuzzy Systems, Pergamon Press, 1237-1240.
137. H.-J. Zimmermann, H.-J. Sebastian [1994] Fuzzy Design- Integration of Fuzzy Theory with Knowledge-Based System Design, Proc. of the Third IEEE International Conference on Fuzzy Systems, Vol. I, 352-357.
138. H.-J. Zimmermann, H.-J. Sebastian [1995] Intelligent System Design Support by Fuzzy-Multi-Criteria Decision Making and/or Evolutionary Algorithms, Proc. of the Fourth IEEE, Vol. I, 367-374. International Conference on Fuzzy Systems,
139. H.-J. Zimmermann, P.Zysno [1980] Latent Connectives in Human Decision Making, *Fuzzy Sets and Systems* **4**, 37–51.
140. H.-J. Zimmermann, P.Zysno [1983] Decisions and evolutions by hierarchical aggregation of information, *Fuzzy Sets and Systems* **10**, 243–260.

Aggregation of Interacting Criteria by Means of the Discrete Choquet Integral

Jean-Luc Marichal

Department of Mathematics
348 TMCB
Brigham Young University
Provo, UT 84602
U.S.A.
E-mail: marichal@math.byu.edu

Abstract. The most often used operator to aggregate criteria in decision making problems is the classical weighted arithmetic mean. In many problems however, the criteria considered interact, and a substitute to the weighted arithmetic mean has to be adopted. Under rather natural conditions, the discrete Choquet integral is proved to be an adequate aggregation operator that extends the weighted arithmetic mean by the taking into consideration of the interaction among criteria. The axiomatic that supports the Choquet integral is presented and some subfamilies are studied.

Index Terms: multicriteria decision making; interacting criteria; Choquet integral.

1 Introduction

Let us consider a finite set of *alternatives* $A = \{a, b, c, \ldots\}$ and a finite set of *criteria* $N = \{1, \ldots, n\}$ in a multicriteria decision making problem. Each alternative $a \in A$ is associated with a *profile* $x^a = (x_1^a, \ldots, x_n^a) \in \mathbb{R}^n$, where, for any $i \in N$, x_i^a represents the partial score of a related to criterion i. We assume that all the partial scores are defined according to the same interval scale, that is, they are defined up to the same positive linear transformation.

From the profile of any alternative a, one can compute a global score $M(x^a)$ by means of an aggregation operator $M : \mathbb{R}^n \to \mathbb{R}$ which takes into account the weights of the criteria. Once the global scores are computed, they can be used to rank the alternatives or select an alternative that best satisfies the given criteria. For instance the optimal alternative $a^* \in A$ could be selected such that

$$M(x^{a^*}) = \max_{a \in A} M(x^a).$$

Until recently, the most often used aggregation operators were the weighted arithmetic means, that is, operators of the form

$$M_\omega(x) = \sum_{i=1}^{n} \omega_i \, x_i \, ,$$

with $\sum_i \omega_i = 1$ and $\omega_i \geq 0$ for all $i \in N$. However, since these operators are not able to model in any understandable way an interaction among criteria, they can be used only in the presence of independent criteria. They are not appropriate for the aggregation of interacting criteria.

In order to have a flexible representation of complex interaction phenomena between criteria (e.g. positive or negative synergy between some criteria), it is useful to substitute to the weight vector ω a non-additive set function on N allowing to define a weight not only on each criterion, but also on each subset of criteria. For this purpose the concept of *fuzzy measure* [29] has been introduced.

Now, a suitable aggregation operator, which generalizes the weighted arithmetic mean, is the discrete Choquet integral (see Definition 2 below), whose use in multicriteria decision making was proposed by many authors (see e.g. [7,11] and the references therein). This integral, which is constructed from the concept of fuzzy measure, is able to take into account the interaction existing among criteria. When criteria are independent, it identifies with the weighted arithmetic mean.

The main aim of this paper is to present the Choquet integral as an appropriate extension to the weighted arithmetic mean for the aggregation of criteria. This operator offers indeed a large flexibility while keeping in some sense a linear form. Although its definition is not very intuitive, we will see that the Choquet integral can be characterized axiomatically by means of rather natural properties.

The outline of this paper is as follows. In Section 2 we present the Choquet integral and we propose an axiomatic characterization. In Section 3 we introduce the importance and interaction indices which enable us to interpret the behavior of aggregation. Finally, in Section 4 we investigate particular Choquet integrals such as the weighted arithmetic mean and the ordered weighted averaging operator.

In order to avoid a heavy notation, cardinality of subsets $S, T, \ldots$ will be denoted whenever possible by the corresponding lower case letters $s, t, \ldots$, otherwise by the standard notation $|S|, |T|, \ldots$. Moreover, we will often omit braces for singletons, e.g. writing $a(i)$, $N \setminus i$ instead of $a(\{i\})$, $N \setminus \{i\}$. Also, for pairs, we will often write ij instead of $\{i, j\}$, as for example $a(ij)$.

For any subset $S \subseteq N$, e_S will denote the characteristic vector of S in $\{0,1\}^n$, i.e., the vector of $\{0,1\}^n$ whose ith component is 1 if and only if $i \in S$.

Finally, $\wedge$ and $\vee$ will denote the minimum and maximum operations, respectively.

2 The Discrete Choquet Integral

In this section we give the definition of the Choquet integral in the discrete case. We also present an axiomatic characterization that motivates the use of this operator in applications.

2.1 The use of the fuzzy measures

It is known that the fuzzy measures are able to model the dependence between criteria in many situations, whatever the nature of the dependence [18]. In fact, they were proposed by Sugeno in 1974 [29] to generalize additive measures. It seems widely accepted that additivity is not suitable as a required property of set functions in many real situations, due to the lack of additivity in many facets of human reasoning. To be able to express human subjectivity, Sugeno proposed to replace the additivity property by a weaker one: *monotonicity*, and he called these non-additive monotonic measures *fuzzy measures*. Let us recall the definition in the discrete case.

Definition 1. A *fuzzy measure* on N is a set function $v : 2^N \to [0,1]$ satisfying the following conditions:
i) $v(\emptyset) = 0, v(N) = 1$,
ii) $S \subseteq T \Rightarrow v(S) \leq v(T)$.

The set of all fuzzy measures on N will be denoted by $\mathcal{F}_N$ as we continue.

For any $S \subseteq N$, $v(S)$ can be interpreted as the weight or the degree of importance of the combination S of criteria, or better, its power to make the decision alone (without the remaining criteria). Thus, in addition to the usual weights on criteria taken separately, weights on any combination of criteria are also defined. Monotonicity then means that adding a new element to a combination cannot decrease its importance. Obviously $v(N)$ has the maximal value, being one by convention.

We will assume that the weights are numerical values defined on a cardinal scale. In particular, expressions like $v(S) + v(T)$ or $v(T \cup i) - v(T)$ can be interpreted.

A fuzzy measure $v \in \mathcal{F}_N$ is said to be *additive* if $v(S \cup T) = v(S) + v(T)$ whenever $S \cap T = \emptyset$. In that case it suffices to define the n coefficients (weights) $v(1), \ldots, v(n)$ to define the measure entirely.

2.2 Definition and alternative representation

The concept of Choquet integral was first introduced in capacity theory [3]. Its use as a (fuzzy) integral with respect to a fuzzy measure was then proposed by Höhle [13] and rediscovered later by Murofushi and Sugeno [21,22], see also [4]. Since this integral is viewed here as an n-place aggregation operator, we will adopt a connective-like notation instead of the usual integral form, and the integrand will be a set of n values $x_1, \ldots, x_n$ of $\mathbb{R}$.

Definition 2. Let $v \in \mathcal{F}_N$. The *Choquet integral* of $x : N \to \mathbb{R}$ with respect to v is defined by

$$\mathcal{C}_v(x) := \sum_{i=1}^{n} x_{(i)} \, [v(A_{(i)}) - v(A_{(i+1)})], \tag{1}$$

where $(\cdot)$ indicates a permutation on N such that $x_{(1)} \leq \ldots \leq x_{(n)}$. Also $A_{(i)} = \{(i), \ldots, (n)\}$, and $A_{(n+1)} = \emptyset$.

For instance, if $x_3 \leq x_1 \leq x_2$, we have

$$\begin{aligned} \mathcal{C}_v(x_1, x_2, x_3) = \; & x_3 \, [v(3,1,2) - v(1,2)] \\ & + x_1 \, [v(1,2) - v(2)] \\ & + x_2 \, v(2). \end{aligned}$$

Thus the discrete Choquet integral is a linear expression up to a reordering of the elements. It is closely related to the discrete Lebesgue integral (weighted arithmetic mean), since both coincide when the measure is additive:

$$\mathcal{C}_v(x) = \sum_{i=1}^{n} v(i) \, x_i, \qquad x \in \mathbb{R}^n.$$

In this sense, the Choquet integral is a generalization of the Lebesgue integral.

In the Appendix we establish a connection between the Choquet integral and the Lovász extension of a pseudo-Boolean function, a concept used in combinatorial optimization [28]. Such a connection enables us to have a geometrical interpretation of the graph of the Choquet integral.

Now, the Möbius transform of a given fuzzy measure $v \in \mathcal{F}_N$ is a set function $a : 2^N \to \mathbb{R}$ defined by

$$a(S) = \sum_{T \subseteq S} (-1)^{s-t} \, v(T), \qquad S \subseteq N, \tag{2}$$

where $s = |S|$ and $t = |T|$. The transformation is invertible and we have (see e.g. [25])

$$v(S) = \sum_{T \subseteq S} a(T), \qquad S \subseteq N. \tag{3}$$

The Möbius transform often makes it possible to express a function of v in a simpler form. For example, the Choquet integral is written as (see Appendix)

$$\mathcal{C}_v(x) = \sum_{T \subseteq N} a(T) \bigwedge_{i \in T} x_i \, , \qquad x \in \mathbb{R}^n.$$

Using Eq. (2), we then have

$$\mathcal{C}_v(x) = \sum_{T \subseteq N} v(T) \, f_T(x), \qquad x \in \mathbb{R}^n,$$

where

$$f_T(x) = \sum_{S \supseteq T} (-1)^{s-t} \bigwedge_{i \in S} x_i \,, \qquad T \subseteq N.$$

2.3 Axiomatic characterization

Although the Choquet integrals have become popular in the field of fuzzy sets and multicriteria decision making, there exist few axiomatic characterizations of this family in the literature. The most representative one is given by Schmeidler [26], using the concept of comonotonic additivity. However, such a characterization is not very attractive in the context of multicriteria decision making.

In [18] the author proposed an axiomatic characterization of the class of all the Choquet integrals with n arguments. The statement is the following.

Theorem 1. *The operators $M_v : \mathbb{R}^n \to \mathbb{R}$ $(v \in \mathcal{F}_N)$ are*

- *linear w.r.t. the fuzzy measures, that is, there exist 2^n functions $f_T : \mathbb{R}^n \to \mathbb{R}$ $(T \subseteq N)$, such that*

$$M_v = \sum_{T \subseteq N} v(T)\, f_T, \qquad v \in F_N.$$

- *non decreasing in each argument,*
- *stable for the admissible positive linear transformations, that is,*

$$M_v(r\,x_1 + s, \ldots, r\,x_n + s) = r\,M_v(x_1, \ldots, x_n) + s$$

for all $x \in \mathbb{R}^n$, $r > 0$, $s \in \mathbb{R}$.

- *properly weighted by v, that is,*

$$M_v(e_S) = v(S), \qquad S \subseteq N,\ v \in F_N,$$

if and only if $M_v = \mathcal{C}_v$ for all $v \in \mathcal{F}_N$.

The axioms presented in the previous characterization are natural enough in the context of multicriteria decision making. The first one is proposed to keep the aggregation model as simple as possible. The second axiom says that increasing a partial score along any criterion cannot decrease the global score. The third axiom only demands that the aggregated value is stable with respect to any change of scale. Finally, assuming that the partial score scale is embedded in $[0,1]$, the fourth axiom suggests that the weight of importance of any subset S of criteria is defined as the global evaluation of the alternative that completely satisfies criteria S and totally fails to satisfy the others.

The fourth axiom is fundamental. It gives an appropriate definition of the weights of subsets of criteria, interpreting them as global evaluation of particular profiles.

3 Behavioral Analysis of Aggregation

Now that we have a tool for a suitable aggregation, an important question arises: How can we interpret the behavior of the Choquet integral or that of its associated fuzzy measure ? Of course the meaning of the values $v(T)$ is not always very clear for the decision maker. These values do not give immediately the global importance of the criteria, nor the degree of interaction among them.

In fact, from a given fuzzy measure, it is possible to derive some indices or parameters that will enable us to interprete the behavior of the fuzzy measure. These indices constitute a kind of *identity card* of the fuzzy measure. In this section, we present two types of indices: importance and interaction. Other indices, such as tolerance and dispersion, were proposed and studied by the author in [15,16].

3.1 Importance indices

The overall importance of a criterion $i \in N$ into a decision problem is not solely determined by the number $v(i)$, but also by all $v(T)$ such that $i \in T$. Indeed, we may have $v(i) = 0$, suggesting that element i is unimportant, but it may happen that for many subsets $T \subseteq N$, $v(T \cup i)$ is much greater than $v(T)$, suggesting that i is actually an important element in the decision.

Shapley [27] proposed in 1953 a definition of a coefficient of importance, based on a set of reasonable axioms. The *importance index* or *Shapley value* of criterion i with respect to v is defined by

$$\phi(v,i) := \sum_{T \subseteq N \setminus i} \frac{(n-t-1)!\,t!}{n!} [v(T \cup i) - v(T)]. \tag{4}$$

The Shapley value is a fundamental concept in game theory expressing a power index. It can be interpreted as a weighted average value of the marginal contribution $v(T \cup i) - v(T)$ of element i alone in all combinations. To make this clearer, it is informative to rewrite the index as follows:

$$\phi(v,i) = \frac{1}{n} \sum_{t=0}^{n-1} \frac{1}{\binom{n-1}{t}} \sum_{\substack{T \subseteq N \setminus i \\ |T|=t}} [v(T \cup i) - v(T)].$$

Thus, the average value of $v(T \cup i) - v(T)$ is computed first over the subsets of same size t and then over all the possible sizes. Consequently, the subsets containing about $n/2$ criteria are the less important in the average, since they are numerous and each criterion is very often involved into them.

The use of the Shapley value in multicriteria decision making was proposed in 1992 by Murofushi [19]. It is worth noting that a basic property of the Shapley value is

$$\sum_{i=1}^{n} \phi(v,i) = 1.$$

Note also that, when v is additive, we clearly have $v(T \cup i) - v(T) = v(i)$ for all $i \in N$ and all $T \subseteq N \setminus i$, and hence

$$\phi(v,i) = v(i), \qquad i \in N. \tag{5}$$

If v is non-additive then some criteria are dependent and (5) generally does not hold anymore. This shows that it is sensible to search for a coefficient of overall importance for each criterion.

In terms of the Möbius representation, the Shapley value takes a very simple form [27]:

$$\phi(v,i) = \sum_{T \ni i} \frac{1}{t}\, a(T).$$

Now, since the Shapley indices are non-negative and sum up to one, it would be interesting to consider the weighted arithmetic mean having these indices as weights. We call this operator the Shapley integral [18].

Definition 3. Let $v \in \mathcal{F}_N$. The *Shapley integral* of $x : N \to \mathbb{R}$ with respect to v is defined by

$$\mathrm{Sh}_v(x) = \sum_{i \in N} \phi(v,i)\, x_i.$$

This operator will play an important role in Theorem 2 (Section 4.3).

Now, the concept of importance index can be easily generalized to subsets of criteria. The *importance index* of subset $S \subseteq N$ with respect to v is defined by

$$\phi(v,S) := \sum_{T \subseteq N \setminus S} \frac{(n-t-s)!\, t!}{(n-s+1)!} [v(T \cup S) - v(T)].$$

This index, introduced by the author in [17] as the *influence index* of criteria S, measures the overall importance of subset S of criteria.

In terms of the Möbius representation, it is given by

$$\phi(v,S) = \sum_{\substack{T \subseteq N \\ T \cap S \neq \emptyset}} \frac{1}{|T \setminus S| + 1}\, a(T).$$

It was shown [17] that this expression is also the average amplitude of the range of C_v that criteria S may control when assigning partial scores in $[0,1]$ to the criteria not in S at random. That is,

$$\begin{aligned}
\phi(v,S) &= \int_0^1 \cdots \int_0^1 \Big[\lim_{\substack{x_j \to 1 \\ j \in S}} C_v(x) - \lim_{\substack{x_j \to 0 \\ j \in S}} C_v(x) \Big]\, \mathrm{d}x_{i_1} \cdots \mathrm{d}x_{i_{n-s}} \\
&= \int_{[0,1]^n} \Big[\lim_{\substack{x_j \to 1 \\ j \in S}} C_v(x) - \lim_{\substack{x_j \to 0 \\ j \in S}} C_v(x) \Big]\, \mathrm{d}x\,,
\end{aligned}$$

where $N \setminus S = \{i_1, \ldots, i_{n-s}\}$.

3.2 Interaction indices

Another interesting concept is that of *interaction* among criteria. We have seen that when the fuzzy measure is not additive then some criteria interact. Of course, it would be interesting to appraise the degree of interaction among any subset of criteria.

Consider first a pair $\{i, j\} \subseteq N$ of criteria. It may happen that $v(i)$ and $v(j)$ are small and at the same time $v(ij)$ is large. Clearly, the number $\phi(v, i)$ merely measures the average contribution that criterion i brings to all possible combinations, but it gives no information on the phenomena of interaction existing among criteria.

Clearly, if the marginal contribution of j to every combination of criteria that contains i is greater (resp. less) than the marginal contribution of j to the same combination when i is excluded, then the expression

$$[v(T \cup ij) - v(T \cup i)] - [v(T \cup j) - v(T)]$$

is positive (resp. negative) for any $T \subseteq N \setminus ij$. We then say that i and j positively (resp. negatively) interact.

This latter expression is called the *marginal interaction* between i and j, conditioned to the presence of elements of the combination $T \subseteq N \setminus ij$. Now, an interaction index for $\{i, j\}$ is given by an average value of this marginal interaction. Murofushi and Soneda [20] proposed in 1993 to calculate this average value as for the Shapley value. Setting

$$(\Delta_{ij}\, v)(T) := v(T \cup ij) - v(T \cup i) - v(T \cup j) + v(T),$$

the *interaction index* of criteria i and j related to v is then defined by

$$I(v, ij) := \sum_{T \subseteq N \setminus ij} \frac{(n-t-2)!\, t!}{(n-1)!} (\Delta_{ij}\, v)(T). \tag{6}$$

It should be mentioned that, historically, the interaction index (6) was first introduced in 1972 by Owen (see Eq. (28) in [24]) in game theory to express a degree of complementarity or competitiveness between elements i and j.

The interaction index among a combination S of criteria was introduced by Grabisch [8] as a natural extension of the case $s = 2$. The *interaction index* of S $(s \geq 2)$ related to v, is defined by

$$I(v, S) := \sum_{T \subseteq N \setminus S} \frac{(n-t-s)!\, t!}{(n-s+1)!} (\Delta_S\, v)(T),$$

where we have set

$$(\Delta_S v)(T) := \sum_{L\subseteq S} (-1)^{s-l} v(L\cup T).$$

In terms of the Möbius representation, this index is written [8]

$$I(v,S) = \sum_{T\supseteq S} \frac{1}{t-s+1}\, a(T), \qquad S\subseteq N. \tag{7}$$

Viewed as a set function, it coincides on singletons with the Shapley value (4).

In terms of the Choquet integral, we have [10, Proposition 4.1]

$$\begin{aligned} I(v,S) &= \int_0^1 \cdots \int_0^1 (\Delta_S\, C_v)(x)\, \mathrm{d}x_{i_1} \cdots \mathrm{d}x_{i_{n-s}} \\ &= \int_{[0,1]^n} (\Delta_S\, C_v)(x)\, \mathrm{d}x\,, \end{aligned}$$

where $N\setminus S = \{i_1,\ldots,i_{n-s}\}$ and

$$(\Delta_S\, C_v)(x) := \sum_{L\subseteq S} (-1)^{s-l} \lim_{\substack{x_i\to 1\\ i\in L}} \lim_{\substack{x_j\to 0\\ j\in S\setminus L}} C_v(x).$$

It was also proved [10, Proposition 5.1] that the transformation (7) is invertible and its inverse is written as

$$a(S) = \sum_{T\supseteq S} B_{t-s}\, I(v,T), \qquad S\subseteq N, \tag{8}$$

where B_n is the nth Bernoulli number, that is the nth element of the numerical sequence $\{B_n\}_{n\in\mathbb{N}}$ defined recursively by

$$\begin{cases} B_0 = 1, \\ \sum\limits_{k=0}^{n} \binom{n+1}{k} B_k = 0, \qquad n\in\mathbb{N}\setminus\{0\}. \end{cases}$$

4 Particular Choquet Integrals

We have seen that the Choquet integral with respect to an additive measure is a weighted arithmetic mean. Of course, they are also other interesting particular Choquet integrals, such as symmetric Choquet integrals. In this section we investigate some of them.

Any vector $\omega\in[0,1]^n$ such that $\sum_i \omega_i = 1$ will be called a *weight vector* as we continue.

4.1 The weighted arithmetic mean

Definition 4. For any weight vector $\omega \in [0,1]^n$, the *weighted arithmetic mean* operator WAM_ω associated to ω is defined by

$$\mathrm{WAM}_\omega(x) = \sum_{i=1}^{n} \omega_i \, x_i.$$

We have seen that WAM_ω is a Choquet integral $\mathcal{C}_v$ with respect to an additive fuzzy measure:

$$v(S) = \sum_{i \in S} \omega_i, \qquad S \subseteq N.$$

The corresponding Möbius representation is given by

$$\begin{cases} a(i) = \omega_i, & \forall i \in N, \\ a(S) = 0, & \forall S \subseteq N, s \geq 2. \end{cases}$$

Conversely, the weights associated to WAM_ω are given by

$$\omega_i = v(i) = a(i), \qquad i \in N.$$

Furthermore the importance and interaction indices of v are written

$$\phi(v,i) = \omega_i \qquad \forall i \in N,$$

$$\phi(v,S) = \sum_{i \in S} \omega_i \ \forall S \subseteq N,$$

$$I(v,S) = 0 \qquad \forall S \subseteq N, s \geq 2.$$

The class of weighted arithmetic mean WAM_ω includes two important special cases, namely:

- the arithmetic mean
$$\mathrm{AM}(x) = \frac{1}{n} \sum_{i=1}^{n} x_i,$$
when $\omega_i = 1/n$ for all $i \in N$. In this case, we have $v(S) = \phi(v,S) = s/n$ for all $S \subseteq N$ and $a(i) = \phi(v,i) = 1/n$ for all $i \in N$.
- the kth projection
$$\mathrm{P}_k(x) = x_k,$$
when $\omega_k = 1$ for some $k \in N$. In this case, we have $v(S) = \phi(v,S) = 1$ if $S \ni k$ and 0 otherwise. Moreover, we have $a(i) = \phi(v,i) = 1$ if $i = k$ and 0 otherwise.

4.2 The ordered weighted averaging

Yager [30] defined in 1988 the ordered weighted averaging operators (OWA) as follows.

Definition 5. For any weight vector $\omega \in [0,1]^n$, the *ordered weighted averaging* operator OWA_ω associated to ω is defined by

$$\mathrm{OWA}_\omega(x) = \sum_{i=1}^{n} \omega_i \, x_{(i)}$$

with the convention that $x_{(1)} \leq \cdots \leq x_{(n)}$.

It was proved in [23, §4] and [6] that the OWA operators coincides with the subclass of discrete Choquet integrals that are symmetric. We give below a rather short proof of this result.

Proposition 1. *Let $v \in \mathcal{F}_N$. Then the following assertions are equivalent.*
i) $v(S) = v(S')$ whenever $|S| = |S'|$.
ii) $\exists \omega \in [0,1]^n$ such that $C_v = \mathrm{OWA}_\omega$.
iii) C_v is a symmetric function.

Proof. *i)* $\Rightarrow$ *ii)* Since v depends only on the cardinality of subsets, the difference

$$v(A_{(i)}) - v(A_{(i+1)})$$

depends only on i. Denoting this difference by ω_i, we have by (1),

$$C_v(x) = \sum_{i=1}^{n} \omega_i \, x_{(i)}, \qquad x \in \mathbb{R}^n,$$

with $\sum_i \omega_i = 1$ and $\omega_i \geq 0$ for all $i \in N$.

ii) $\Rightarrow$ *iii)* Trivial.

iii) $\Rightarrow$ *i)* Fix $S, S' \subseteq N$ such that $|S| = |S'|$. Of course, there exists a permutation π of N such that $S' = \pi(S)$, where $\pi(S) := \{\pi(i) \mid i \in S\}$. Now, by symmetry, we have

$$v(S) = C_v(e_S) = C_v(e_{\pi(S)}) = v(\pi(S)) = v(S'),$$

which completes the proof. □

The fuzzy measure v associated to OWA_ω is given by

$$v(S) = \sum_{i=n-s+1}^{n} \omega_i, \qquad S \subseteq N, \; S \neq \emptyset,$$

and its Möbius representation by [9, Theorem 1]

$$a(S) = \sum_{j=0}^{s-1} \binom{s-1}{j} (-1)^{s-1-j} \, \omega_{n-j}, \qquad S \subseteq N, \; S \neq \emptyset.$$

Conversely, the weights associated to OWA_ω are given by

$$\omega_{n-s} = v(S \cup i) - v(S) = \sum_{T \subseteq S} a(T \cup i), \qquad i \in N,\ S \subseteq N \setminus i. \tag{9}$$

Now, let us show that the Möbius representation can take a very simple form in terms of ω. Consider the difference operator

$$\Delta_k\, x_k := x_{k+1} - x_k$$

for sequences $(x_k)_{k\in\mathbb{N}}$. Denoting by Δ_k^s the sth iterate of Δ_k, it is well-known that we have (cf. Berge [1, Chap. 1, Sect. 8])

$$\Delta_k^s\, x_k = \sum_{j=0}^{s} \binom{s}{j} (-1)^{s-j}\, x_{k+j}\,, \qquad s \in \mathbb{N}. \tag{10}$$

We then have

$$a(S) = \left[\Delta_k^{s-1}\, \omega_{n-k}\right]_{k=0}, \qquad S \subseteq N,\ S \neq \emptyset. \tag{11}$$

According Proposition 1, the use of a symmetric Choquet integral means that all the criteria have the same importance. More precisely, all the combinations of criteria of the same size have the same importance. Notice that this not imply that criteria do not interact. The following result, proved in [9, Theorem 1], shows that symmetric criteria can be dependent. We give below a more concise and precise proof of it.

Proposition 2. *Let $\omega \in [0,1]^n$ be a weight vector and let $v \in \mathcal{F}_N$ such that $C_v = \mathrm{OWA}_\omega$. Then*

$$\phi(v,i) = 1/n, \qquad i \in N, \tag{12}$$

and, for all $S \subseteq N$ such that $s \geq 2$,

$$I(v,S) = \frac{1}{n-s+1} \sum_{j=0}^{s-2} \binom{s-2}{j} (-1)^{s-j} (\omega_{s-j-1} - \omega_{n-j}).$$

Proof. By Proposition 1, we can set $v_t := v(T)$ for all $T \subseteq N$. Next, for any $S \subseteq N$, $S \neq \emptyset$, and any $T \subseteq N \setminus S$, we have, by (9) and (10),

$$\begin{aligned}
\sum_{L \subseteq S} (-1)^{s-l}\, v(L \cup T) &= \sum_{l=0}^{s} \binom{s}{l} (-1)^{s-l}\, v_{l+t} \\
&= \Delta_t^s v_t \\
&= \Delta_t^{s-1} (v_{t+1} - v_t) \\
&= \Delta_t^{s-1}\, \omega_{n-t},
\end{aligned}$$

and hence, by definition of $I(v,S)$,

$$I(v,S) = \sum_{T \subseteq N \setminus S} \frac{(n-t-s)!\,t!}{(n-s+1)!} \Delta_t^{s-1} \omega_{n-t}$$
$$= \frac{1}{n-s+1} \sum_{t=0}^{n-s} \Delta_t^{s-1} \omega_{n-t}.$$

If $S = \{i\}$ for some $i \in N$, we have

$$\phi(v,i) = I(v,i) = \frac{1}{n} \sum_{t=0}^{n-1} \omega_{n-t} = \frac{1}{n}.$$

If $s \geq 2$, we can set $\lambda_t := \Delta_t^{s-2} \omega_{n-t}$ and we have

$$I(v,S) = \frac{1}{n-s+1} \sum_{t=0}^{n-s} \Delta_t \lambda_t$$
$$= \frac{1}{n-s+1} \sum_{t=0}^{n-s} (\lambda_{t+1} - \lambda_t)$$
$$= \frac{1}{n-s+1} (\lambda_{n-s+1} - \lambda_0).$$

Another application of (10) then leads to the result. □

The previous result shows that $I(v,S)$ depends only on s, so that the interaction among criteria is constant on all subsets of the same size. For example, for any pair $\{i,j\} \subseteq N$, we have

$$I(v,ij) = \frac{\omega_1 - \omega_n}{n-1}. \tag{13}$$

Regarding the importance of subsets of criteria, we have the following formula [17, Proposition 6.2]:

$$\phi(v,S) = \sum_{i=1}^{n} \omega_i \, \frac{i \wedge s \wedge (n-i+1) \wedge (n-s+1)}{n-s+1}, \qquad S \subseteq N.$$

The class of ordered weighted averaging operators OWA_ω includes some important special cases, namely:

- the arithmetic mean

$$\mathrm{AM}(x) = \frac{1}{n} \sum_{i=1}^{n} x_i,$$

when $\omega_i = 1/n$ for all $i \in N$.

- the kth order statistic

$$\mathrm{OS}_k(x) = x_{(k)},$$

when $\omega_k = 1$ for some $k \in N$. In this case, we have

$$v(S) = \begin{cases} 1 & \text{if } s \geq n-k+1, \\ 0 & \text{otherwise,} \end{cases}$$

$$a(S) = \begin{cases} (-1)^{s-n+k-1}\binom{s-1}{n-k} & \text{if } s \geq n-k+1, \\ 0 & \text{otherwise.} \end{cases}$$

Moreover, we have

$$\phi(v,S) = \frac{k \wedge s \wedge (n-k+1) \wedge (n-s+1)}{n-s+1} \qquad \forall S \subseteq N,$$

$$I(v,S) = \frac{(-1)^{k+1}}{n-s+1}\left[\binom{s-2}{n-k}(-1)^{n-s} + \binom{s-2}{s-k-1}\right] \forall S \subseteq N, s \geq 2,$$

with the convention that $\binom{p}{q} = 0$ whenever $p < q$ or $q < 0$.
In particular, $I(v, ij) = 0$ whenever $k \neq 1$ and $k \neq n$. Thus there exist non-additive fuzzy measures that present no interaction between pairs of criteria.

- the min operator

$$\min(x) = \min_{i \in N} x_i,$$

when $\omega_1 = 1$. In this case, we have $v(S) = a(S) = 1$ if $S = N$ and 0 otherwise. Moreover,

$$\phi(v,S) = I(v,S) = \frac{1}{n-s+1}, \qquad S \subseteq N, \ S \neq \emptyset.$$

- the max operator

$$\max(x) = \max_{i \in N} x_i,$$

when $\omega_n = 1$. In this case, we have $v(S) = 1$ and $a(S) = (-1)^{s-1}$ for all $S \neq \emptyset$. Moreover,

$$\phi(v,S) = \frac{1}{n-s+1} \text{ and } I(v,S) = \frac{(-1)^{s-1}}{n-s+1}, \qquad S \subseteq N, \ S \neq \emptyset.$$

4.3 The 2-order Choquet integrals

We know that a problem involving n criteria requires 2^n coefficients in $[0,1]$ in order to define the fuzzy measure v on every subset. Of course, a decision maker is not able to give such an amount of information. To overcome this problem, Grabisch [8] proposed to approximate v by a *k-order fuzzy measure*.

Definition 6. $v \in \mathcal{F}_N$ is said to be of order k ($k \leq n$) if its corresponding Möbius representation a fulfils $a(T) = 0$ for all T such that $t > k$, and there exists at least one subset K of k elements such that $a(K) \neq 0$.

This concept allows us to range freely between purely additive measures ($k = 1$), defined by n coefficients, and general fuzzy measures ($k = n$), defined by 2^n coefficients. Indeed, when varying k from 1 to n, we recover all possible fuzzy measures. Looking at Eq. (7)–(8), we see that considering k-order fuzzy measures merely amounts to assuming that the interactions $I(v,T)$ for $t > k$ are zero.

Although the additive model is very simple to handle (only n coefficients are needed), it is restrictive and leads to a very poor modeling tool for application in multicriteria decision making. Grabisch [8] then suggested to consider the 2-order case, which seems to be the most interesting in practical applications, since it permits to model somehow interaction among criteria while remaining very simple. Indeed, only $n + \binom{n}{2} = \frac{n(n+1)}{2}$ coefficients are required to define the fuzzy measure, namely the coefficients

$$\begin{aligned} v(i) &= a(i), \qquad i \in N, \\ v(ij) &= a(i) + a(j) + a(ij), \qquad \{i,j\} \subseteq N. \end{aligned}$$

The other coefficients are then given by

$$\begin{aligned} v(S) &= \sum_{i \in S} a(i) + \sum_{\{i,j\} \subseteq S} a(ij) \\ &= \sum_{\{i,j\} \subseteq S} v(ij) - (s-2) \sum_{i \in S} v(i), \end{aligned}$$

for all $S \subseteq N$ such that $s \geq 2$. Moreover, we have immediately

$$\phi(v,i) = a(i) + \frac{1}{2} \sum_{j \in N \setminus i} a(ij), \qquad i \in N, \tag{14}$$

$$\phi(v,S) = \sum_{i \in S} \phi(v,i), \qquad S \subseteq N, \tag{15}$$

$$I(v,ij) = a(ij), \qquad i,j \in N, \tag{16}$$

and $I(v,S) = 0$ for all $S \subseteq N$ such that $s > 2$.

The Choquet integral with respect to a 2-order fuzzy measure is then written as

$$C_v(x) = \sum_{i \in N} a(i)\, x_i + \sum_{\{i,j\} \subseteq N} a(ij)\, (x_i \wedge x_j).$$

The following result presents this Choquet integral in terms of the interaction indices. This will allow us to interpret the aggregation process more easily.

Theorem 2. *Let $v \in \mathcal{F}_N$ of order 2. Then the best weighted arithmetic mean* WAM_ω *that minimizes*

$$\int_{[0,1]^n} \left[\mathcal{C}_v(x) - \mathrm{WAM}_\omega(x)\right]^2 dx$$

is given by the Shapley integral Sh_v. *Moreover, we have*

$$\mathcal{C}_v(x) = \mathrm{Sh}_v(x) - \frac{1}{2} \sum_{\{i,j\}\subseteq N} I(v,ij)\,|x_i - x_j|, \qquad x \in \mathbb{R}^n. \tag{17}$$

Proof. The first part has been proved in [15, Sect. 7.3].

Now, by using Eq. (8), we have, for all $x \in \mathbb{R}^n$,

$$\begin{aligned}
\mathcal{C}_v(x) &= \sum_{\substack{T\subseteq N\\ t\geq 1}} a(T) \bigwedge_{i\in T} x_i \\
&= \sum_{\substack{T\subseteq N\\ t\geq 1}} \Big[\sum_{\substack{K\supseteq T\\ k\geq 1}} B_{k-t}\, I(v,K)\Big] \bigwedge_{i\in T} x_i \\
&= \sum_{\substack{K\subseteq N\\ k\geq 1}} \Big[\sum_{\substack{T\subseteq K\\ t\geq 1}} B_{k-t} \bigwedge_{i\in T} x_i\Big] I(v,K).
\end{aligned}$$

Since $I(v,K) = 0$ whenever $k \geq 3$, we simply have

$$\begin{aligned}
\mathcal{C}_v(x) &= \mathrm{Sh}_v(x) + \sum_{\{i,j\}\subseteq N} \big[-\frac{1}{2}x_i - \frac{1}{2}x_j + x_i \wedge x_j\big]\, I(v,ij) \\
&= \mathrm{Sh}_v(x) - \frac{1}{2} \sum_{\{i,j\}\subseteq N} I(v,ij) \big[(x_i \vee x_j) - (x_i \wedge x_j)\big],
\end{aligned}$$

as expected. □

Equation (17) shows a decomposition of the Choquet integral $\mathcal{C}_v$ into a linear part and a non-linear part. The linear part, namely the Shapley integral, appears to be a first order approximation of $\mathcal{C}_v$. The non-linear part brings some correction to the Shapley integral that we can interpret in terms of interaction as follows. Consider a pair of criteria $\{i,j\} \subseteq N$.

- A negative $I(v,ij)$, that is $I(v,ij) \in [-1,0]$, implies a disjunctive behavior between i and j. If only one criterion of the two is satisfied then the linear average should be improved by a positive amount that is proportional to the gap between the scores x_i and x_j but also to the amplitude of the interaction $|I(v,ij)|$.
- A positive $I(v,ij)$, that is $I(v,ij) \in [0,1]$, implies a conjunctive behavior. In case of unilateral satisfaction, the linear average should be decreased by the same quantity as explained in the first case.

Now, regarding OWA operators, we also have the following.

Proposition 3. *The operator* OWA_ω *is a Choquet integral defined from a 2-order fuzzy measure if and only if* ω_s *is linear in* s*, that is,* $\omega_s = \alpha\, s + \beta$ *for some* $\alpha, \beta \in \mathbb{R}$.

Proof. (Necessity) By (9), we simply have $\omega_s = a(i) + (n-s)\,a(ij)$ for all $i, j, s \in N$.

(Sufficiency) By (11), we have

$$a(S) = \left[\Delta_k^{s-1} \alpha\,(n-k) + \beta\right]_{k=0}, \qquad S \subseteq N,\ S \neq \emptyset,$$

and hence $a(S) = 0$ whenever $s > 2$. □

Theorem 3. *Any* OWA_ω *of order 2 is of the form*

$$\mathrm{OWA}_\omega(x) = \frac{1}{n}\sum_{i=1}^{n} x_i + \frac{1}{2}\,\frac{\omega_1 - \omega_n}{n-1}\sum_{i=1}^{n}(n+1-2\,i)\,x_{(i)}, \tag{18}$$

for all $x \in \mathbb{R}^n$.

Proof. By Eq. (12)–(16), we immediately have

$$a(i) = \frac{1}{n} - \frac{1}{2}(\omega_1 - \omega_n), \qquad i \in N,$$

$$a(ij) = \frac{\omega_1 - \omega_n}{n-1}, \qquad i, j \in N,$$

and by (9), we have

$$\begin{aligned} \omega_s &= a(i) + (n-s)\,a(ij) \\ &= \frac{1}{n} + \frac{\omega_1 - \omega_n}{2(n-1)}(n+1-2\,s), \end{aligned}$$

for all $i, j, s \in N$, which is sufficient. □

Interestingly enough, comparing Eq. (17) and (18) leads to the following remarkable identity:

$$\sum_{\{i,j\}\subseteq N} |x_i - x_j| = \sum_{i=1}^{n}(2i-n-1)\,x_{(i)}, \qquad x \in \mathbb{R}^n.$$

5 Concluding Remarks

The Choquet integral is an appropriate substitute to the weighted arithmetic mean to aggregate dependent decision criteria. The motivation is based mainly on an axiomatic characterization of the class of Choquet integrals, but also on an intuitive interpretation of its expression, especially in the 2-order case.

We hope that this will encourage people of the multicriteria decision making community to use this innovative technique of aggregation, which seems very promising.

References

1. C. Berge, *Principes de combinatoire* (Dunod, Paris, 1968). English translation: *Principles of combinatorics* (Academic Press, New York, 1971).
2. A. Chateauneuf and J.Y. Jaffray, Some characterizations of lower probabilities and other monotone capacities through the use of Möbius inversion, *Mathematical Social Sciences* **17** (1989) 263–283.
3. G. Choquet, Theory of capacities, *Annales de l'Institut Fourier* **5** (1953) 131–295.
4. L.M. de Campos, M.T. Lamata and S. Moral, A unified approach to define fuzzy integrals, *Fuzzy Sets and Systems* **39** (1991) 75–90.
5. S. Fujishige, Characterization of subdifferentials of submodular functions and its relation to Lovász extension of submodular functions. Report No. 82241-OR, Inst. für Ökonometrie und Operations Research, Bonn, 1982.
6. M. Grabisch, On equivalence classes of fuzzy connectives : the case of fuzzy integrals, *IEEE Trans. Fuzzy Systems* **3** (1) (1995) 96–109.
7. M. Grabisch, The application of fuzzy integrals in multicriteria decision making, *European Journal of Operational Research* **89** (1996) 445-456.
8. M. Grabisch, *k*-order additive discrete fuzzy measures and their representation, *Fuzzy Sets and Systems* **92** (1997) 167–189.
9. M. Grabisch, Alternative representations of OWA operators, in: R.R. Yager and J. Kacprzyk (eds.), *The Ordered Weighted Averaging Operators: Theory and Applications*, (Kluwer Academic Publisher, 1997) 73–85.
10. M. Grabisch, J.-L. Marichal, and M. Roubens, Equivalent representations of set functions, *Mathematics of Operations Research* **25** (2000) 157–178.
11. M. Grabisch, T. Murofushi, and M. Sugeno (eds.), *Fuzzy Measures and Integrals - Theory and Applications.* Series: Studies in Fuzziness and Soft Computing (Physica Verlag, Heidelberg, 2000).
12. P.L. Hammer and S. Rudeanu, *Boolean methods in operations research and related areas* (Springer, Berlin, 1968).
13. U. Höhle, Integration with respect to fuzzy measures, *Proc. IFAC Symposium on Theory and Applications of Digital Control*, New Delhi, January 1982, 35–37.
14. L. Lovász, Submodular function and convexity, in: A. Bachem, M. Grötschel, B. Korte (eds.), *Mathematical programming. The state of the art. Bonn 1982*, (Springer-Verlag, Berlin-Heidelberg-New York-Tokyo, 1983) 235–257.
15. J.-L. Marichal, *Aggregation operators for multicriteria decision aid*, Ph.D. thesis, Institute of Mathematics, University of Liège, Liège, Belgium, 1998.
16. J.-L. Marichal, Behavioral analysis of aggregation in multicriteria decision aid, in: J. Fodor, B. De Baets, and P. Perny (eds.), *Preferences and Decisions under Incomplete Knowledge.* Series: Studies in Fuzziness and Soft Computing Vol. 51 (Physica Verlag, Heidelberg, 2000) 153–178.
17. J.-L. Marichal, The influence of variables on pseudo-Boolean functions with applications to game theory and multicriteria decision making, *Discrete Applied Mathematics* **107** (2000) 139–164.
18. J.-L. Marichal, An axiomatic approach of the discrete Choquet integral as a tool to aggregate interacting criteria, *IEEE Transactions on Fuzzy Systems* **8** (6) (2000) 800–807.
19. T. Murofushi, A technique for reading fuzzy measures (I): the Shapley value with respect to a fuzzy measure, *2nd Fuzzy Workshop*, Nagaoka, Japan, October 1992, 39–48. In Japanese.

20. T. Murofushi and S. Soneda, Techniques for reading fuzzy measures (III): interaction index, *9th Fuzzy System Symposium*, Sapporo, Japan, May 1993, 693–696. In Japanese.
21. T. Murofushi and M. Sugeno, An interpretation of fuzzy measure and the Choquet integral as an integral with respect to a fuzzy measure, *Fuzzy Sets and Systems* **29** (1989) 201–227.
22. T. Murofushi and M. Sugeno, A theory of fuzzy measures. Representation, the Choquet integral and null sets, *Journal of Mathematical Analysis and Applications* **159** (2) (1991) 532–549.
23. T. Murofushi and M. Sugeno, Some quantities represented by the Choquet integral, *Fuzzy Sets & Systems* **56** (1993) 229–235.
24. G. Owen, Multilinear extensions of games, *Management Sciences* **18** (1972) 64–79.
25. G.C. Rota, On the foundations of combinatorial theory I. Theory of Möbius functions, *Zeitschrift für Wahrscheinlichkeitstheorie und Verwandte Gebiete* **2** (1964) 340–368.
26. D. Schmeidler, Integral representation without additivity, *Proc. Amer. Math. Soc.* **97** (1986) 255–261.
27. L.S. Shapley, A value for n-person games, in: H.W. Kuhn and A.W. Tucker (eds.), *Contributions to the Theory of Games*, Vol. II, Annals of Mathematics Studies, 28, (Princeton University Press, Princeton, NJ, 1953) 307–317.
28. I. Singer, Extensions of functions of 0-1 variables and applications to combinatorial optimization, *Numerical Functional Analysis and Optimization* **7** (1) (1984-85) 23–62.
29. M. Sugeno, *Theory of fuzzy integrals and its applications*, Ph.D. Thesis, Tokyo Institute of Technology, Tokyo, 1974.
30. R.R. Yager, On ordered weighted averaging aggregation operators in multicriteria decision making, *IEEE Trans. on Systems, Man and Cybernetics* **18** (1988) 183–190.

Appendix: Pseudo-Boolean Functions and Their Lovász Extensions

Any set function $v : 2^N \to \mathbb{R}$ can be assimilated unambiguously with a pseudo-Boolean function $f : \{0,1\}^n \to \mathbb{R}$. The correspondence is straightforward: We have

$$f(x) = \sum_{T \subseteq N} v(T) \prod_{i \in T} x_i \prod_{i \notin T} (1 - x_i), \qquad x \in \{0,1\}^n,$$

and $v(S) = f(e_S)$ for all $S \subseteq N$.

In particular, any pseudo-Boolean function that corresponds to a fuzzy measure is increasing in each variable and fulfils the boundary conditions: $f(e_\emptyset) = 0$ and $f(e_N) = 1$.

Hammer and Rudeanu [12] showed that any pseudo-Boolean function has a unique expression as a multilinear polynomial in n variables:

$$f(x) = \sum_{T \subseteq N} a(T) \prod_{i \in T} x_i, \qquad x \in \{0,1\}^n,$$

with $a(T) \in \mathbb{R}$ for all $T \subseteq N$. These coefficients $a(T)$ correspond to the Möbius transform of v, see (3). Indeed, we have

$$v(S) = f(e_S) = \sum_{T \subseteq S} a(T)$$

for every $S \subseteq N$.

Lovász [14, Sect. 3] observed that any $x \in (\mathbb{R}^+)^n \setminus \{e_\emptyset\}$ can be written uniquely in the form

$$x = \sum_{i=1}^{k} \lambda_i \, e_{S_i} \tag{19}$$

where $\lambda_1, \ldots, \lambda_k > 0$ and $\emptyset \neq S_1 \subsetneq \cdots \subsetneq S_k \subseteq N$. For example, we have

$$\begin{aligned}(1,5,3) &= 2\,(0,1,0) + 2\,(0,1,1) + 1\,(1,1,1)\\ (0,5,3) &= 2\,(0,1,0) + 3\,(0,1,1).\end{aligned}$$

Hence any function $f : \{0,1\}^n \to \mathbb{R}$ with $f(e_\emptyset) = 0$ can be extended to $\hat{f} : (\mathbb{R}^+)^n \to \mathbb{R}$, by $\hat{f}(e_\emptyset) = 0$ and

$$\hat{f}(x) = \sum_{i=1}^{k} \lambda_i \, f(e_{S_i}) \qquad \Big(x = \sum_{i=1}^{k} \lambda_i \, e_{S_i} \in (\mathbb{R}^+)^n \setminus \{e_\emptyset\}\Big).$$

Indeed, $\hat{f}$ is well defined (due to the uniqueness of (19)) and $\hat{f}(x) = f(x)$ for all $x \in \{0,1\}^n$. The function $\hat{f}$ is called [5] the *Lovász extension* of f.

Now, the Lovász extension of an arbitrary function $f : \{0,1\}^n \to \mathbb{R}$ is defined by

$$\hat{f}(x) = f(e_\emptyset) + \hat{f}_0(x), \qquad x \in (\mathbb{R}^+)^n,$$

where $\hat{f}_0$ is the Lovász extension of $f_0 = f - f(e_\emptyset)$.

The hypercube $[0,1]^n$ can be subdivided into $n!$ simplices $\mathcal{B}_\pi$ of the form

$$\mathcal{B}_\pi := \{x \in [0,1]^n \mid x_{\pi(1)} \leq \cdots \leq x_{\pi(n)}\}, \qquad \pi \in \Pi,$$

where Π is the set of all permutations of N.

It is well known [28] that $\mathcal{B}_\pi$ is a polytope with vertices $\varepsilon_i^\pi = e_{\{\pi(i),\ldots,\pi(n)\}}$ $(i = 1, \ldots, n+1)$. Singer [28, Sect. 2] showed that $\hat{f}$ is defined on each cone $\mathcal{K}_\pi = \{\lambda \, \mathcal{B}_\pi \mid \lambda \geq 0\}$ as the unique affine function that coincides with f at the $n+1$ vertices of $\mathcal{B}_\pi$. More formally, $\hat{f}$ can be written as

$$\hat{f}(x) = f(e_\emptyset) + \sum_{i=1}^{n} x_{\pi(i)} \, [f(\varepsilon_i^\pi) - f(\varepsilon_{i+1}^\pi)], \qquad x \in \mathcal{K}_\pi. \tag{20}$$

Geometrically, on each simplex $\mathcal{B}_\pi$, the graph of $\hat{f}$ is the portion of the unique hyperplane passing through $(\varepsilon_1^\pi, f(\varepsilon_1^\pi)), \ldots, (\varepsilon_{n+1}^\pi, f(\varepsilon_{n+1}^\pi))$ (which is, actually, their convex hull). According to Singer [28], the restriction of $\hat{f}$ to $[0,1]^n$

is called the *tight extension* of f associated to the standard triangulation $\{\mathcal{B}_\pi \mid \pi \in \Pi\}$ of $[0,1]^n$.

A practical form of $\hat{f}$ is given by the following result.

Proposition 4. *The Lovász extension of $f : \{0,1\}^n \to \mathbb{R}$ is given by*

$$\hat{f}(x) = \sum_{T \subseteq N} a(T) \bigwedge_{i \in T} x_i, \qquad x \in (\mathbb{R}^+)^n, \tag{21}$$

where the coefficients $a(T)$ are those of the unique multilinear expression of f.

Proof. The function (21) agrees with f at all the vertices of $[0,1]^n$, and identifies with an affine function on each cone $\mathcal{K}_\pi$, which is sufficient. □

Let $v \in \mathcal{F}_N$. By (20), we immediately see that the Choquet integral $\mathcal{C}_v$, defined on $(\mathbb{R}^+)^n$, is nothing else than the Lovász extension of the pseudo-Boolean function f which represents v:

$$\mathcal{C}_v = \hat{f} \qquad \text{on } (\mathbb{R}^+)^n.$$

Moreover, since the expression in (21) is stable on $\mathbb{R}^n$ for the positive linear transformations (cf. Theorem 1), it corresponds to the Choquet integral. Indeed, any function from $\mathbb{R}^n$ to $\mathbb{R}$ fulfilling this property is completely defined by its restriction on $[0,1]^n$, see Proposition 3.1 in [18]. We then have the following result (see also [2]).

Proposition 5. *Any Choquet integral $\mathcal{C}_v : \mathbb{R}^n \to \mathbb{R}$ can be written as*

$$\mathcal{C}_v(x) = \sum_{T \subseteq N} a(T) \bigwedge_{i \in T} x_i, \qquad x \in \mathbb{R}^n,$$

where a is the Möbius representation of v.

Thus, the Choquet integral $\mathcal{C}_v$ is a piecewise affine function on $\mathbb{R}^n$, which extends the pseudo-Boolean function representing v:

$$\mathcal{C}_v(e_S) = v(S), \qquad S \subseteq N.$$

Moreover, we clearly see that $\mathcal{C}_v$ is an increasing function if and only if v is also an increasing function.

Representation Models for Aggregating Linguistic Information: Issues and Analysis

F. Herrera[1], E. Herrera-Viedma[1], and L. Martínez[2]

[1] Dept. of Computer Science and Artificial Intelligence.
University of Granada, 18071 - Granada, Spain
[2] Dept. of Computer Science.
University of Jaén, 23071 - Jaén, Spain

e-mail: herrera,viedma@decsai.ugr.es, martin@ujaen.es

Abstract. The linguistic information has been used successfully in many areas. The aggregation of linguistic information is a crucial aspect. In the literature we can find different linguistic computational models that present linguistic aggregation operators as: (i) The computational model based on the Extension Principle, which operates over the fuzzy numbers that supports the semantics of the linguistic labels. (ii) The symbolic one makes the computations directly over the order index of the linguistic labels. And, (iii) the model based on the linguistic 2-tuple representation, which uses the symbolic translation to make the linguistic computations.

Depending upon the linguistic aggregation model, different results can be obtained. In this chapter we shall make a comparative analysis of the aggregation approaches according to the results obtained in a decision-making application.

Keywords: linguistic variables, linguistic aggregation.

1 Introduction

Usually, the most of the problems we solve deal with quantitative aspects. These are easily assessed by means of precise numerical values that specify in a simply manner the importance or preference of these aspects in the problem. However, in practice we find a lot of these models are difficult to apply because they fail to assess the aspects which describe the problem under consideration. This is due to the fact that these problems deal with qualitative aspects that are complex to assess by means of precise values. In order to overcome this drawback the use of the fuzzy linguistic approach [26] has provided very good results in different areas, as information retrieval [3,25], economics [13,23], planning [1], consensus reaching processes [4,12], decision-making [11,22], etc.

We focus our study in those linguistic problems that deal with multiple criteria or experts manage linguistic preferences and are solved following a common resolution scheme with the two following processes [21]:

- *Aggregation process.* The individual linguistic preference values are combined to obtain collective preference values.

- *Exploitation process.* The collective preference values are analysed to obtain the best alternative/s for the problem.

This resolution scheme implies processes of computing with words, that is, processes of aggregation of linguistic information. In the literature we can find different linguistic computational models to accomplish these processes:

- *The approximative computational model based on the Extension Principle* [2,6]. This model uses fuzzy arithmetic based on the Extension Principle to make computations over the linguistic variables. This model can present the results in two ways:
  1. By means of the fuzzy numbers obtained from the fuzzy arithmetic computations based on the Extension Principle.
  2. Or by means of linguistic labels computed from the fuzzy numbers obtained using a linguistic approximation process.
- *The ordinal linguistic computational model [7].* This symbolic model makes direct computations on labels, using the ordinal structure of the linguistic term sets. Its results are inherently linguistic labels due to either the operators used, basicaly max and min operators [24] or because in the computations on the order index there exist an approximation by means of the *round operator* [10].
- *The 2-tuple linguistic computational model.* It uses the 2-tuple linguistic representation model and its characteristics to make linguistic computations [14,17], obtaining as results linguistic 2-tuples. A linguistic 2-tuple is a pair of values, where the first one is a linguistic label and the second one is a real number that represents the value of the symbolic translation, that it is the basic concept of the linguistic 2-tuple representation model that will be introduced in the following section.

The aim of this contribution is to make a comparative analysis among the linguistic computational models for the aggregation of linguistic information from different points of view. To do so, we shall solve a simple decision-making problem and compare the results obtained by each model. This study will help us to decide what linguistic computational model is most adequated to make computations over linguistic information.

This contribution is structured as follows: in Section 2 we shall make a brief review of the fuzzy linguistic approach and introduce a simple decision-making problem. In Section 3 we shall review the different linguistic computational models and solve the above decision-making problem with each one. In Section 4 a comparative study of the different linguistic computational models is presented. Finally some concluding remarks are pointed out.

2 Preliminaries

In this section we review the fuzzy linguistic approach and present a decision-making problem that will be solved along this contribution using different linguistic methods.

2.1 Fuzzy linguistic approach

Usually, we work in a quantitative setting, where the information is expressed by means of numerical values. However, many aspects of different activities in the real world cannot be assessed in a quantitative form, but rather in a qualitative one, i.e., with vague or imprecise knowledge. In that case a better approach may be to use linguistic assessments instead of numerical values. The fuzzy linguistic approach represents qualitative aspects as linguistic values by means of linguistic variables [26]. This approach is adequated in some situations, for example, when attempting to qualify phenomena related to human perception, we are often led to use words in natural language. This may arise for different reasons. There are some situations where the information may be unquantifiable due to its nature, and thus, it may be stated only in linguistic terms (e.g., when evaluating the "comfort" or "design" of a car, terms like "bad", "poor", "tolerable", "average", "good" can be used [19]). In other cases, precise quantitative information may not be stated because either it is not available or the cost of its computation is too high, then an "approximate value" may be tolerated (e.g., when evaluating the speed of a car, linguistic terms like "fast", "very fast", "slow" are used instead of numerical values).

We have to choose the appropriate linguistic descriptors for the term set and their semantics. In order to accomplish this objective, an important aspect to analyze is the *"granularity of uncertainty"*, i.e., the level of discrimination among different counts of uncertainty. Typical values of cardinality used in the linguistic models are odd ones, such as 7 or 9 [20], where the mid term represents an assessment of "approximately 0.5", and with the rest of the terms being placed symmetrically around it [2].

One possibility of generating the linguistic term set consists of directly supplying the term set by considering all terms distributed on a scale on which a total order is defined [24]. For example, a set of seven terms S, could be given as follows:

$$S = \{s_0 = N, s_1 = VL, s_2 = L, s_3 = M, s_4 = H, s_5 = VH, s_6 = P\}$$

Usually, in these cases, it is required that in the linguistic term set there exist:

1. A negation operator: Neg(s_i) $= s_j$ such that $j = g$-i (g+1 is the cardinality of S).
2. $s_i \leq s_j \iff i \leq j$. Therefore, there exists a minimization and a maximization operator.

The semantics of the linguistic terms is given by fuzzy numbers defined in the [0,1] interval. A way to characterize a fuzzy number is to use a representation based on parameters of its membership function [2]. The linguistic assessments given by the users are just approximate ones, some authors [2,6,8,11] consider that linear trapezoidal membership functions are good enough to capture the vagueness of those linguistic assessments. The parametric representation is achieved by the 4-tuple (a, b, d, c), where b and d indicate the interval in which the membership value is 1, with a and c indicating the left and right limits of the definition domain of the trapezoidal membership function. A particular case of this type of representation are the linguistic assessments whose membership functions are triangular, i.e., $b = d$, then we represent this type of membership functions by a 3-tuple (a, b, c). An example may be the following (Figure 1) :

$$N = (0, 0, .25) \;\; VL = (0, .2, .4) \;\; L = (.25, .4, .50)$$
$$M = (.4, .5, .6) \;\; H = (.5, .6, .75) \;\; VH = (.6, .8, 1)$$
$$P = (.75, 1, 1).$$

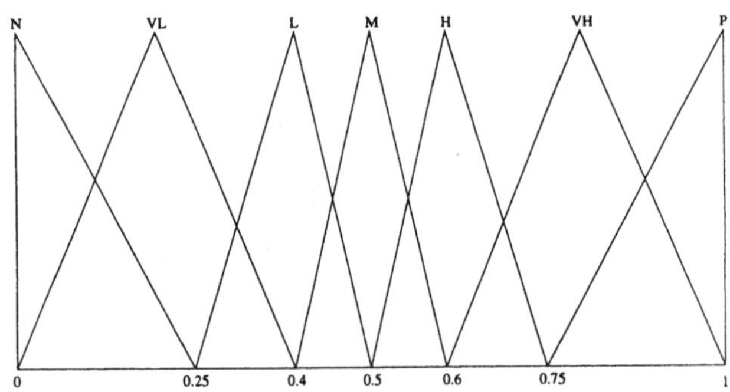

Fig. 1. A set of seven linguistic terms with its semantics

Other authors use a non-trapezoidal representation, e.g., Gaussian functions [3].

2.2 A decision-making problem

Here, we propose a simple decision-making problem to be solved with different linguistic computational models following the resolution scheme presented in the introduction.

- **Description**

  A distribution company needs to renew its computing system, so it contracts a consulting company to carry out a survey of the different possibilities existing on the market, to decide which is the best option for its needs. The alternatives are the following:

x_1	x_2	x_3	x_4
LINUX	WINDOWS-2000	VMS	SOLARIS

The consulting company has a group of four consultancy departments

p_1	p_2	p_3	p_4
Cost analysis	System analysis	Risk analysis	Techonology analysis

- **Initialization**

  Each department provides a performance vector expressing its preferences for each alternative. The preferences are assessed in the term set $S = \{N, VL, L, M, H, VH, P\}$ (see Figure 1).

		alternatives			
	L_{ij}	x_1	x_2	x_3	x_4
	p_1	L	M	M	M
experts	p_2	L	L	VL	M
	p_3	VL	VL	M	M
	p_4	H	VL	L	M

  whose membership functions, L_{ij}, are assumed to be of the triangular type $L_{ij} = (a_{ij}, b_{ij}, c_{ij})$.

3 Linguistic modelling in decision-making problems

In this section we apply the different linguistic computational models to the linguistic decision problem presented just above, following the resolution scheme presented in the introduction.

3.1 Solution based on the extension principle

The Extension Principle is a basic concept in the fuzzy sets theory [8] which is used to generalize crisp mathematical concepts to fuzzy sets, i.e., it allows non-fuzzy function to be fuzzified in the sense that if the function arguments are made into fuzzy sets, then the function value is also a fuzzy set. One of the several formulations of this concept is the following [26]:

Definition 1. Let X be the cartesian product of the universes $X_1, ..., X_r$ and $\tilde{A}_1, ..., \tilde{A}_r$ be fuzzy sets in $X_1, ..., X_r$, respectively. Let f be a function defined from X, $(X = X_1 \times ... \times X_r)$ to Y, $y = f(x_1, ..., x_r)$. The Extension Principle allows us to define a fuzzy set $\tilde{B}$ in Y, from the fuzzy sets $\tilde{A}_1, ..., \tilde{A}_r$ representing its value from f, according to the following expression:

$$\tilde{B} = \{(y, \mu_{\tilde{B}}(y))/y = f(x_1, ..., x_r), (x_1, ..., x_r) \in X\}$$

where

$$\mu_{\tilde{B}}(y) = \begin{cases} \sup_{(x_1,...,x_r) \in f^{-1}(y)} \min\{\mu_{\tilde{A}_1}(x_1), ..., \mu_{\tilde{A}_r}(x_r)\}, & if \;\; f^{-1}(y) \neq \emptyset \\ 0, & otherwise. \end{cases}$$

For $r = 1$, the extesion principle is:

$$\tilde{B} = f(A) = \{(y, \mu_{\tilde{B}}(y))/y = f(x), x \in X)$$

where

$$\mu_{\tilde{B}}(y) = \begin{cases} \sup_{x \in f^{-1}(y)} \mu_{\tilde{A}}(x), & if \;\; f^{-1}(y) \neq \emptyset \\ 0, & otherwise. \end{cases}$$

The resolution scheme for linguistic problems, presented in the introduction, is divided in two processes: *aggregation* and *exploitation*. Depending on the process, this computational model can deal with information from different nature.

- *Aggregation process.* The approximative model based on the Extension Principle uses the membership functions of the linguistic terms, μ_{s_i} where $s_i \in S$, for computing collective (aggregated) values that are expressed by means of fuzzy numbers. A formal scheme can be:

$$S^n \xrightarrow{\tilde{F}} F(\mathcal{R})$$

  where S^n symbolizes the n cartesian product of S, $\tilde{F}$ is an aggregation operator based on the Extension Principle, and $F(\mathcal{R})$ the set of fuzzy sets over the Real line $\mathcal{R}$.
  Now, we apply this model to the decision-making problem. There exist a lot of aggregation operators, an usual one is the mean operator, that will be used in this paper.

$$L_j = \tilde{x}(L_{ij}) = (\frac{1}{m}\sum_{i=1}^{m} a_{ij}, \frac{1}{m}\sum_{i=1}^{m} b_{ij}, \frac{1}{m}\sum_{i=1}^{m} c_{ij})$$

  being m the number of experts, $\tilde{x}$ the fuzzy mean operator based on the extension principle and L_j the collective value for x_j. We obtain the following collective preference vector:

L_1	L_2	L_3	L_4
(.25,.4,.53)	(.16,.32,.47)	(.26,.4,.52)	(.4,.5,.6)

- *Exploitation process.* The objective of this process is to order the collective values obtained in the aggregation process. To do so, this computational model can act on two different ways:
  1. *Ranking fuzzy numbers obtained on the aggregation process.* Therefore a fuzzy ranking function must be applied to them, obtaining an alternatives order according to the real numbers computed by the fuzzy ranking:

$$F(\mathcal{R}) \xrightarrow{\tilde{O}} \mathcal{R}$$

In the specialized literature we can find different fuzzy ranking functions [27], in this paper we shall use the following two different ones.

(a) The González fuzzy ranking [9]. Let F_j (j=1,...,m) be m fuzzy numbers whose membership functions are $F_j = (a_j, b_j, c_j)$. Define the following fuzzy ranking function:

$$V^{\alpha}(F_j) = \alpha(c + \frac{b}{2}) + (1 - \alpha)(c - \frac{a}{2})$$

The value α is an index of rating attitude.

Applying this fuzzy ranking to the collective fuzzy numbers with $\alpha = 0.1$ we obtain the following ranking values:

$V^{0.1}(L_1)$	$V^{0.1}(L_2)$	$V^{0.1}(L_3)$	$V^{0.1}(L_4)$
.442	.414	.423	.445

Therefore, the best alternative for the problem according to this fuzzy ranking is $\{x_4\}$, i.e., the *Solaris based System.*

(b) Another fuzzy ranking was presented by Chang in [5], where:

$$Ch(\tilde{F}_j) = \frac{1}{6}(c_i - a_i)(a_i + b_i + c_i)$$

This fuzzy ranking obtains the following ranking for the collective values:

$Ch(L_1)$	$Ch(L_2)$	$Ch(L_3)$	$Ch(L_4)$
.055	.049	.051	.05

The solution set of alternatives in this case is $\{x_1\}$, i.e., *the Linux based system*

2. *Applying a linguistic approximation process to the fuzzy numbers obtained in the aggregation process.* This approximation process obtains the linguistic term closest to the unlabelled fuzzy number in the initial term set. Once the aggregated values are expressed by means of linguistic terms these have an inner order defined by the structure of the linguistic term set:

$$F(\mathcal{R}) \stackrel{app_1(\cdot)}{\longrightarrow} S$$

Different linguistic approximation functions, $app_1(\cdot)$, can be used. In this paper we shall use the following one. Let $F_j = (a_j, b_j, c_j)$, $(j = 1, ..., m)$ be the collective fuzzy numbers obtained in the aggregation process the linguistic approximation process, $app_1(\cdot)$, comes back the closest linguistic term, $s_l \in S$, to the unlabelled fuzzy number, F_j.

$$d(s_l, F_j) = \sqrt{P_1(a_l - a_j)^2 + P_2(b_l - b_j)^2 + P_3(c_l - c_j)^2}$$

being (a_l, b_l, c_l) and (a_j, b_j, c_j) the membership functions of "s_l" and "F_j" respectively, with P_1, P_2, P_3 being the weights that represent the importance of a, b *and* c. Therefore, $app_1(\cdot)$ chooses s_l^* $(app_1(F_j) = s_l^*)$, such that, $d(s_l^*, F_j) \leq d(s_l, F_j)$ $\forall s_l \in S$, where S is the linguistic term set used as expression domain.
Applying this linguistic approximation process to the collective vector, we shall obtain a collective linguistic vector that is easy to order and thus to choose the best alternative/s.
This linguistic approximation process is applied to the collective values, with $P_1 = 0.2$, $P_2 = 0.6$, $P_3 = 0.2$. We select these values because of the parameter "b_i" is the most representative of the membership function and "a_i", "c_i" are equally representative.
The collective preference vector obtained after the linguistic approximation is:

$app_1(L_1)$	$app_1(L_2)$	$app_1(L_3)$	$app_1(L_4)$
M	L	L	M

The exploitation process applies a choice degree to the collective performance vector for obtaining the solution set of alternatives. In this case we obtain as solution set:

$$\{x_1, x_4\}$$

that is a solution set with multiple alternatives.

3.2 Solution based on the linguistic computational symbolic model

Here, we shall apply the same decision process for solving the problem, but in this case we shall deal with the symbolic approach presented in [7].

- *Aggregation process.* The symbolic operator we shall use to aggregate linguistic variables is the Convex Combination [7]:

  Definition 2.[7]*Let $A = \{a_1, ..., a_m\}$ be a set of linguistic terms to be aggregated, the Convex Combination is defined in a recursive way as:*
  For m=2:

  $$C^2\{\{w_1, 1-w_1\}, \{b_1, b_2\}\} = w_1 \odot s_j \oplus (1-w_1) \odot s_i = s_k, \ \ s_j, \ s_i \in S, \ (j \geq i)$$

  such that,

  $$k = \min\{g, i + round(w_1 \cdot (j - i))\},$$

where $g+1$ is the cardinality of S, round$(\cdot)$ is the usual round operation, and $b_1 = s_j,\ b_2 = s_i$.

For m > 2:

$C^m\{w_k, b_k, k = 1,...,m\} = w_1 \odot b_1 \oplus (1 - w_1) \odot C^{m-1}\{\eta_h, b_h, h = 2,\ldots,m\} = C^2\{\{w_1, 1 - w_1\}, \{b_1, C^{m-1}\{\eta_h, b_h, h = 2,\ldots,m\}\}\}$

where $W = [w_1,\ldots,w_m]$ is a weighting vector associated to A, such that, (i) $w_i \in [0,1]$, and (ii) $\sum_i w_i = 1$; and $B = \{b_1,\ldots,b_m\}$ is a vector, such that, $B = \{a_{\sigma(1)},\ldots,a_{\sigma(m)}\}$, where, $a_{\sigma(j)} \leq a_{\sigma(i)}\ \forall\ i \leq j$, with σ being a permutation over the values a_i. $\eta_h = w_h/\Sigma_2^m w_k, h = 2,\ldots,m$. With $\odot$ and $\oplus$ being the product of a number by a label and the addition of two labels respectively.

In our example the weighting vector is $\{.25,.25,.25,.25\}$, then the collective performance values obtained are:

L_1	L_2	L_3	L_4
M	M	L	M

- *Exploitation process.*The alternatives with the highest collective performance are:

$$\{x_1, x_2, x_4\}$$

Here again we find a multiple alternative solution set.

3.3 Solution based on linguistic 2-tuples

This model is based on the representation model presented in [14,17] and it is based on the concept of symbolic translation and uses it for representing the linguistic information by means of 2-tuples, (s_i, α), where s is a linguistic term and α is a numerical value representing the symbolic translation.

Let $S = \{s_0,...,s_g\}$ be a linguistic term set, and $\beta \in [0,g]$ a numerical value in its interval of granularity (e.g.: let β be a value obtained from a symbolic aggregation operation [2,7]).

Definition 3. *Let β be the result of an aggregation of the indexes of a set of labels assessed in a linguistic term set S, i.e., the result of a symbolic aggregation operation. $\beta \in [0,g]$, being $g+1$ the cardinality of S. Let $i =$ round(β) and $\alpha = \beta - i$ be two values, such that, $i \in [0,g]$ and $\alpha \in [-.5,.5)$ then α is called a Symbolic Translation.*

Roughly speaking, the symbolic translation of a linguistic term, s_i, is a numerical value assessed in $[-.5,.5)$ that supports the "difference of information" between a counting of information $\beta \in [0,g]$ obtained after a symbolic aggregation operation and the closest value in $\{0,...,g\}$ that indicates the index of the closest linguistic term in S ($i =$ round(β)).

From this concept we shall develop a linguistic representation model which represents the linguistic information by means of 2-tuples (s_i, α_i), $s_i \in S$ and $\alpha_i \in [-.5, .5)$:

- s_i represents the linguistic label of the information, and
- α_i is a numerical value expressing the value of the translation from the original result β to the closest index label, i, in the linguistic term set (s_i), i.e., the Symbolic Translation.

This model defines a set of transformation functions between linguistic terms and 2-tuples, and between numeric values and 2-tuples.

Definition 4. *Let $S = \{s_0, ..., s_g\}$ be a linguistic term set and $\beta \in [0, g]$ a value representing the result of a symbolic aggregation operation, then the 2-tuple that expresses the equivalent information to β is obtained with the following function:*

$$\Delta : [0, g] \longrightarrow S \times [-0.5, 0.5)$$

$$\Delta(\beta) = (s_i, \alpha), \ with \begin{cases} s_i & i = round(\beta) \\ \alpha = \beta - i & \alpha \in [-.5, .5) \end{cases}$$

where round(·) is the usual round *operation, s_i has the closest index label to "β" and "α" is the value of the symbolic translation.*

Proposition 1. *Let $S = \{s_0, ..., s_g\}$ be a linguistic term set and (s_i, α) be a 2-tuple. There is always a Δ^{-1} function, such that, from a 2-tuple it returns its equivalent numerical value $\beta \in [0, g] \subset \mathcal{R}$.*

Proof.
It is trivial, we consider the following function:

$$\Delta^{-1} : S \times [-.5, .5) \longrightarrow [0, g]$$

$$\Delta^{-1}(s_i, \alpha) = i + \alpha = \beta$$

Remark: From definitions 2 and 3 and from proposition 1, it is obvious that the conversion of a linguistic term into a linguistic 2-tuple consist of adding a value 0 as symbolic translation:

$$s_i \in S \Longrightarrow (s_i, 0)$$

This representation model has a computational technique based on 2-tuples presented in [14,17], that it will be used in the aggregation and exploitation processes:

1. Aggregation of 2-tuples

The aggregation of information consists of obtaining a value that summarizes a set of values, therefore, the result of the aggregation of a set of 2-tuples must be a 2-tuple. In [14] we can find some 2-tuple aggregation operators, that are based on classical aggregation operators.

2. Comparison of 2-tuples

The comparison of linguistic information represented by 2-tuples is carried out according to an ordinary lexicographic order.

Let (s_k, α_1) and (s_l, α_2) be two 2-tuples, with each one representing a counting of information:

- if $k < l$ then (s_k, α_1) is smaller than (s_l, α_2)
- if $k = l$ then
  1. if $\alpha_1 = \alpha_2$ then (s_k, α_1), (s_l, α_2) represents the same information
  2. if $\alpha_1 < \alpha_2$ then (s_k, α_1) is smaller than (s_l, α_2)
  3. if $\alpha_1 > \alpha_2$ then (s_k, α_1) is bigger than (s_l, α_2)

Here, we shall use the linguistic representation model with 2-tuples to solve the decision making problem presented in Subsection 2.4. To do so, we follow the same selection model.

The performance vectors of the experts are transformed into 2-tuples:

		alternatives			
		x_1	x_2	x_3	x_4
experts	p_1	$(P, 0)$	$(M, 0)$	$(M, 0)$	$(M, 0)$
	p_2	$(L, 0)$	$(L, 0)$	$(VL, 0)$	$(M, 0)$
	p_3	$(VL, 0)$	$(VL, 0)$	$(M, 0)$	$(M, 0)$
	p_4	$(H, 0)$	$(H, 0)$	$(L, 0)$	$(M, 0)$

Now we apply the resolution scheme using linguistic 2-tuples.

1. *Aggregation process.* We aggregate these 2-tuples, using the 2-tuple arithmetic mean [14].

   Definition 6. *Let $x = \{(r_1, \alpha_1), \ldots, (r_m, \alpha_m)\}$ be a set of 2-tuples, the 2-tuple arithmetic mean $\overline{x}^e$ is computed as,*

   $$\overline{x}^e = \Delta(\sum_{i=1}^{m} \frac{1}{m} \Delta^{-1}(r_i, \alpha_i)) = \Delta(\frac{1}{m} \sum_{i=1}^{m} \beta_i)$$

   Obtaining the following collective values:

$\overline{x}_1^e(L_1)$	$\overline{x}_2^e(L_2)$	$\overline{x}_3^e(L_3)$	$\overline{x}_4^e(L_4)$
(M,.25)	(M,-.5)	(L,.25)	(M,0)

2. *Exploitation process.* We obtain as solution set of alternatives:

   $$\{x_1\}.$$

4 Analysis of the linguistic computational models

Here, we shall analyse the linguistic computational models from different points of view studying the results obtained in the above decision-making problem and which can be observed in the following table:

	Ranking Fuzzy Numbers		Ling. Approx.	Symbolic Model	2-tuples
	González	Chang	S	S	S
x_1	.442	**.055**	**M**	**M**	**(M,.25)**
x_2	.414	.049	*L*	M	(M,-.5)
x_3	.423	.051	*L*	L	(L,.25)
x_4	**.445**	.050	**M**	**M**	(M,.0)

Table I. Results

The bold typed values in each column represent the solution set of alternatives correspondent to each model.

The different points of view we shall use in this comparative analysis are:

1. *Linguistic description.* This point of view analyses the expression domain used to express the results. This criterion focus in the closeness of the results to the initial linguistic expression domain.
2. *Consistency.* This criterion measures if from the same inputs the linguistic computational models obtaining the same final results.
3. *Accuracy.* This point of view studies the precision of the linguistic operations and the granularity of uncertainty of the results obtained by the different linguistic computational models.

We have chosen these points of view for analysing the different linguistic computational models becuase, they are critical criteria to decide if the reached solution is good and easy to understand by the users that want to solve a problem by means of linguistic information.

To carry out this analysis we shall make a table in which we describe the score of each linguistic computational model according to the different points of view. This score is expressed in the following linguistic term set $S = \{LOW, MEDIUM, HIGH\}$.

	Approx. model (Fuzzy Numbers)	*Approx. model (Labels)*	*Ordinal model*	*2-tuple model*
Ling. Descrip.	LOW	HIGH	HIGH	HIGH
Consistency	MEDIUM	HIGH	HIGH	HIGH
Accuracy	HIGH	LOW	LOW	HIGH

Table II. Analysis of the linguistic computational models

Now we explain the results of the Table II by models (columns):

- *Linguistic approximative model with fuzzy numbers.* This model has a LOW score in the linguistic description because its results are expressed by means of numerical values, a value MEDIUM for consistency due to although it is not usual sometimes it is possible that with the same inputs obtain different results due to the fact that fuzzy numbers have not an inner order, and must be ordered by fuzzy ranking functions. Finally, it has a HIGH value for accuracy because it does not lose information in its computation processes.
- *Linguistic approximative model with linguistic labels.* In this case, the linguistic description is HIGH because the results are expressed in the initial linguistic domain, the consistency has an score of HIGH because from the same input always obtains the same results, while the accuracy is LOW due to this model loses information in the linguistic approximation processes.
- *Ordinal model.* Its linguistic description is HIGH because it expresses the results in the initial linguistic term set and also it has a HIGH value for consistency due to from the same input always obtains the same results, while its accuracy is LOW because its computation processes have loss of information.
- *2-tuple model.* This model has all the scores HIGH because its results are expressed in the initial linguistic domain, from the same inputs always obtains the same results and its computations have not loss of information.

5 Concluding remarks

The above analysis shows that the 2-tuple computational model is always so good as the other linguistic computational models. In linguistic description, accuracy and consistency to be used as the base of representation for expressing aggregation operators. Therefore, we can say the 2-tuple linguistic computational model is the most adequated for dealing with linguistic information in processes of computing with words.

We have just seen that the linguistic 2-tuple model is very adequated for computing with words processes, besides this model provides tools for dealing with multigranularity linguistic contexts [15], linguistic and numerical contexts [16] easily.

References

1. G.I. Adamopoulos, G.P. Pappis, A Fuzzy Linguistic Approach to a Multicriteria-sequencing Problem, *European Journal of Operational Research* **92** (1996) 628-636.
2. P.P. Bonissone and K.S. Decker, *Selecting Uncertainty Calculi and Granularity: An Experiment in Trading-off Precision and Complexity*, in: L.H. Kanal and

J.F. Lemmer, Eds., Uncertainty in Artificial Intelligence (North-Holland, 1986) 217-247.

3. G. Bordogna and G. Passi, A Fuzzy Linguistic Approach Generalizing Boolean Information Retrieval: A Model and Its Evaluation, *J. of the American Society for Information Science* **44** (1993) 70-82.
4. G. Bordogna, G. Passi, A Linguistic Modeling of Consensus in Group Decision Making Based on OWA operators, *IEEE Trans. on Systems, Man and Cybernetics, Part A: Systems and Humans* **27** (1997) 126-132.
5. P. Chang, Y. Chen, Fuzzy Number in Business Conditions Monitoring Indicators: Fuzzy Set Methodologies in Economic Condition, in:*Fuzzy Engineeting toward Human Friendly Systems* (Proceedings of the International Fuzzy Engineering Symposium 91, Yokohama, Japan) **2** (1991) 1091-1100.
6. R. Degani and G. Bortolan, The problem of Linguistic Approximation in Clinical Decision Making, *Int. Journal of Approximate Reasoning* **2** (1988) 143-162
7. M. Delgado, J.L. Verdegay and M.A. Vila, On Aggregation Operations of Linguistic Labels, *Int. J. of Intelligent Systems* **8** (1993) 351-370.
8. D. Dubois and H. Prade, *Fuzzy Sets and Systems: Theory and Applications*, Academic, New York, 1980.
9. A. González, A study of the ranking function approach through mean values, *Fuzzy Sets and Systems* **35** (1990) 29-40.
10. F. Herrera, E. Herrera-Viedma, J.L. Verdegay, Direct Approach Processes in Group Decision Making Using Linguistic OWA Operators, *Fuzzy Sets and Systems* **79** (1996) 175-190.
11. F. Herrera, E. Herrera-Viedma, J.L. Verdegay, A Model of Consensus in Group Decision Making Under Linguistic Assessments, *Fuzzy Sets and Systems* **79** (1996) 73-87.
12. F. Herrera, E. Herrera-Viedma, J.L. Verdegay, *Consensus Based on Fuzzy Coincidence for Group Decision Making in Linguistic Setting*, in: J. Kacprzyk, H. Nurmi and M. Fedrizzi, Eds., Consensus Under Fuzziness (Kluwer Academic Publishers, 1997) 121-146.
13. F.Herrera, E. López, C. Mendaña, M.A. Rodríguez, A Linguistic Decision Model for Personnel Management Solved with a Linguistic Biobjective Genetic Algorithm, *Fuzzy Sets and Systems* **118** (2001) 47-64.
14. F. Herrera and L. Martínez, A 2-tuple Fuzzy Linguistic Representation Model for Computing with Words, *IEEE Transactions on Fuzzy Systems* **8:6** (2000) 746-752.
15. F. Herrera and L. Martínez, A Fusion Method for Multi-Granularity Linguistic Information based on the 2-tuple Fuzzy Linguistic Representation Model, Dept. Computer Sciences and A.I., Granada University, Technical Report#DECSAI-990107, 1999. (http://decsai.ugr.es/~herrera/public.html)
16. F. Herrera and L. Martínez, An Approach for Combining Linguistic and Numerical Information Based on 2-tuple Fuzzy Representation Model in Decision-Making, Int. J. of Uncertainty, *Fuzziness and Knowledge-Based Systems* **8:5** (2000) 539-562.
17. F. Herrera and L. Martínez, A 2-tuple Fuzzy Linguistic Representation Model based on a Symbolic Translation, *In Proc. EUROFUSE-SIC'99 Conference*, Budapest (1999) 25-28.
18. K. Kim, K.S. Park, Ranking Fuzzy Numbers with Maximizing Set and Minimizing Set, *Fuzzy Sets and Systems* **35** (1990) 143-150.

19. E. Levrat, A.Voisin, S. Bombardier and J. Bremont, Subjective Evaluation of Car Seat Comfort with Fuzzy Techniques, *Int. Journal of Intelligent Systems* **12** (1997) 891-913.
20. G.A. Miller, The Magical Number Seven or Minus Two: Some limits On Our Capacity of Processing Information, *Psychological Rev.* **63** (1956) 81-97.
21. M. Roubens, Fuzzy Sets and Decision Analysis, *Fuzzy Sets and Systems* **90** (1997) 199-206.
22. M. Tong, P. P. Bonissone, Linguistic Solutions to Fuzzy Decision Problems, *Studies in the Management Sciences* **20** (1984) 323-334.
23. R.R. Yager, On Weighted Median Aggregation, *Int. J. Uncertainty, Fuzziness and Knowledge-Based Systems* **2** (1994) 101-113.
24. R.R. Yager, An Approach to Ordinal Decision Making, *Int. J. of Approximate Reasoning*, **12** (1995) 237-261.
25. R.R. Yager, *Protocol for Negotiations among Multiple Intelligent Agents*, in: J. Kacprzyk, H. Nurmi and M. Fedrizzi Eds., Consensus Under Fuzziness. (Kluwer Academic Publishers, 1996) 165-174.
26. L.A. Zadeh, The Concept of a Linguistic Variable and Its Applications to Approximate Reasoning. Part I, *Information Sciences* **8** (1975) 199-249, Part II, *Information Sciences* **8** (1975) 301-357, Part III, *Information Sciences* **9** (1975) 43-80.
27. Q. Zhu and E.S. Lee,*Comparison and Ranking of Fuzzy Numbers.* , in: J. Kacprzyk and M. Fedrizzi Eds., Fuzzy Regression Analysis. (Physica-Verlag, 1992) 21-44.

Aggregation Techniques for Statistical Confidentiality

Josep Domingo-Ferrer[1] and Vicenç Torra[2]

[1] Dept. Enginyeria Informàtica i Matemàtiques,
Universitat Rovira i Virgili,
Autovia de Salou, s/n, E-43006 Tarragona (Catalunya, Spain),
e-mail:jdomingo@etse.urv.es, http://www.etse.urv.es/~jdomingo

[2] Institut d'Investigació en Intel.ligència Artificial - CSIC,
Campus UAB s/n, E-08193 Bellaterra (Catalunya, Spain)
e-mail: vtorra@iiia.csic.es, http://www.iiia.csic.es/~vtorra

Abstract. This chapter describes microaggregation, a technique for statistical confidentiality that uses aggregation operators. We describe the goals of statistical confidentiality and its application to continuous and categorical data. We show the application of the method to a small publicly available data set. The chapter finishes by reviewing some of the practical problems of the application of microaggregation to statistical disclosure control.

1 Introduction

National Statistical Offices disseminate information that allows researchers, media and general public to perform their own analyses and studies. This information is collected beforehand from individual and corporate respondents. Although released information should be as detailed as possible from the users' viewpoint, dissemination is in conflict with respondents' privacy. Disclosure risk is defined as the risk of re-identification of particular respondents. That is, if some sensitive and confidential data that have been released are subsequently linked to a particular respondent, confidentiality is lost.

Statistical offices must avoid disclosure to protect the confidentiality of the information provided by respondents. The two usual approaches to disclosure protection (*i.e.* re-identification avoidance) are either to partially suppress data (data suppression or sampling methods) or to perturb them before publication. In this way, the disclosure risk decreases. However, data have to maintain the so-called analytical validity [20], *i.e.* similar results should be obtained when the same statistical analysis is performed on disclosure-protected data and on the original confidential data. Therefore, as [14] point out, "Statistical offices are confronted with the problem of ensuring that the risk of a breach of confidentiality (a disclosure) is acceptably low, while at the same time preserving as much as possible the information content of the data to be released."

Statistical Disclosure Control (SDC) is the discipline that seeks to modify statistical data so that they can be published without giving away the identity of any respondent behind the data. SDC methods for microdata[1] (sets of records, each containing information about an individual respondent such as a person, household, ...) are usually known as *masking methods*. At present, there is a wide range of masking methods for microdata [6]. For example: additive noise, global recoding, record swapping, microaggregation, resampling, PRAM, etc). This chapter focuses on microaggregation, a perturbative method based on aggregation operators.

In addition to masking methods, other tools are required by Statistical Offices to ensure data protection. We underline the following:

- Measures to assess the level of information loss due to a masking method (information loss measures).
- Measures to assess the disclosure risk associated to the masked data (disclosure risk measures).

Information loss measures quantify the data suppression/perturbation introduced by an SDC method in the masked dataset. Disclosure risk measures evaluate the re-identification risk inherent to the masked dataset. Both measures are useful to make the decisions and trade-offs when releasing the data. For perturbative masking (which does not involve data suppression) information loss measures are based on the differences between some computations on the original data and on the masked data, *e.g.* differences between correlations, differences between contingency tables, distance between original and masked records. Disclosure risk measures for perturbative masking are based on re-identification algorithms.

Re-identification algorithms take as input an original microdata set and a masked version of it and try to link masked records with the original ones. For example, record linkage [15] is used to link records in separate files that relate to the same respondent. Re-identification algorithms in [12,20,15] are based on the presence of a set of common variables in both files. The more records are correctly linked, the more re-identification and the higher the disclosure risk. Algorithms have been designed so that not only exact matches are considered but also partial matches. This is so because, as pointed out in [20], "the normal situation in record linkage is that identifiers in pairs of records that are truly matches disagree by small or large amounts and that different combinations of the non-unique, error-filled identifiers need to be used in correctly matching different pairs of records". In particular, this is the case with masked data.

In this chapter, we consider the use of aggregation methods for statistical confidentiality. The structure of this chapter is as follows. In Section 2 the general approach to microaggregation is described. Section 3 discusses

[1] National Statistical Offices also release *tabular data* (tables containing aggregated data), but these will not be dealt with in this paper.

microaggregation for quantitative variables. Section 4 deals with microaggregation for qualitative variables. Section 5 contains the chapter conclusions, as well as some comments on open research issues.

2 Microaggregation

Microaggregation is a perturbative masking method and as such consists on the replacement of the variable values of each respondent by perturbed values. In the case of microaggregation, perturbed values correspond to the aggregation of the values of some similar respondents.

Microaggregation relies on a confidentiality rule that is very often applied by Statistical Offices: publication of a data vector is not allowed if it corresponds to a group of less than k respondents. To allow for publication, at least k respondents should share the same values. To satisfy this condition, groups of at least k similar respondents are formed and the individual value for each respondent is replaced by the average value of the group. According to this, microaggregation is defined by the following two steps:

1. Partition of the respondents into a set of disjoint groups so that the number of respondents in each group is at least k.
2. For each respondents, the value of each variable is replaced by the aggregation of the corresponding values of all respondents in the same group.

According to this, formalization of the method is based on the notion of "similarity" / "distance" and the use of an aggregation operator:

1. Let $X = \{x_1, x_2, \cdots, x_n\}$ be a set of given data for n respondents for which p variables are observed. Thus, x_i is a p-dimensional vector $x_i = (x_{i1}, \cdots, x_{ip})$. Let $d(x, y)$ be a p-dimensional distance. Then, if we denote a partition of X into g groups by $G = \{G_1, \cdots, G_g\}$, with $|G_i|$ being the cardinality of groups G_i and $\hat{x}_i$ being the average data vector of G_i, the optimal partition G for microaggregation is the one that minimizes:
$$\sum_{i=1}^{g} \sum_{x_j \in G_i} d(x_j, \hat{x}_i) \tag{1}$$
while satisfying $|G_i| \geq k$ for all $G_i \in G$.
2. Given an optimal solution G, x_j is replaced by $\hat{x}_i$ if x_j belongs to group G_i ($\hat{x}_i$ is the average of the elements in G_i).

Note that usual clustering methods (e.g. k-means) do not directly apply to this problem because they do not consider group cardinality contraints.

Exactly solving the above minimization problem has been shown to be NP-hard when data are numerical and $d(\cdot, \cdot)$ is the squared Euclidean distance [13]. This is enough to understand why existing practical methods are of heuristic nature [5]. We distinguish below two types of heuristics, depending on whether they deal with univariate or multivariate data:

Univariate microaggregation: A single variable is considered in the masking process (i.e., $p = 1$) or, if several ones are considered, they are masked in successive and independent processes. Fixed-size microaggregation and variable-size microaggregation methods exist. The former yield partitions where all groups except perhaps one have the same size k. The remaining group can contain more than k elements. In variable-size microaggregation the only restriction is that all groups have at least k elements. In this case, several groups can exist with more than k elements. Figure 1 illustrates the advantages of variable-sized groups. If fixed-size microaggregation with $k = 3$ is used, we obtain a partition of the data into three groups, which looks rather unnatural for the data distribution given. On the other hand, if variable-sized groups are allowed then the five data on the left can be kept in a single group and the four data on the right in another group; such a variable-size grouping yields more homogeneous groups, which results in lower information loss. The following approaches have been considered for group building:

- Elements are sorted either in ascending or descending order. Groups are defined with k consecutive elements. This approach is followed in [2].
- Genetic algorithms [10,5]. The fitness function is proportional to Expression (1).
- Modified Ward's algorithm [10,5]. Ward's hierarchical agglomerative clustering has been modified to accomodate group cardinality constraints.

Multivariate microaggregation: In this case, several variables are considered and masked simultaneously (not necessarily all variables at a time, see Section 3). This is, a single partition is used to mask several variables. Two alternatives can be considered:

- *Projected data.* A new variable is defined as the projection of the multivariate data into a single axis. The optimal partition is computed on the new variable. Aggregated values are computed using that partition and the original values.
- *Unprojected data.* Clustering is performed directly on the multivariate data. Modified Ward's algorithm can again be used here.

3 The numerical case

When variables are numerical the usual distance $d(\cdot,\cdot)$ in Expression (1) is the squared Euclidean distance. Therefore, the function to minimize is the within-groups sum of squares and, thus, Expression (1) becomes:

$$\sum_{i=1}^{g} \sum_{x_j \in G_i} ||x_j - \hat{x}_i||^2 \qquad (2)$$

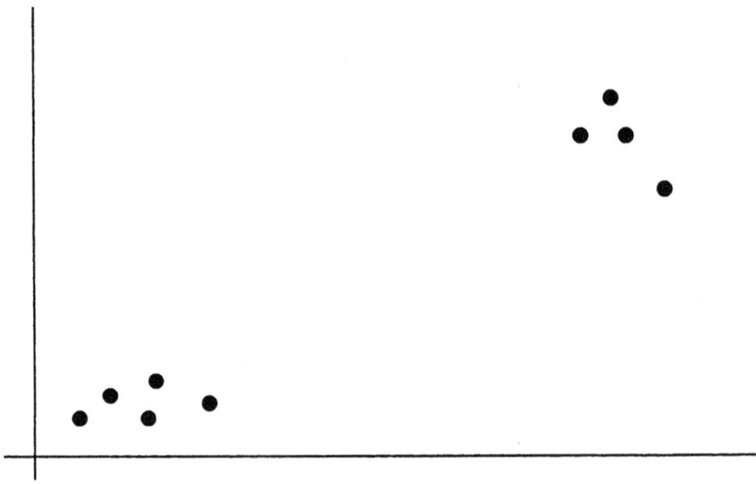

Fig. 1. Variable-sized groups versus fixed-sized groups

At present [10,5,2], the aggregation of the values of all respondents in the same group is computed using the arithmetic mean. There are several results worth mentioning when the arithmetic mean is used as $\hat{x}_i$ in Expression (2):

- Modified Ward's algorithm is attractive in this case because the underlying Ward hierarchical clustering chooses at each step two groups for joining which minimize the increase in the within-group sum of squares.
- The use of heuristics is theoretically justified, because it has been shown in [13] that minimizing Expression (2) with $\hat{x}_i$ being the arithmetical mean is NP-hard.
- It is reported in [7] that multivariate microaggregation on unprojected data taking two or three variables at a time (rather than all variables) is the microaggregation algorithm offering the best tradeoff between information loss and disclosure risk.

It has to be said that aggregation operators other than the arithmetical mean could also be applied, the most straightforward being the OWA operator [21,22]. This is so because this operator can model low influence of extreme values[2]. This is an attractive property in statistical confidentiality, where no individual value should dominate (*i.e.* contribute too much to) the group aggregated value published in the masked dataset.

[2] An OWA operator with a non-decreasing fuzzy quantifier Q such that its derivative is large for values near to 0.5 and small for values near 0 and 1. This is, a fuzzy quantifier corresponding to "about one half".

4 The categorical case

Two aspects have to be addressed in the categorical case: the definition of a distance and the definition of an aggregation procedure. The degree of difficulty of both definitions is similar.

In general, there exist two main approaches to deal with ordinal scales and define operators on these scales. On the one hand there are sets of methods that rely on an underlying semantics defined for each term (e.g. an interval or a fuzzy set). Operators are then defined according to this semantics. On the other hand there exist methods that try to circumvent the explicit definition of a semantics by directly operating over the terms. The latter approach can be implemented by first establishing the desired properties of the operator and then considering all possible operators that satisfy these properties (*e.g.*, t-conorms in [11]); another implementation possibility is first assuming a set of basic operators in the ordinal scale and then building complex operators from the basic ones. This second possibility is the one chosen in [3] to define the convex combination operator (assumption of a negation function on the ordinal scale) and in [8] to define aggregation (assumption of a dual pair of t-norm / t-conorm on the ordinal scale). However, the basic operator is sometimes associated with an implicit semantics. This is the case when negation is considered.

In order to define qualitative aggregation for statistical disclosure control, we consider two approaches: (i) define operations based on user-defined semantics and (ii) define operations based on semantics inferred from basic operators. In particular, we consider the negation function as the basic operator. As shown below, negation implicitly induces a semantics [16] from which aggregation operators and distances can be defined. We do not consider the definition of aggregation operators from properties because aggregation is not associative and the definition of a non-associative m-dimensional operator on a n-dimensional domain through tables is far from practical: n^m values are required (and consistency has to be checked).

We take negation as our basic operator instead of other alternatives (*e.g.*, t-norms in [8]) because the meaning of negation is clear for non-experienced users. As [4] and [18] point out, the negation of a term can be understood as its antonym. From a practical point of view, classical negation (an involutive operator over the domain) restricts the semantics too much because it implies equal informativeness for all terms. To overcome this difficulty, we use the negation introduced in [16]. Both negations are reviewed below.

Classical negation functions are defined over an ordinal scale $L = \{l_0, \cdots, l_n\}$ (with $<$ defined such that $l_0 < \cdots < l_n$) as functions from L to L that satisfy the following two properties [11,9]:

N1: if $l < l'$ then $N(l) > N(l')$ for all l,l' in L
N2: $N(N(l)) = l$ for all l in L

For a given L, conditions N1 and N2 determine the negation function N. This is formally stated by the following proposition proven in [1]:

Proposition 1. *For each set of ordered linguistic labels* $L = \{l_0, \cdots, l_n\}$ *there exists only one negation function that satisfies conditions N1 and N2. This negation is of the form:* $N(l_i) = l_{n-i}$ *for all* l_i *in* L.

This proposition implies that each term in the pair $< l_i, l_{n-i} >$ carries the same information or, in other words, that l_i and l_{n-i} are equally informative. Moreover, this proposition shows that as soon as conditions N1 and N2 are fulfilled, implicit assumptions are made for the underlying semantics of the labels.

However, equal informativeness is not adequate in some situations because the *amount* of information should be different for different labels. To overcome equal informativeness, [16] introduced a new negation function over linguistic labels. The new negation is a function from L to parts of L (*i.e.*, $\wp(L)$). [16] also introduced a method to induce a semantics from that negation. Here semantics is understood as a mapping from the set of labels L into intervals in $[0, 1]$.

Definition 1. [16] A function N from L to $\wp(L)$ is a negation function if it satisfies:

C0: $N(l)$ is a non-empty interval in L
C1: if $l < l'$ then $\max N(l) \geq \min N(l')$ for all l,l' in L
C2: if $l \in N(l')$, then $l' \in N(l)$

In Definition 1, C1 and C2 are generalizations, respectively, of N1 and N2, and C0 is a technical condition.

Note that the negation defined here is, following [4], an inner operator. An alternative definition would be to consider a membership function for each l_i and the negation of each membership function. This would be an external operator following [4]. Both definitions of negation correspond to the idea of antonyms.

A semantics is a mapping of each label into subintervals of the unit interval. Thus, assuming that there exists a negation function N_I in $[0, 1]$ that is the counterpart of the function N in L, it is possible to define a semantics consistent with N_I. In other words, for each label l_i an interval $I(l_i) \subseteq [0, 1]$ is defined, where the set of intervals constitute a partition of $[0, 1]$ consistent with N_I. Consistence is defined as follows (see Definition 3.8 in [16]):

- The negation of all the elements in the interval of label l_i is in the intervals of the negation of l_i. This is,

$$\text{for all } l_i \in L, \text{for all } x \in I(l_i), N_I(x) \in I^*(N(l_i))$$

  where $I^*(\Lambda)$ is defined as $I^*(\Lambda) = \cup_{l \in \Lambda} I(l)$.

- No label in $N(l_i)$ is "superfluous". This condition needs only to be formulated for the extreme labels in $N(l_i)$: if $N(l_i) = \{l_{i_0}, \cdots, l_{i_n}\}$, then:
  1. there exists a value $x \in I(l_i)$ such that $N_I(x) \in I_R(l_{i_0})$
  2. there exists a value $x \in I(l_i)$ such that $N_I(x) \in I_L(l_{i_n})$

  where I_R and I_L mean intervals open in the upper limit and the lower limit, respectively.

An example of semantics that is consistent with the usual negation in the unit interval $N_I(x) = 1 - x$ is the following one given in [16]:

Proposition 2. *Let N be a negation function from L to $\wp(L)$ according to Definition 1. Then, the following expression defines a semantics for L into $[0,1]$ consistent with $N_I(x) = 1 - x$.*

$$I(l_i) = \left[\frac{\sum_{l<l_i} |N(l)|}{\sum_{l\in L} |N(l)|}, \frac{\sum_{l\leq l_i} |N(l)|}{\sum_{l\in L} |N(l)|}\right]$$

where $|\cdot|$ is the cardinality operator.

When the negation function N is such that $|N(l)| = 1$ for all $l \in L$, the intervals reduce to the one with maximal neutrality (this is, maximal similarity on their imprecision). In other words, all intervals have the same measure.

Using the mapping of Proposition 2 it is possible to define a distance and an aggregation operator over L by considering the intervals induced by N. An example is given after the next definition.

Definition 2. Let L be an ordinal scale and N be its negation function. Then the distance $d : L \times L \to [0,1]$ is defined as

$$d(l_i, l_j) = (center(I(l_i)) - center(I(l_j)))^2$$

where $center(\cdot)$ is the center of the interval (if $I(l_i) = [min, max]$ we define $center(I(l_i)) = (min + max)/2$). Similarly, an m-dimensional aggregation operator $\mathbb{C} : L^m \to L$ is defined as

$$\mathbb{C}(a_1, \cdots, a_m) = I^{-1}((1/m)\sum_{i=1}^{m} center(I(a_i)))$$

where $I^{-1}(x) = l$ if $x \in I(l)$.

4.1 Example

This approach has been applied to mask 21 records of the *American Housing Survey 1993* (data publicly available from the U. S. Bureau of the Census through the *Data Extraction System* [19]). Two variables have been considered: *BUILT* (year the household structure was built) and *DEGREE* (long term average heating degree days). Both are defined on an ordinal scale.

Variable *BUILT* has an explicit semantics: its range is $(01-09, 80-93)$ where 09 corresponds to years earlier than or equal to 1919, 08 to the interval $(1920-1929)$, 07 to the interval $(1930-1939)$, and in a similar way for categories $06, 05, 04, 03$ and 02; 01 corresponds to year 1979, and categories from $80-93$ correspond to years $1980-1993$. Distance and aggregation are computed for this variable according to these intervals and using expressions in Definition 2.

The range of variable *DEGREE* is {*coldest, cold, cool, mild, mixed, hot*}. In this case, distance and aggregation have been defined using negation functions. We have used the one given in Table 1. Note that the term *mild* overlaps with *hot* and *mixed* as its negation is defined as {*cool, cold, coldest*}. This term is not considered when building the intervals. Using Proposition 2, the negation in Table 1 induces the intervals given in Table 2 (a graphical interpretation is given in Figure 2).

The original records to be masked and the masked records are given in Table 3 (left and right columns, respectively). Computing the information loss of this method for the records in Table 3 with a distance-based information loss (based on the distance between the old and the new record [17]) yields an average loss of 0.144 (the average is computed over the set of records).

$N(coldest) = \{hot\}$	$N(mixed) = \{cool, cold\}$
$N(cold) = \{hot, mixed\}$	$N(hot) = \{cold, coldest\}$
$N(cool) = \{mixed\}$	$N(mild) = \{cool, cold, coldest\}$

Table 1. Negation function for variable *DEGREE*

$I(coldest) = [7/8, 8/8]$	$I(mixed) = [2/8, 4/8]$
$I(cold) = [5/8, 7/8]$	$I(hot) = [0/8, 2/8]$
$I(cool) = [4,8, 5/8]$	$I(mild) = [0/8, 4/8]$

Table 2. Induced intervals for variable *DEGREE*

5 Conclusions and open research issues

This chapter has described microaggregation and its application to disclosure protection of continuous and categorical variables. While quantitative

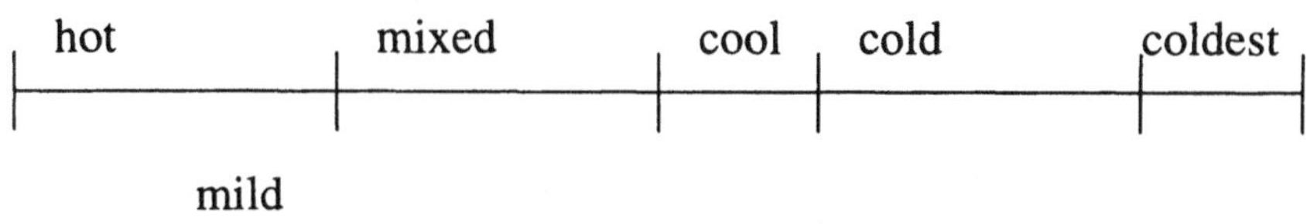

Fig. 2. Intervals induced for variable *DEGREE*

Built	Degree	Built'	Degree'
80	mild	81	mixed
81	cool	81	mixed
81	cool	81	mixed
84	mild	84	cool
84	cold	84	cool
84	cold	84	cool
84	cool	85	cool
85	mixed	85	cool
85	cool	85	cool
85	cool	86	mixed
86	cool	86	mixed
86	mild	86	mixed
86	mild	86	mixed
86	mild	86	mixed
87	cool	86	mixed
87	coldest	88	cold
88	cool	88	cold
89	cool	88	cold
92	cold	92	cold
92	cold	92	cold
93	cold	92	cold

Table 3. Original variables BUILT and DEGREE and their masked versions. Groups displayed in this table have been built with $k = 3$.

microaggregation has been a research topic for about ten years, qualitative microaggregation is quite new. Due to this, the research problems in both areas are quite different.

For continuous variables, the squared Euclidean distance is the natural similarity measure to use and the arithmetic mean is the most straightforward aggregation operator. With these choices, it has very recently been shown that exact optimal microaggregation of continuous variables is NP-hard. Practical microaggregation methods such as those described in [5] are heuristic. An open research issue is to design heuristics which guarantee that the information loss they cause is at most $p\%$ percent above the minimal information loss, where p is an input parameter. A second open research issue is to find the microaggregation heuristic that achieves an optimal tradeoff between information loss and disclosure risk. Indeed, in [7] it is shown that multivariate microaggregation on unprojected data taking only two or three variables at a time is a good option; it remains to be seen which is the best option. A third open issue is whether aggregation operators alternative to the arithmetic mean (*e.g.* OWA) could improve the aforementioned tradeoff and/or reduce the complexity of exact optimal microaggregation.

In ordinal scales, research should be focused on the definition of methods for microaggregation. However, not all existing aggregation operators are suitable. Note that operations for statistical confidentiality should have a clear semantics and be defined in an easy way (the smaller the number of parameters, the best), because their potential users are seldom experts in soft computing methods. Moreover, operators have to be tested to determine their information loss and the disclosure risk when data is released. It is clear that methods with high information loss and disclosure risk are of little interest.

In this work, we have proposed the use of negation functions to define a semantics for ordinal scales when such a semantics does not exist beforehand. Further work is needed to exhaustive analyze its implications for information loss and disclosure risk of the resulting methods.

Acknowledgments

Partial support of the European Commission under project IST-2000-25069 "CASC" and of the U.S. Census Bureau under contracts no. OBLIG-2000-29158-0-0 and OBLIG-2000-29144-0-0 is gratefully acknowledged.

References

1. Agustí, J., Esteve, F., Garcia, P., Godo, L., Sierra., C., (1991), Combining multiple-valued logics in modular expert systems, in *Proc. 7th Conference on Uncertainty in AI*, Los Angeles, July.
2. Defays, D., Nanopoulos, P., (1993), Panels of enterprises and confidentiality: the small aggregates method, in *Proc. of the 1992 Symposium on Design and Analysis of Longitudinal Surveys*, Ottawa: Statistics Canada, 195-204.
3. Delgado, M., Verdegay, J. L., Vila, M. A., (1993), On aggregation operations of linguistic labels, *Int. J. of Intel. Syst.*, 8:351-370.
4. De Soto, A. R., Trillas, E., (1999), On antonym and negate in fuzzy logic, *Int. J. of Intel. Syst.*, 14:295-303.
5. Domingo-Ferrer, J., Mateo-Sanz, J. M., (2000), Practical data-oriented microaggregation for statistical disclosure control, *IEEE T. on Knowledge and Data Engineering* (to appear).
6. Domingo-Ferrer, J., Torra, V., (2000), Disclosure methods and information loss for microdata, in *Confidentiality, Disclosure and Data Access: Theory and Practical Applications for Statistical Agencies*, North-Holland, forthcoming.

7. Domingo-Ferrer, J., Mateo-Sanz, J. M.,(2001), An empirical comparison of SDC methods for continuous microdata in terms of information loss and disclosure risk, in *Second Eurostat-UN/ECE Joint Work Session on Statistical Data Confidentiality*, Skopje, Macedonia, March.
8. Godo, L., Torra, V., (2000), On aggregation operators for ordinal qualitative information, *IEEE T. on Fuzzy Systems*, 8:143-154.
9. Herrera, F., Herrera-Viedma, E., Verdegay, J. L., (1995), A sequential selection process in group decision making with a linguistic assessment approach, *Int. J. Information Science*, 80:223-239.
10. Mateo-Sanz, J. M., Domingo-Ferrer, J., (1998), A comparative study of microaggregation methods, *Qüestiió*, 22:511-526.
11. Mayor, G., Torrens, J., (1993), On a class of operators for expert systems, *Int. J. of Intel. Syst.*, 8:771-778.
12. Newcombe, H. B., Kennedy, J. M., Axford, S. J., James, A. P., (1959), Automatic linkage of vital records, *Science*, 130:954-959.
13. Oganian, A., Domingo-Ferrer, J., (2001), On the complexity of microaggregation, in *Second Eurostat-UN/ECE Joint Work Session on Statistical Data Confidentiality*, Skopje, Macedonia, March.
14. Pannekoek, J., Willenborg, L., (1999), Preface to the special issue on statistical disclosure control, *Netherlands Official Statistics*, 14:4-5.
15. Robinson-Cox, J. F., (1998), A record-linkage approach to imputation of missing data: analyzing tag retention in a tag-recapture experiment, *J. of Agricultural, Biological and Environmental Statistics*, 3:48-61.
16. Torra, V., (1996), Negation functions based semantics for ordered linguistic labels, *Int. J. of Intel. Syst.*, 11:975-988.
17. Torra, V., (2000), *On information loss measures for categorical variables*, Deliverable 3 of contract no. OBLIG-2000-29144-0-0 "Optimizing the Tradeoff Between Information Loss and Disclosure Risk for Categorical Microdata", submitted to U. S. Bureau of the Census.
18. Torra, V., (2001), Aggregation of Linguistic Labels when Semantics is Based on Antonyms, , *Int. J. of Intel. Syst.*, (in press).
19. U. S. Bureau of the Census (2000), *The Data Extraction System*, http://www.census.gov/DES/www/welcome.html
20. Winkler, W.E., (1995), Matching and record linkage, in *Business Survey Methods*, New York: Wiley, 355-384.
21. Yager, R. R., (1988), On ordered weighted averaging aggregation operators in multi-criteria decision making, *IEEE Trans. on SMC*, 18:183-190.
22. Yager, R. R., (1996), Quantifier guided aggregation using OWA operators, *Int. J. of Intel. Syst.*, 11:49-73.

Quantifier Guided Aggregation of Fuzzy Criteria with Associated Importances

I. Blanco, M. Delgado, M.J. Martín-Bautista, D. Sánchez, and M.A. Vila

Department of Computer Science and Artificial Intelligence
University of Granada, Avda. Andalucia 38, 18071 Granada, Spain

Abstract. The evaluation of quantified sentences of the form "Q of D are A" is recognized as a suitable tool for quantifier guided aggregation of fuzzy criteria with associated importances. In this paper we discuss the properties any good evaluation method should verify. We study a new method to evaluate quantified sentences. Our new method is shown to be an extension of the quantified aggregation via the Choquet integral when all the criteria are equally important and a monotone increasing quantifier is employed.

Keywords: Aggregation of fuzzy criteria, fuzzy quantifiers, quantified sentences, fuzzy integrals.

1 Introduction

Aggregation plays a central role in many human reasoning processes. In particular, the aggregation of fuzzy valuations (defined on a set of objects) into a single $[0,1]$ value has received a lot of attention. Applications of aggregation processes to flexible database querying are described in [4,5,13,19]. Other fields where aggregation has been applied are inductive learning [9], control [10] and expert systems [12], among others.

Aggregation has been extensively applied in multicriteria decision problems, where the set of objects is actually a set of criteria [1]. For each possible solution, every criterion has associated a value in $[0,1]$ that represents its satisfaction. The problem is to aggregate the satisfaction of the individual criteria for a given solution, in order to obtain the goodness of the solution. A more general problem is to consider that each criterion has associated a degree of importance. In this case the objective is to aggregate the fulfillment of the *important* criteria [11].

In some situations it is interesting to aggregate not all but some of the values [15]. For example, a given solution in a multicriteria decision problem could be one that verifies, say, *most* of the criteria, and not necessarily all of them. If an importance value is associated to each object, a solution could be good if, for instance, *most* of the *important* criteria are satisfied. Here, *most* is a linguistic quantifier, and we say that the aggregation is guided by a quantifier.

Quantifiers other than "most" can be used. Following Zadeh [21], two important classes of quantifiers are the *absolute* and the *relative* ones. The former are fuzzy natural quantities, represented by fuzzy sets on $\mathbb{N}$. The latter are fuzzy proportions, and they are represented by fuzzy sets on the real interval $[0, 1]$. The quantifier "most" is an example of this kind. Usually, relative quantifiers are the ones employed in aggregation.

Zadeh [21] and Yager [18] distinguished three categories of relative quantifiers:

1. Regular Increasing Monotone (RIM). A quantifier Q is a RIM quantifier if it verifies: a) $Q(0) = 0$. b) $Q(1) = 1$. c) $Q(x) \geq Q(y)$ if $x > y$.
2. Regular Decreasing Monotone (RDM). A quantifier Q is a RDM quantifier if it verifies: a) $Q(0) = 1$. b) $Q(1) = 0$. c) $Q(y) \geq Q(x)$ if $x > y$.
3. Regular Unimodal (RUM). A quantifier Q is a RUM quantifier if it verifies: a) $Q(0) = 0$. b) $Q(1) = 0$. c) $\exists a, b \in [0, 1]$ with $a \leq b$ such that
   - For $y < a$, $Q(x) \leq Q(y)$ if $x < y$
   - For $x \in [a, b]$, $Q(x) = 1$
   - For $x > b$, $Q(x) \geq Q(y)$ if $x < y$

In most of the common problems where aggregation is a needed task, it is usual to employ only some of the categories just described. For example, in multicriteria decision making, it seems natural that the more criteria satisfied, the better the solution under evaluation. Hence, RIM quantifiers are commonly used for this problem. However, it is possible to use RDM and RUM quantifiers for the aggregation, as it is suggested in [20].

A natural way to perform quantifier guided aggregation with associated importances is the use of type II quantified sentences [22]. This technique has been employed for example in [8,17,20]. Quantified sentences take the form "Q of D are A", where Q is a fuzzy quantifier, and both D and A are fuzzy sets defined on a set X with $D \neq \emptyset$. We assume X to be finite, i.e. $X = \{x_1, \ldots, x_n\}$. Both D and A are possibility distributions induced in X by imprecise properties that can be attributed to the objects pertaining to X. In multicriteria aggregation with importances, X is the set of criteria. The fuzzy set D represents the possibility distribution induced in X by the property *to be important*. On his turn, $A(x_i)$ is the degree to which the criteria x_i is satisfied by the solution we are evaluating at this moment.

In [7] we introduced a set of "coherence" properties that any "reasonable" method to evaluate quantified sentences should verify. In the same paper we also analyzed several existing methods to show whether they verify such coherence conditions, and we concluded that no method fulfills all the properties. Then in [7] we introduced both a possibilistic and a probabilistic method to perform the evaluation. Both of them verify all the proposed properties. It is to be remarked that for type I sentences (i.e. when there are no importances associated to the objects, and hence $D = X$) involving RIM quantifiers, the new methods become the well-known aggregation methods based on the Sugeno and the Choquet integral, respectively.

In this paper we extend the set of properties that any good method should verify, and we propose a new method that verifies those properties. The paper is organized as follows. In section 2 we discuss on the properties for evaluation methods. In section 3 we describe some existing approaches and methods. Section 4 is devoted to the definition of a new method and the study of the set of properties that it verifies. Finally, section 5 contains some remarks and a sketch of the future work.

2 Desirable properties for reasonable evaluation methods

Let us consider $X = \{x_1, \ldots, x_n\}$ a finite set, $D \neq \emptyset$ and A two fuzzy sets defined over X, and Q a relative quantifier. Let $Eval_Q(A/D)$ be the evaluation of the sentence "Q of D are A". We consider that any suitable method to assess $Eval_Q(A/D)$ should verify the following properties.

1. Crisp case: If A and D are crisp subsets of X, then the result of the evaluation must be
$$Q\left(\frac{|A \cap D|}{|D|}\right)$$
2. When $D = X$, the evaluation method must be suitable for the evaluation of type I sentences.
3. Evaluation must be time-efficient (as much as possible). We consider efficient a complexity that doesn't exceed $O(nlogn)$.
4. If $D \subseteq A$ the result of the evaluation must be $Q(1)$ (in such case, the percentage of elements in D that pertain to A is 100%).
5. If $D \cap A = \emptyset$ the result of the evaluation must be $Q(0)$ (in such case, the percentage of elements in D that pertain to A is 0%).
6. Quantifiers $\exists$ and $\forall$.
   (a) The evaluation of "$\exists$ of D are A" must return a degree of intersection of D and A. Coherency with fuzzy logic holds if there exist a t-norm $\otimes$ and a t-conorm $\oplus$ such that the evaluation is
$$\bigoplus_{x_i \in X} (A(x_i) \otimes D(x_i))$$
   (b) The evaluation of "$\forall$ of D are A" must return a degree of inclusion of D in A. Coherency with fuzzy logic holds if there exist a t-norm $\otimes$ and a fuzzy implication I such that the evaluation is
$$\bigotimes_{x_i \in X} I\left(D(x_i), A(x_i)\right)$$
7. Evaluation must allow us to use any quantifier.

8. Evaluation is not to be too "strict", i.e., given a quantifier Q defined over the rational interval $[0,1]$ with $Q \neq \emptyset$ and $Q \neq H = \{p/q \mid p \in \{0,\dots,n\}$ and $q \in \{1,\dots,n\}$ and $p \leq q\}$, we must be able to find fuzzy sets A and D such that the evaluation of the sentence "Q of D are A" is not in $\{0,1\}$.
9. If $Q \subseteq Q'$, then $Eval_Q(A/D) \leq Eval_{Q'}(A/D)$.
10. $Eval_Q(A/D) = Eval_Q((A \cap D)/D)$, i.e., the evaluation of "Q of D are A" must return the same value that the evaluation of "Q of D are $A \cap D$".
11. Let Q be a quantifier and let $antQ(x) = Q(1-x)$ be its antonym. Let $\neg A(x_i) = 1 - A(x_i)$. Then $Eval_Q(A/D) = Eval_{antQ}(\neg A/D)$, i.e., the evaluation of "Q of D are A" must return the same value that the evaluation of "$antQ$ of D are $\neg A$".
12. $Eval_Q(A/D) = 1 - Eval_{\neg Q}(A/D)$, where $\neg Q(x) = 1 - Q(x)$, i.e., the evaluation of "Q of D are A" must return the negation of the evaluation of "$\neg Q$ of D are A".
13. De Morgan's laws:
    (a) $1 - Eval_\forall(A/D) = Eval_\exists(\neg A/D)$
    (b) $1 - Eval_\exists(A/D) = Eval_\forall(\neg A/D)$

Properties 1-9 were introduced in [7]. Property 9 applies only for type I sentences, but we consider it to be desirable also for type II sentences. Property 10 is very intuitive. Properties 11 and 12 have been employed as a tool to evaluate type I sentences involving only RDM and RUM quantifiers, previously studied for different evaluation methods in [2,3,12,21]. Property 13 was studied in [2] for the Sugeno and Choquet fuzzy integrals, that are well-known methods for the evaluation of type I sentences ($D = X$). Properties 11-13 ensure that the evaluation method is suitable for quantifier-guided aggregation. In fact, it is sufficient to require properties 11 and 12 because of the following result:

Proposition 1. *Any evaluation method verifying properties 11 and 12 verifies property 13.*

Proof. Inmediate since

$$\exists(x) = \begin{cases} 1 & x > 0 \\ 0 & x = 0 \end{cases} \qquad \forall(x) = \begin{cases} 1 & x = 1 \\ 0 & x < 1 \end{cases}$$

and hence $\forall = \neg ant\exists$ and $\exists = \neg ant\forall$

3 Existing methods; the cardinality approach

We studied in [7] whether some of the existing methods for type II sentences verify the properties. No method verifies all the properties, particularly the very used ones by Zadeh [21] and Yager [16,17,19]. To fill this gap, we introduced in the same paper a method, we called GD, that verifies all the properties.

Method GD is based on what we called the cardinality approach. Following this approach, the evaluation of the sentence is obtained as the compatibility degree between a) the relative cardinality of A with respect to D, and b) the quantifier. This approach was previously used for example by Zadeh [21]. Zadeh's method employs an scalar measure of the relative cardinality of A with respect to D, based on σ-counts. The compability of the relative cardinality and the quantifier is obtained by evaluating the quantifier in the point given by the cardinality, i.e.

$$Z_Q(A/D) = Q\left(\frac{\sum_{x_i \in X}(A(x_i) \wedge D(x_i))}{\sum_{x_i \in X} D(x_i)}\right)$$

For the definition of GD we introduced a new relative cardinality measure, called ER, as a random set over $\mathbb{N}$. Next we introduce some preliminary concepts and the definition of ER.

Let A and D be fuzzy sets over X. We will call $M(A)$ and $M(A/D)$ the following subsets of $[0,1]$:

$$M(A) = \{A(x_i) \mid x_i \in X \text{ and } A(x_i) > 0\} \cup \{1\} \tag{1}$$

i.e., the level set of A plus $\{1\}$, and

$$M(A/D) = M(A \cap D) \cup M(D) \tag{2}$$

where the interesection is performed by means of the minimum. It is always possible to write: $M(A/D) = \{\alpha_1, \ldots, \alpha_m\}$ where we assume $1 = \alpha_1 > \alpha_2 > \ldots > \alpha_m > \alpha_{m+1} = 0$. Also we call $C(A/D, \alpha_i)$ the ratio

$$C(A/D, \alpha_i) = \frac{|(A \cap D)_{\alpha_i}|}{|D_{\alpha_i}|} \tag{3}$$

Obviously, $C(A/D, \alpha_i)$ is defined only when $|D_{\alpha_i}| > 0$. We call $CR(A/D)$ the following subset of $[0,1] \cap \mathbb{Q}$:

$$CR(A/D) = \left\{C(A/D, \alpha_i) \,\middle|\, \alpha_i \in M(A/D) \text{ and } |D_{\alpha_i}| > 0\right\} \tag{4}$$

The following is our definition of the fuzzy cardinality ER:

Definition 1. Given A and D fuzzy subsets of X, the fuzzy relative cardinality ER(A/D), $\forall q \in \mathbb{Q}$ is to be the fuzzy set

$$ER(A/D, q) = \sum_{\{\alpha_i \mid C(A/D, \alpha_i) = q\}} (\alpha_i - \alpha_{i+1}) \tag{5}$$

Here, $ER(A/D, q)$ is the degree with which q pertains to $ER(A/D)$. In the above definition we assume D is normalized (i.e. $\exists x_i \in X$ such that $D(x_i) = 1$). Otherwise, there is a certain probability (specifically $1 - \max_{x_i \in X} D(x_i)$)

that the relative cardinality keeps undefined. To avoid such situation we previously normalize D by using $nf(D) = \max_{x_i \in X} D(x_i)$. We apply $nf(D)$ to $A \cap D$ as well because, that way, the possible relative cardinalities does not change. Obviously, the probability assigned by ER to each possible relative cardinality is scaled by the normalization factor. But this is right since otherwise $\sum_{q \in CR(A/D)} ER(A/D, q) = nf(D)$ and, with the scaling, we obtain the desirable result $\sum_{q \in CR(A/D)} ER(A/D, q) = 1$.

Since the normalization of D and $A \cap D$ is in fact the normalization of the values in $M(A/D)$, and hence a normalization of the differences $\alpha_i - \alpha_{i+1}$, we can avoid the previous normalization of D and $A \cap D$ by using the expression:

$$ER(A/D, q) = \frac{1}{nf(D)} \sum_{\{\alpha_i \mid C(A/D, \alpha_i) = q\}} (\alpha_i - \alpha_{i+1})$$

From ER we define the evaluation method GD as follows:

Definition 2. Given A and D fuzzy subsets of X, and Q a quantifier, the evaluation $GD_Q(A/D)$ for the sentence "Q of D are A" is to be the compatibility between the quantifier Q and the fuzzy relative cardinality $ER(A/D)$,

$$GD_Q(A/D) = \sum_{c \in CR(A/D)} ER(A/D, c) Q(c) \qquad (6)$$

An interesting property of method GD is that in the particular case of type I sentences with RIM quantifiers, GD reduces to the Choquet integral. In [7] we showed that GD verifies properties 1-8. Looking at (6) it is easy to see that GD verifies property 9 as well. It is easy to show that GD can be expressed as

$$GD_Q(A/D) = \frac{1}{nf(D)} \sum_{\alpha_i \in M(A/D)} (\alpha_i - \alpha_{i+1}) Q\left(\frac{|(A \cap D)_{\alpha_i}|}{|D_{\alpha_i}|}\right)$$

As $((A \cap D) \cap D) = (A \cap D)$, it is obvious that $GD_Q(A/D) = GD_Q((A \cap D)/D)$. Hence, GD verifies property 10. The following proposition shows that GD verifies property 12.

Proposition 2. $GD_{\neg Q}(A/D) = 1 - GD_Q(A/D)$

Proof.

$$GD_{\neg Q}(A/D) = \sum_{q \in CR(A/D)} ER(A/D, q) \neg Q(q) =$$

$$= \sum_{q \in CR(A/D)} ER(A/D, q)(1 - Q(q)) =$$

$$\left(\sum_{q \in CR(A/D)} ER(A/D, q) \right) - GD_Q(A/D) =$$

$$= 1 - GD_Q(A/D)$$

Method GD does not verify the property 11 in general. However, the following proposition holds:

Proposition 3. $GD_Q(A/X) = GD_{antQ}(\neg A/X)$

Proof. It is obvious that $CR(A/X) = \{p/n \mid p \in \{0,\ldots,n\},\ n = |X|\}$. In [6] we showed that

$$ER(A/X, p/n) = ED(A,p)\ \forall p \in \{0,\ldots,n\}$$

and

$$ED(A,p) = ED(\neg A, n-p)$$

where $ED(A,p)$ is a measure of the fuzzy cardinality of A, given by

$$ED(A,p) = \begin{cases} \alpha_i - \alpha_{i+1} & \exists \alpha_i \in M(A) \cup \{1\} \text{ such that } |A_{\alpha_i}| = p \\ 0 & \text{otherwise} \end{cases}$$

Therefore

$$ER(A/X, p/n) = ED(A/p) = ED(\neg A, n-p) = ER(\neg A/X, 1-p/n)$$

Moreover it is easy to show that

$$p/n \in M(A/X) \Leftrightarrow (1-p/n) \in M(\neg A/X)$$

and thus

$$GD_Q(A/X) = \sum_{p/n \in CR(A/X)} ER(A/X, p/n)Q(p/n) =$$

$$= \sum_{(1-p/n) \in CR(\neg A/X)} ED(\neg A/X, 1-p/n) antQ(1-p/n) =$$

$$= GD_{antQ}(\neg A/X)$$

Hence, if $D = X$ (in general, when D is crisp), the method GD verifies properties 11-13. But in the general case, property 11 is not verified by GD. Since we consider that we need the mentioned properties in order to aggregate, we introduce in the next section a modification of GD called $GD2$ that verifies, among others, properties 11-13.

4 A new method for quantifier-guided aggregation

To start we introduce a new relative cardinality measure of fuzzy sets, called ER2. Let $M2(A) = M(A)$, let $M2(D) = \{\alpha \in M(D) \mid |D_\alpha| > 0\}$ and let $M2(A/D) = M2(A) \times M2(D)$.

Definition 3. Given A and D fuzzy subsets of X, the relative cardinality of A with respect to D is to be

$$ER2(A/D,q) = \frac{1}{nf(D)} \sum_{\left\{\substack{<\alpha_i,\beta_j>\in M2(A/D)\ \mid \\ q=C2(A/D,\alpha_i,\beta_j)}\right\}} (\alpha_i - \alpha_{i+1})(\beta_j - \beta_{j+1}) \quad (7)$$

where

$$C2(A/D,\alpha_i,\beta_j) = \frac{|A_{\alpha_i} \cap D_{\beta_j}|}{|D_{\beta_j}|}$$

The normalization factor $1/nf(D)$ is introduced because when D is not normalized, we obtain

$$\sum_{\left\{\substack{<\alpha_i,\beta_j>\in M2(A/D)\ \mid \\ q=C2(A/D,\alpha_i,\beta_j)}\right\}} (\alpha_i - \alpha_{i+1})(\beta_j - \beta_{j+1}) = nf(D)$$

Now we introduce the method $GD2$.

Definition 4. The evaluation $GD2_Q(A/D)$ of the sentence "Q of D are A" will be the compatibility between the quantifier Q and the fuzzy relative cardinality $ER2(A/D)$, i.e.:

$$GD2_Q(A/D) = \sum_{c\in CR2(A/D)} ER2(A/D,c)Q(c) \quad (8)$$

where

$$CR2(A/D) = \left\{C2(A/D,\alpha_i,\beta_j) \;\middle|\; <\alpha_i,\beta_j>\in M2(A/D)\right\}$$

Proposition 4. *An alternative formulation of our $GD2_Q(A/D)$ is*

$$\frac{1}{nf(D)} \sum_{<\alpha_i,\beta_j>\in M2(A/D)} (\alpha_i - \alpha_{i+1})(\beta_j - \beta_{j+1})Q\left(C2(A/D,\alpha_i,\beta_j)\right) \quad (9)$$

Proof.

$$nf(D)GD2_Q(A/D) = \sum_{c\in CR2(A/D)} ER2(A/D,c)Q(c) =$$

$$= \sum_{c\in CR2(A/D)} \left(\sum_{c=C2(A/D,\alpha_i,\beta_j)} (\alpha_i - \alpha_{i+1})(\beta_j - \beta_{j+1}) \right) Q(C2(A/D,\alpha_i,\beta_j))$$

Since for every $c \in CR2(A/D)$ at least one pair $< \alpha_i, \beta_j > \in M2(A/D)$ such that $c = C2(A/D, \alpha_i, \beta_j)$ exists, then

$$nf(D)GD2_Q(A/D) =$$

$$= \sum_{<\alpha_i,\beta_j>\in M2(A/D)} (\alpha_i - \alpha_{i+1})(\beta_j - \beta_{j+1})Q\left(C2(A/D, \alpha_i, \beta_j)\right)$$

Example 1. Let $X = \{x_1, x_2, x_3, x_4\}$ and let A and D be the following fuzzy sets.

$$A = \{0.5/x_1 + 0.8/x_2 + 1/x_3 + 0.3/x_4\}$$
$$D = \{1/x_1 + 0.7/x_2 + 0.2/x_3 + 0.5/x_4\}$$

We shall obtain the evaluation of the sentence "Q of D are A" for the quantifiers $\exists$, $\forall$ and M, this last given by $M(x) = x$. From the definition of A and D it follows than

$$M2(A) = \{1, 0.8, 0.5, 0.3\}$$
$$M2(D) = \{1, 0.7, 0.5, 0.2\}$$

and $M2(A/D) = M2(A) \times M2(D)$. Since D is normalized, $nf(D) = 1$. Table 1 shows the set $A_{\alpha_i} \cap D_{\beta_j}$ for each $< \alpha_i, \beta_j > \in M2(A/D)$.

α_i \ β_j	1	0.7	0.5	0.2
1	$\emptyset$	$\emptyset$	$\emptyset$	$\{x_3\}$
0.8	$\emptyset$	$\{x_2\}$	$\{x_2\}$	$\{x_2, x_3\}$
0.5	$\{x_1\}$	$\{x_1, x_2\}$	$\{x_1, x_2\}$	$\{x_1, x_2, x_3\}$
0.3	$\{x_1\}$	$\{x_1, x_2\}$	$\{x_1, x_2, x_4\}$	X

Table 1. Set $A_{\alpha_i} \cap D_{\beta_j}$ for each $< \alpha_i, \beta_j > \in M2(A/D)$

Table 2 shows the value $C2(A/D, \alpha_i, \beta_j)$ for each $< \alpha_i, \beta_j > \in M2(A/D)$.

α_i \ β_j	1	0.7	0.5	0.2
1	0	0	0	1/4
0.8	0	1/2	1/3	2/4
0.5	1	1	2/3	3/4
0.3	1	1	1	1

Table 2. Value $C2(A/D, \alpha_i, \beta_j)$ for each $< \alpha_i, \beta_j > \in M2(A/D)$

From the table 2 it is easy to obtain

$$CR2(A/D) = \{0, \frac{1}{4}, \frac{1}{3}, \frac{1}{2}, \frac{2}{3}, \frac{3}{4}, 1\}$$

The fuzzy relative cardinality $ER2(A/D)$ is calculated as follows:

$$\begin{aligned}
ER2(A/D,0) &= (1-0.8)(1-0.7)+(1-0.8)(0.7-0.5)+\\
&\quad +(1-0.8)(0.5-0.2)+(0.8-0.5)(1-0.7)=\\
&= 0.25\\
ER2(A/D,1/4) &= (1-0.8)(0.2-0)=0.04\\
ER2(A/D,1/3) &= (0.8-0.5)(0.5-0.2)=0.09\\
ER2(A/D,1/2) &= (0.8-0.5)(0.7-0.5)+(0.8-0.5)(0.2-0)=\\
&= 0.12\\
ER2(A/D,2/3) &= (0.5-0.3)(0.5-0.2)=0.06\\
ER2(A/D,3/4) &= (0.5-0.3)(0.2-0)=0.04\\
ER2(A/D,1) &= (0.5-0.3)(1-0.7)+(0.5-0.3)(0.7-0.5)+\\
&\quad +(0.3-0)(1-0.7)+(0.3-0)(0.7-0.5)+\\
&\quad +(0.3-0)(0.5-0.2)+(0.3-0)(0.2-0)=\\
&= 0.4
\end{aligned}$$

Hence,

$$\begin{aligned}
GD2_{\forall}(A/D) &= \sum_{q\in CR2(A/D)} ER2(A/D,q)\forall(q) = 0.25\forall(0)+\\
&\quad +0.04\forall(1/4)+0.09\forall(1/3)+0.12\forall(1/2)+\\
&\quad +0.06\forall(2/3)+0.04\forall(3/4)+0.4\forall(1)=0.4\\
GD2_{\exists}(A/D) &= \sum_{q\in CR2(A/D)} ER2(A/D,q)\exists(q) = 0.25\exists(0)+\\
&\quad +0.04\exists(1/4)+0.09\exists(1/3)+0.12\exists(1/2)+\\
&\quad +0.06\exists(2/3)+0.04\exists(3/4)+0.4\exists(1)=0.75\\
GD2_{M}(A/D) &= \sum_{q\in CR2(A/D)} ER2(A/D,q)M(q) = 0.25M(0)+\\
&\quad +0.04M(1/4)+0.09M(1/3)+0.12M(1/2)+\\
&\quad +0.06M(2/3)+0.04M(3/4)+0.4M(1)=0.57
\end{aligned}$$

In the following we will show some interesting properties of $GD2$.

Proposition 5. *Let* $M2^{+\alpha}(A) = M2(A)\cup\{\alpha\}$ *with* $\alpha\in(0,1]$ *and* $\alpha\notin M2(A)$. *Let* $M2^{A+\alpha}(A/D) = M2^{+\alpha}(A)\times M2(D)$. *Then*

$$\frac{1}{nf(D)} \sum_{<\alpha_i,\beta_j>\in M2^{A+\alpha}(A/D)} (\alpha_i-\alpha_{i+1})(\beta_j-\beta_{j+1})Q\left(C2(A/D,\alpha_i,\beta_j)\right) = $$

$$= GD2_Q(A/D)$$

Proof. Since for any $\alpha \notin M2(A)$, $\alpha_{i_0} \in M2(A)$ exists such that $\alpha_{i_0} > \alpha > \alpha_{i_0+1}$, and $A_{\alpha_{i_0}} = A_\alpha$. Then $Q(C(A/D, \alpha_{i_0}, \beta_j)) = Q(C(A/D, \alpha, \beta_j))$ and

$$\sum_{<\alpha_i,\beta_j>\in M2^{A+\alpha}(A/D)} (\alpha_i - \alpha_{i+1})(\beta_j - \beta_{j+1}) Q\left(C2(A/D, \alpha_i, \beta_j)\right) =$$

$$= GD2_Q(A/D) - \frac{1}{nf(D)} \left(\left(\sum_{\beta_j \in M2(D)} (\beta_j - \beta_{j+1}) Q(C(A/D, \alpha_{i_0}, \beta_j)) \right) \right.$$

$$(\alpha_{i_0} - \alpha_{i_0+1}) + \left(\sum_{\beta_j \in M2(D)} (\beta_j - \beta_{j+1}) Q(C(A/D, \alpha_{i_0}, \beta_j)) \right) (\alpha_{i_0} - \alpha) +$$

$$\left. + \left(\sum_{\beta_j \in M2(D)} (\beta_j - \beta_{j+1}) Q(C(A/D, \alpha, \beta_j)) \right) (\alpha - \alpha_{i_0+1}) \right) =$$

$$= GD2_Q(A/D) - \left(\frac{1}{nf(D)} \sum_{\beta_j \in M2(D)} (\beta_j - \beta_{j+1}) Q(C(A/D, \alpha_{i_0}, \beta_j)) \right)$$

$$((\alpha_{i_0} - \alpha) + (\alpha - \alpha_{i_0+1}) - (\alpha_{i_0} - \alpha_{i_0+1})) = GD2_Q(A/D)$$

Proposition 6. *Let $M2^{+\beta}(D) = M2(D) \cup \{\beta\}$ with $\beta \in (0,1]$ and $\beta \notin M2(D)$ and such that $|D_\beta| > 0$. Let $M2^{D+\beta}(A/D) = M2(A) \times M2^{+\beta}(D)$. Then*

$$\frac{1}{nf(D)} \sum_{<\alpha_i,\beta_j>\in M2^{D+\beta}(A/D)} (\alpha_i - \alpha_{i+1})(\beta_j - \beta_{j+1}) Q\left(C2(A/D, \alpha_i, \beta_j)\right) =$$

$$= GD2_Q(A/D)$$

Proof. Analogous to proposition 5.

Corollary 1. *Let $M2^{+\alpha}(A) = M2(A) \cup \{\alpha\}$ and $M2^{+\beta}(D) = M2(D) \cup \{\beta\}$, with $\alpha, \beta \in (0,1]$ and $\alpha \notin M2(A)$ and $\beta \notin M2(D)$, and such that $|D_\beta| > 0$. Let us define $M2^{A+\alpha,D+\beta}(A/D) = M2^{+\alpha}(A) \times M^{+\beta}(D)$. Then*

$$\frac{1}{nf(D)} \sum_{<\alpha_i,\beta_j>\in M2^{A+\alpha,D+\beta}(A/D)} (\alpha_i - \alpha_{i+1})(\beta_j - \beta_{j+1}) Q\left(C2(A/D, \alpha_i, \beta_j)\right) =$$

$$= GD2_Q(A/D)$$

By this corollary, the evaluation by means of $GD2$ can be performed by using any two sets of α-cuts of A and D, provided that those sets contain $M2(A)$ and $M2(D)$ respectively. This will help us to design an efficient algorithm for the evaluation.

The following proposition relates method $GD2$ to method GD.

Proposition 7.

$$GD2_Q(A/D) = \sum_{\beta_j \in M2(D)} \frac{(\beta_j - \beta_{j+1})}{nf(D)} GD_Q(A/D_{\beta_j})$$

Proof.

$$GD2_Q(A/D) =$$

$$= \frac{1}{nf(D)} \sum_{<\alpha_i,\beta_j> \in M2(A/D)} (\alpha_i - \alpha_{i+1})(\beta_j - \beta_{j+1}) Q\left(C2(A/D, \alpha_i, \beta_j)\right) =$$

$$= \frac{1}{nf(D)} \sum_{\beta_j \in M2(D)} (\beta_j - \beta_{j+1}) \sum_{\alpha_i \in M2(A)} ((\alpha_i - \alpha_{i+1}) Q(C2(A/D, \alpha_i, \beta_j))) =$$

$$= \sum_{\beta_j \in M2(D)} \frac{(\beta_j - \beta_{j+1})}{nf(D)} \sum_{\alpha_i \in M2(A)} ((\alpha_i - \alpha_{i+1}) Q(C(A/D_{\beta_j}, \alpha_i))) =$$

$$= GD2_Q(A/D) = \sum_{\beta_j \in M2(D)} \frac{(\beta_j - \beta_{j+1})}{nf(D)} GD_Q(A/D_{\beta_j})$$

Now, we shall discuss the method $GD2$ on the basis of properties 1-13.

Proposition 8. *The method $GD2$ verifies properties 1-9 and 11-13, and it verifies property 10 when D is crisp.*

Proof. **Property 1:** Since D and A are crisp,

$$GD_Q(A/D_{\beta_j}) = GD_Q(A/D) = \frac{|A \cap D|}{|D|} \quad \forall \beta_j \in M2(D)$$

On the other hand

$$\sum_{\beta_j \in M2(D)} \frac{(\beta_j - \beta_{j+1})}{nf(D)} = 1$$

Hence

$$GD2_Q(A/D) = \sum_{\beta_j \in M2(D)} \frac{(\beta_j - \beta_{j+1})}{nf(D)} \frac{|A \cap D|}{|D|} = \frac{|A \cap D|}{|D|}$$

Property 2: If D is crisp then $GD_Q(A/D_{\beta_j}) = GD_Q(A/D)\ \forall \beta_j \in M2(D)$. Hence

$$GD2_Q(A/D) = \sum_{\beta_j \in M2(D)} \frac{(\beta_j - \beta_{j+1})}{nf(D)} GD_Q(A/D_{\beta_j}) = GD_Q(A/D)$$

Particularly, if $D = X$ then $GD2_Q(A/X) = GD_Q(A/X)$. As GD is a valid method for the evaluation of type I sentences, $GD2$ verifies property 2.

As a consequence, when Q is a RIM quantifier, method $GD2$ reduces to the Choquet fuzzy integral, since in that case $GD_Q(A/X)$ reduces to the Choquet fuzzy integral.

Property 3: We shall prove that when the possible values of $M2(A/D)$ are restricted to a fixed set of k equidistant degrees in $(0,1]$, the method $GD2$ has complexity $O(n)$. In general the complexity of $GD2$ is $O(n^2)$, because the number of possible distinct pairs in $M2(A/D)$ is n^2. When the possible degrees with which an object $x_i \in X$ is in D (or A) are restricted to k equidistant values (i.e. $\{1/k, 2/k, \ldots, k/k = 1\}$, we can define a matrix $V[kk]$ whose (i,j) entry contains the number of objects in X that pertain to $A_{\alpha_i} \cap D_{\beta_j}$. This matrix can be calculated in $O(n)$. Then, the evaluation by $GD2$ can be obtained from V by an algorithm with complexity $O(k^2) = O(1)$. This technique has been successfully used in [14] for the evaluation of GD with $k = 100$. Smaller values for k can be considered guaranteeing a good enough representation of the original sets by α-cuts.

Property 4: We shall prove that when $D \subset A$ in the sense that $D_{\beta_j} \subseteq A$ $\forall \beta_j \in M2(D)$, then the method $GD2$ verifies property 4. In the case $D = A$ with D being not crisp, the property does not hold. If $D_{\beta_j} \subseteq A \; \forall \beta_j \in M2(D)$ then $GD_Q(A/D_{\beta_j}) = Q(1) \; \forall \beta_j \in M2(D)$. By proposition 7 and given that

$$\sum_{\beta_j \in M2(D)} \frac{(\beta_j - \beta_{j+1})}{nf(D)} = 1$$

it follows that $GD2_Q(A/D) = Q(1)$.

Property 5: If $D \cap A = \emptyset$ then $A_{\alpha_i} \cap D_{\beta_j} = \emptyset \; \forall < \alpha_i, \beta_j > \in M2(A/D)$. Hence $C2(A/D, \alpha_i, \beta_j) = 0 \; \forall < \alpha_i, \beta_j > \in M2(A/D)$. By proposition 4 and given that

$$\sum_{<\alpha_i, \beta_j> \in M2(A/D)} (\alpha_i - \alpha_{i+1}) \frac{(\beta_j - \beta_{j+1})}{nf(D)} = 1$$

we obtain $GD2_Q(A/D) = Q(0)$.

Property 6: We shall prove that

1. $$GD2_\exists(A/D) = \sum_{\left\{ \begin{array}{c} <\alpha_i, \beta_j> \in M2(A/D) \mid \\ (A_{\alpha_i} \cap D_{\beta_j}) \neq \emptyset \end{array} \right\}} (\alpha_i - \alpha_{i+1}) \frac{(\beta_j - \beta_{j+1})}{nf(D)}$$

   This expression can be seen as a measure of intersection of A and D.

2. $$GD2_\forall(A/D) = \sum_{\left\{ \begin{array}{c} <\alpha_i, \beta_j> \in M2(A/D) \mid \\ D_{\beta_j} \subseteq A_{\alpha_i} \end{array} \right\}} (\alpha_i - \alpha_{i+1}) \frac{(\beta_j - \beta_{j+1})}{nf(D)}$$

   This expression can be seen as a measure of strict inclusion of D in A.

1. From proposition 4 and since $\exists(C2(A/D, \alpha_i, \beta_j)) = 1 \Leftrightarrow A_{\alpha_i} \cap D_{\beta_j} \neq \emptyset$ and 0 otherwise.
2. From proposition 4 and since $\forall(C2(A/D, \alpha_i, \beta_j)) = 1 \Leftrightarrow D_{\beta_j} \subseteq A_{\alpha_i}$ and 0 otherwise.

Property 7: The method $GD2$ doesn't restrict the quantifiers that can be used.
Property 8: We must find fuzzy sets A and D so that the evaluation of the sentence "Q of D are A" is not in $\{0, 1\}$.

1. If Q is not crisp, then $\exists p, q \in \mathbb{N}$ such that $p < q$ and $0 < Q(p/q) < 1$. Let $c = p/q$. We define A and D as follows:
$$A = \{1/x_1 + \cdots + 1/x_p + 0/x_{p+1} + \cdots + 0/x_n\}$$
$$D = \{1/x_1 + \cdots + 1/x_p + 1/x_{p+1} + \cdots + 1/x_q + 0/x_{q+1} + \cdots + 0/x_n\}$$
Then $M2(A/D) = \{< 1, 1 >\}$ and $CR2(A/D) = \{c\}$. Also $ER2(A/D) = \{1/c\}$ and hence $GD2_Q(A/D) = Q(c)$. Since $0 < Q(c) < 1$, the property holds.
2. If Q is crisp, then let $Q(0) = w \in \{0, 1\}$. Since $Q \neq \emptyset$ and $Q \neq H$, $\exists p, q \in \mathbb{N}$ such that $p < q$ and $Q(p/q) = 1 - w$. Let $0 < a < 1$ and let $c = p/q$. Then, we define A and D as follows:
$$A = \{a/x_1 + \cdots + a/x_p + 0/x_{p+1} + \cdots + 0/x_n\}$$
$$D = \{1/x_1 + \cdots + 1/x_p + 1/x_{p+1} + \cdots + 1/x_q + 0/x_{q+1} + \cdots + 0/x_n\}$$
Then $M2(A/D) = \{< 1, 1 >, < a, 1 >\}$, and $CR2(A/D) = \{0, c\}$, and, $ER2(A/D) = \{(1-a)/0 + a/c\}$, and hence $GD2_Q(A/D) = w(1-a) + (1-w)a$. If $w = 1$ then $GD2_Q(A/D) = (1-a)$. If $w = 0$ then $GD2_Q(A/D) = a$. Since $0 < a < 1$ then $0 < 1 - a < 1$, and the property holds.

Property 9: Trivial, regarding the alternative formulation of $GD2$ given by proposition 4.
Property 10: If D is crisp then, as we showed before, $GD2_Q(A/D) = GD_Q(A/D)$. As method GD verifies property 10, the proposition holds.
Property 11: By proposition 3, method GD verifies property 11 when D is crisp. Hence, $GD_Q(A/D_{\beta_j}) = GD_{antQ}(\neg A/D_{\beta_j})\ \forall \beta_j \in (0, 1]$. Then

$$GD2_Q(A/D) = \sum_{\beta_j \in M2(D)} \frac{(\beta_j - \beta_{j+1})}{nf(D)} GD_Q(A/D_{\beta_j}) =$$
$$\sum_{\beta_j \in M2(D)} \frac{(\beta_j - \beta_{j+1})}{nf(D)} GD_{antQ}(\neg A/D_{\beta_j}) = GD2_{antQ}(\neg A/D)$$

Property 12: By proposition 2, method GD verifies property 12. Hence, $GD_Q(A/D_{\beta_j}) = GD_{\neg Q}(A/D_{\beta_j})\ \forall \beta_j \in (0, 1]$. Then

$$GD2_Q(A/D) = \sum_{\beta_j \in M2(D)} \frac{(\beta_j - \beta_{j+1})}{nf(D)} GD_Q(A/D_{\beta_j}) =$$

$$\sum_{\beta_j \in M2(D)} \frac{(\beta_j - \beta_{j+1})}{nf(D)} GD_{\neg Q}(A/D_{\beta_j}) = GD2_{\neg Q}(A/D)$$

Property 13: By proposition 1 and given that $GD2$ verifies properties 11 and 12.

5 Concluding remarks

In this paper we have introduced a new method for the evaluation of type II quantified sentences. This method can be employed to perform the aggregation of fuzzy criteria with associated importances. The new method verifies most of the properties that we consider suitable for the evaluation. These properties provide a set of intuitive constraints that support the method as a valid tool. A future work will be the study of a possibilistic method based on the same set of crisp representatives that $GD2$ (the set of α-cuts $\{A_{\alpha_i} \mid \alpha_i \in M2(A)\}$ and $\{D_{\beta_j} \mid \beta_j \in M2(D)\}$). This should be an extension of the method based on the Sugeno fuzzy integral (this last limited to the case of type I sentences and RIM quantifiers). A similar relation between methods GD and ZS was described in [7], as ZS is a possibilistic method defined by us on the basis of the same crisp representatives than GD.

References

1. R.E. Bellman and L.A. Zadeh. Decision-making in a fuzzy environment. *Manage. Sci.*, 17(4):141–164, 1970.
2. P. Bosc and L. Lietard. Monotonic quantified statements and fuzzy integrals. In *Proc. NAFIPS/IFIS/NASA Conference*, pages 8–12, 1994.
3. P. Bosc and L. Lietard. On the comparison of the Sugeno and the Choquet fuzzy integrals for the evaluation of quantified statements. In *Proc. of EUFIT'95*, pages 709–716, 1995.
4. P. Bosc and O. Pivert. Fuzzy querying in conventional databases. In L.A. Zadeh and J. Kacprzyk, editors, *Fuzzy Logic for the Management of Uncertainty*, pages 645–671. Wiley and Sons, 1992.
5. J.C. Cubero, J.M. Medina, O. Pons, and M.A. Vila. Using OWA operator in flexible query processing. In R.R. Yager and J. Kacprzyk, editors, *The Ordered Weighted Averaging Operators: Theory, Methodology and Applications*, pages 258–274. Kluwer Academic Publishers, 1997.
6. M. Delgado, M.J. Martín-Bautista, D. Sánchez, and M.A. Vila. A probabilistic definition of a nonconvex fuzzy cardinality. Fuzzy Sets and Systems, To appear, 2000.
7. M. Delgado, D. Sánchez, and M.A. Vila. Fuzzy cardinality based evaluation of quantified sentences. *International Journal of Approximate Reasoning*, 23:23–66, 2000.
8. J. Kacprzyk, J. Fedrizzi, and M. Nurmi. Group decision making with fuzzy majorities represented by linguistic quantifiers. In J.L. Verdegay and M. Delgado, editors, *Approximate Reasoning Tools for Artificial Intelligence*, pages 126–145. Verlag TÜV Rheinland, 1990.

9. J. Kacprzyk and C. Iwanski. Inductive learning from incomplete and imprecise examples. In B. Bouchon-Meunier, R. Yager, and L.A. Zadeh, editors, *Uncertainty in Knowledge Bases*, volume 521 of *Lecture Notes in Computer Science*, pages 424–430. Springer-Verlag, 1991.
10. J. Kacprzyk and R.R. Yager. Softer optimization and control models via fuzzy linguistic quantifiers. *Information Sciences*, 34:157–178, 1984.
11. J. Kacprzyk and R.R. Yager. Using fuzzy logic with linguistic quantifiers in multiobjective decision making optimization: A step towards more 'human-consistent' models. In R. Slowinski and J. Teghem, editors, *Stochastic Vs. Fuzzy Approaches to Multiobjective Mathematical Programming Under Uncertainty*, pages 331–352. Kluwer Publishing, 1990.
12. Y. Liu and E. Kerre. An overview of fuzzy quantifiers part I (interpretations) and II (reasoning and applications). *Fuzzy Sets and Systems* 95, pp.1-21, 135-146, 1998.
13. D. Rasmussen. SummarySQL - a general purpose fuzzy query language. *Datalogiske Skrifter*, 75, 1997. Universidad de Roskilde.
14. D. Sánchez. *Adquisición de Relaciones Entre Atributos En Bases de Datos Relacionales (in spanish)*. PhD thesis, Department of Computer Science and Artificial Intelligence, University of Granada, December 1999.
15. R.R. Yager. Quantifiers in the formulation of multiple objective decision functions. *Information Sciences*, 31:197–139, 1983.
16. R.R. Yager. General multiple-objective decision functions and linguistically quantified statements. *International Journal of Man-Machine Studies*, 21:389–400, 1984.
17. R.R. Yager. On ordered weighted averaging aggregation operators in multicriteria decisionmaking. *IEEE Transactions on System, Man and Cybernetics*, 18(1):183–190, 1988.
18. R.R. Yager. Connectives and quantifiers in fuzzy sets. *Fuzzy Sets and Systems*, 40:39–76, 1991.
19. R.R. Yager. Fuzzy quotient operators for fuzzy relational data bases. In *Proc. of IFES'91*, pages 289–296, 1991.
20. R.R. Yager. Quantifier guided aggregation using OWA operators. *International Journal of Intelligent Systems*, 11:49–73, 1996.
21. L. A. Zadeh. A computational approach to fuzzy quantifiers in natural languages. *Computing and Mathematics with Applications*, 9(1):149–184, 1983.
22. L.A. Zadeh. PRUF - a meaning representation language for natural languages. *International Journal of Man-Machine Studies*, 10:395–460, 1978.

Part 4

Fuzzy Quantities and Their Aggregation

Verbally Generated Fuzzy Quantities and Their Aggregation

Milan Mareš* and Radko Mesiar**

* Institute of Information Theory and Automation
Academy of Sciences of the Czech Republic
P.O.Box 18, 182 08 Praha 8, Czech Republic
E-mail: mares@utia.cas.cz

** Faculty of Civil Engineering
Slovak Technical University
Radlinského 11, 813 68 Bratislava, Slovak Republic
E-mail: mesiar@cvt.stuba.sk

Foreword

The processing of vague data recently becomes one of attractive topics in the fuzzy set theory and its applications. As the vagueness is usually represented by some verbal expressions, this branch of the fuzzy sets is frequently called "computing with words". Seemingly, but only seemingly, it could be understood as computational processing of fuzzy numbers or fuzzy quantities in the already classical sense. Other authors understand the computing with words rather as a fuzzy logical discipline being near to fuzzy reasoning methods and other related branches. Both approaches are rational and fully acceptable but, in the matter of facts, none of them appears to be complete. Their parallel existence offers a conclusion that the fair approach to computing with words can consist in some kind of their combination. Computing with words has two faces – quantitative and qualitative one – and each of them would be somehow reflected. The fuzzy set theoretical model of verbal variables, their generating and processing suggested in this contribution and in some of the referred papers is intended to offer such combined view on the quantitative – qualitative dualism existing in the "computing with words" and to develop at least elementary methods for manipulation with such dualistic verbal data.

The text of the contribution is divided into two main parts. The first one of them, presents in four chapters the introductory heuristic discussions of the subject, elementary concepts of the fuzzy quantities theory, and the illustrative description of the suggested model in its elementary form. Its basic components and their simple properties are introduced. The second part, presents in two last chapters mathematically more advanced treatment of the above elementary model. It is based on the application of the triangular norms (t-norms) and conorms to the processing of the qualitative component of the verbal data.

Acknowledgements

The authors have to express their thanks to many partners, both institutional and personal. A significant place among the institutional ones have several grant agencies. The research whose results are summarized in the following chapters was supported by the Grant Agency of the Czech Republic, Grant No. 402/99/0032, the Grant Agency of the Academy of Sciences of CR, Key Project No. K 1075601 and grant No. A 1075106, Ministry of Education, Youth and Sports of CR Project No. VS 96063, and VEGA 1/8331/01. Very important was the technical and organizational support given by authors' home institutions, namely the Institute of Information Theory and Automation of the Academy of Sciences of the Czech Republic in Prague and Faculty of Civil Engineering of the Slovak Technical University in Bratislava, Slovak Republic.

There are also many persons with whom particular topics were discussed and whose opinions were inspirative for the research summarized here. Let us mention at least some of them, namely Prof. Vilém Novák from Ostrava, Dr. Didier Dubois from Toulouse, Prof. Bernard De Baets and Prof. Etienne E. Kerre from Gent.

The authors also want to express their deepest thanks to Mrs. Iva Marešová without whose technical contribution and extremally kind assistance the manuscript could never appear.

Introductory Remarks

The processing of verbal variables being treated in this contribution is a rather synthetical process including several more or less independent specific components. For its presentation and for better orientation in its structure it appears rational to start with its simple and more illustrative model, to define the elementary forms of its concepts and to discuss as thoroughly as acceptable its connection to the modelled reality. This is done in the presented part of this contribution. We briefly remember the classical fuzzy quantity concept and usual way of its handling based on so called extension principle. We discuss its properties and certain inconsistency between the common fuzzy quantities theory and practical intuitive processing of fuzzy data, and suggest a heuristic point-of-view on its nature and on a possible strategy how to overcome it. Further, we introduce here the fundamental concepts of an alternative model of fuzzy quantity generated by verbal specification, and describe the possibilities of their processing. This introductory explanation is divided into two degrees. In the first one we deal with rather isolated concept of verbal variable not being connected with any specific source of vague data, and in the second we extend the previous approach and relate it to the individual properties of the source (personal or inpersonal) of the generated quantities. Essential components of the first four chapters are formed by heuristic discussions interpreting the introduced notions with respect to the relation between abstract model and everyday reality of meeting vague quantitative data. The essential properties of these formal concepts are also illustrated by examples showing their practical features. Last two chapters deal with triangular norms and conorms, and the processing of generated fuzzy quantities based on them.

1 Starting Position

There already exists a quite good and generally accepted fuzzy set theoretical model of vague quantitative data, namely the theory of fuzzy numbers [5] and fuzzy quantities [30,31]. Here, we briefly remember the basic notions of the fuzzy quantities theory and discuss their advantages and also disadvantages regarding their applicability as a model of verbal quantitative expressions.

1.1 Fuzzy Quantity and Extension Principle

Due to, e. g. [30] or [31], we call *fuzzy quantity* any fuzzy subset a of the real line R with *membership function* $\mu_a : R \to [0,1]$ such that

$$\text{there exists } x_a \in R \text{ such that } \mu_a(x_a) = 1, \tag{1}$$

$$\text{there exists } x_a^{(1)},\, x_a^{(2)} \in R \text{ such that } x_a^{(1)} < x_a^{(2)} \text{ and} \tag{2}$$
$$\mu_a(x) = 0 \quad \text{for all} \quad x \notin [x_a^{(1)}, x_a^{(2)}].$$

Number x_a in (1) is called the *modal value* of a. The fuzzy quantities can be processed by means of analogies of algebraic operations. The extensions of the deterministic operations to their fuzzy quantitative analogies follow so called *extension principle* (see [53]). There exist several modifications of its formulation. For our purposc wc rcmcmbcr thc following onc.

Let $a_1, a_2, \ldots, a_n$ be fuzzy quantities, let f be an operation with n deterministic real arguments

$$f : \underbrace{R \times R \times \cdots \times R}_{n-\text{times}} \to R,$$

and values $f(x_1, x_2, \ldots, x_n) = x \in R$ for real $x_1, x_2, \ldots, x_n \in R$. Then $a = f(a_1, a_2, \ldots, a_n)$ is a fuzzy quantity constructed from $a_1, a_2, \ldots, a_n$ by means of operation f, and its membership function μ_a is defined by

$$\mu_a(x) = \sup \Big[\min(\mu_{a_1}(x_1), \mu_{a_2}(x_2), \ldots, \mu_{a_n}(x_n)) : \tag{3}$$
$$x_1, x_2, \ldots, x_n \in R,\; x = f(x_1, x_2, \ldots, x_n) \Big].$$

For special cases of simple algebraic operations, the addition of fuzzy quantities a, b (denoted $a \oplus b$) is a fuzzy quantity and its membership function is

$$\mu_{a\oplus b}(x) = \sup_{y \in R} \left[\min(\mu_a(y), \mu_b(x-y)) \right], \tag{4}$$

and their product $a \odot b$ is a fuzzy quantity with membership function

$$\begin{aligned} \mu_{a\odot b}(x) &= \sup_{\substack{y\in R\\ y\neq 0}} [\min(\mu_a(y), \mu_b(x/y))] \quad \text{if } x \neq 0, \\ &= \max(\mu_a(0), \mu_b(0)) \qquad\qquad \text{if } x = 0. \end{aligned} \tag{5}$$

The extension principle can be easily used also for operations combining fuzzy quantities with crisp numbers, where each crisp number x is considered for a degenerated fuzzy quantity $\langle x \rangle$ with

$$\begin{aligned} \mu_{\langle x \rangle}(y) &= 1 \text{ for } y = x, \\ &= 0 \text{ for } y \neq x. \end{aligned} \tag{6}$$

More about the extension principle can be found in [5] and in Section 5.3, but also, e. g., in [30]. For our purpose it is sufficient to remember that operations extended by means of the extension principle from crisp to fuzzy quantities posses some but farely not all algebraical properties of their deterministic counterparts (as summarized, e. g., in [31]). So, for example, if the opposite of the fuzzy quantity a is $-a$ such that

$$\mu_{-a}(x) = \mu_a(-x), \quad x \in R$$

then $a \oplus (-a)$ is not identical with zero in the sense of degenerated fuzzy quantity $\langle 0 \rangle$, and for real numbers r_1, $r_2 \in R$ and fuzzy quantity a the distributive equality between

$$\langle r_1 \rangle \odot a \oplus \langle r_2 \rangle \odot a \quad \text{and} \quad \langle r_1 + r_2 \rangle \odot a$$

does not also generally hold.

In any manipulation with fuzzy quantities we understand the equality $a = b$ between them as a point-wise equality between their membership functions $\mu_a(x) = \mu_b(x)$ for all $x \in R$. More about equalities between fuzzy quantities can be found, e. g., in [30,31,5].

The fuzzy quantities can be also compared by means of an ordering relation. Proper definition of generally acceptable ordering over fuzzy quantities is not simple. In [16] it is possible to find a remarkable list of attempts to find such definition. One of them, mentioned also in [5] and in [30,31], displays an interesting consistency with the extension principle. It is based on the methodological idea that any relation between fuzzy quantities (i. e., quantities with several possible values) is to be also fuzzy – valid only with some possibility depending on the possibilisties of values of the vague quantities on the input. In formula, if a, b are fuzzy quantities with membership functions μ_a, μ_b then the possibility that an ordering relation $a \geq b$ is true is equal to the number

$$\mu_{\geq}(a, b) = \sup\left[\min(\mu_a(x), \mu_b(y)) : x,\, y \in R,\, x \geq y\right],$$

and the "similarity" relation $a \sim b$ is true with the possibility

$$\mu_{\geq}(a,b) = \sup\left[\min(\mu_a(x), \mu_b(x)) : x \in R\right].$$

Even if these actual relations display some interesting properties (see [5,31]), the plurality of alternative definitions of ordering between fuzzy quantities shown in [16] points at certain hesitation regarding the nature of the ordering and comparison between fuzzy quantities. We remember this hesitation when we will analyse the behaviour of the verbal variables introduced in the following chapters.

1.2 Heuristic Discussion

The above model of fuzzy numbers (or fuzzy quantities) and their processing remembered above is generally accepted. The extension principle respecting all re-combinations of possible values of the treated fuzzy numbers is evidently logical and excerping all significant possibilities. In some sense it remembers by its formal structure (and also by its entire logic) the convolution principle used in the probability theory. Up to this degree there is everything excellent. Anyhow, we have already seen that there is "something" in the structure of such fuzzy quantities what prevents us from choosing single ordering relation as the best one – in fact, we do not know what exactly is to be compared.

Moreover, there is another, perhaps more serious, feature of the fuzzy quantities and extension principle which provokes our intuitive feeling about the "fair world". The extension principle, especially if it is repeated many times, enormously extends the range of positively possible values of the result. Especially, if the summation (4) is repeated for a long series of fuzzy quantities $a_1 \oplus a_2 \oplus \cdots \oplus a_n$ (this repetition is correct as the operation (4) is associative, see [30,31]) then the extent of possible results is rather shocking. It, for example, devaluates fuzzy approaches to the critical path method in the network analysis (see, e. g., [33]) even if the natural vagueness of some time-data in the practical applications of network analysis urgently demand the implementation of fuzzy set theoretical methods. Formally, our first reaction could be rather flegmatic. Repeated manipulation with vague numbers simply increases the vagueness of the result. It is acceptable by our intuition and we have to include this fact in our mental world. But, does the extension principle really reflect the speed of the increasing of vagueness? Does it correspond with our everyday experience?

Though the concept of fuzziness was introduced by Lotfi A. Zadeh in 1965 (see [52]), the vagueness belongs to our life since the time immemorial. Our forefathers were used to read and understand vague (i. e., fuzzy) quantitative data or data with exact numbers but related to vague units. Everybody of us has sometimes met historical (even not so very historical) contributions with instructions using such quantities like "pinch", "handful", "two pieces of wood" (small or big?), "3 days riding distance", "hot", "for a moment", "proportional weight", "morally acceptable measure of profit" and many others. In many technological, geographical, medical, pharmacological and other instructions such "fuzzy" quantities were repeatedly combined but the results were not as uncertain as we could expect. It was always the same soup which was cooked, the composition of metals was of stabile quality, and the Pony Express arrived to El Paso quite regulary. Why? Either, the "pinch", "day riding distance", or "high temperature" are not as vague units as we admit, or their processing does not fully respect the extension principle and the fuzzy quantities modelling the verbal variables are of a specific structure deserving special treatment rather different from the extension principle (better said – generalizing it). In the following chapters we attempt to develop the idea hidden in the latter one of these possibilities.

2 Verbal Variable

In this chapter we finally specify one of the main objects of our interest and analyze its structure. We prepare the formal tools by means of which the general verbal quantitative expression can be transformed into fuzzy quantity representing the actual vague number. It is useful to mention one fundamental fact. The vagueness represented by formal fuzzy sets disposes with infinite dense set of real-valued functions – the membership functions. On the other hand, any real language, except names of numbers, has only finite (and quite limited) number of words bearing some information about the structure of quantitative uncertainty. It means that there exists an unavoidable disproportion between a finite set of real objects (verbal expressions) and infinite set of mathematical objects (membership functions) applicable for their formal representation. This disproportion implies the necessity to accept some degree of subjectivity or arbitrarity in connecting verbal expressions with actual membership functions or related structures called below the shapes. Such arbitrarity is necessary and it is one of the arguments for the fact that the application of the fuzzy set theory to the real world situation always is not only a matter of mathematical exactness but also a matter of experience, intuitive proportionality and other similar phenomena outside the formalistic purism.

2.1 Structure of Verbal Variable

By *verbal variable* we understand any verbal expression with quantitative content. For example "approximately 8", "much more than 10 ", "something between 15 and 20", "near to 5", "about 7", and also "many", "few", "several", "acceptably many ", etc.

Among these examples we can find some the structure of which is especially transparent. It regards such verbal variables like "approximately 8", "near to 5", "about 7". We can consider them for in some sense elementary or basic ones and we can also see that the information contained in their formulations can be decomposed into two components. Namely, the quantitative one, represented by crisp numbers, in our examples "8", "5", "7", and the qualitative one describing the structure or uncertainty connected with their validity, represented by words, in our examples by "approximately", "near to", "about". We can see that in our elementary examples the verbal uncertainty connected with the possible values of the verbal variables changes with their distance from the crisp "core" value. This possibility is the smaller the greater is the distance of the considered value from the "core". This property can be considered for essential for the characterization of such elementary verbal variables. Let us remember that their structure of uncertainty need not be symmetrical as illustrated, e. g., by "almost 6" or "not much more than 9".

It is useful to note that there exist also elementary verbal variables with seemingly "anonymous" crisp value, like "many", "several", "a bit" and others. In fact, the crisp numerical values of such variables intuitively exist and correspond with the actual applications. They are accompanied with usually relatively wide component of verbal uncertainty, so that the very crisp numerical value does not appear to be so significant. We know that such "anonymous" verbal variable like "much", e. g., "much money ", means something very different if it is used for family savings or state budget. Some crisp numbers are intuitively percepted behind both expressions, these crisp quantities significantly differ, and the extent of possible values hidden under the words "much money" is so wide that the exact values of these crisp quantities are not felt to be so dogmatically fixed.

The above elementary verbal variables can be used for the construction of more sophisticated verbal quantitative structures. They are, among our examples, represented by expressions like "much more than 10", "something between 15 and 20", and also expressions like "near to or a bit more than 30", "approximately natural multiple of 10" can be considered to belong to this class of verbal expressions. Their possible values are not concentrated to exactly one crisp number but they can be constructed like some kind of combination of the elementary verbal variables. In our examples, "much more than 10" is an algebraical summation (whatever we formally mean by the operation) of verbal variables "10" (which can be considered for crisp), and "much". On the other hand, "something between 15 and 20" is a logical conjunction of "not more than 15" and "not less than 10". Expression "near to or a bit more than 30" is a more complex structure formed by logical disjunction of "near to 30" and "a bit more than 30" where the latter verbal variable is a sum of "30" and "a bit". Following this idea we can see that "approximately natural multiple of 10" is the unlimited sequence of logical disjunctions of the elementary verbal variables "near to 10" or "near to 20" or ... or "near to $n \cdot 10$" or ... for $n = 1, 2, \ldots$. Similarly, "manytimes more than 5" can be decomposed to the product of "many" and "5".

Summarizing the heuristic analysis illustrated by the above examples we can see that usual verbal expressions determining some quantitative values, called here "verbal variables", can be decomposed into two types of components, namely, crisp quantitative values (or some kind of their combinations) and verbal descriptions of the type of their vagueness (or, again, some kind of their combination). This quantitatively–qualitative dualism should be observable even in the structure of fuzzy quantities formally modelling the verbal variables.

2.2 Simple Representation of Verbal Variables

The above heuristics can be formalized in the following way which bridges the distance between vague linguistic expressions and exact mathematical (fuzzy set theoretical) concepts.

Remember that a verbal expression consists of deterministic numerical value and of a component representing the uncertainty connected with it. If we want to represent a verbal variable A by some fuzzy quantity a with membership function μ_a then these two components should be in some way included into its structure.

In the following sections we call the deterministic component of A its *crisp core* and represent it by a real number $x_A \in R$. The set of real numbers and its formal structures offer the possibility to manage the processing of the crisp cores in a well known classical manner.

The qualitative or semantic component of a verbal variable A will be represented by a real-valued function $\varphi_A : R \to [0, 1]$, called the *shape* of A. We suppose that

$$\varphi_A(0) = 1, \tag{1}$$

$$\text{if } x_1 < x_2 \le 0, \text{ then } \varphi_A(x_1) \le \varphi_A(x_2), \tag{2}$$

$$\text{if } 0 \le x_1 < x_2, \text{ then } \varphi_A(x_1) \ge \varphi_A(x_2), \tag{3}$$

$$\text{the set } \{x \in R : \varphi_A(x) > 0\} \text{ is limited.} \tag{4}$$

Conditions (2) and (3) express monotonicity of the shapes valid only for the "possible" arguments. This formulation opens the possibility to define (partwise monotonous) shapes also for the vagueness regarding, e. g., integer values of argument, only.

In our practical examples we index the shapes by the verbal expressions which are to be represented, e. g., $\varphi_{\text{approximately}}$, φ_{about}, φ_{almost}, etc. At least slightly careful reader can note that the shapes formally remember of the membership functions of fuzzy quantities. Really, they represent in certain sense normalized form of the fuzzy quantities modelling the verbal variables, and being combined with the crisp cores (and another component) they can specify the actual membership function of the fuzzy quantity which corresponds with the given verbal variable.

The shapes defined in this manner represent the uncertainty component of the verbal variables which we have in the previous subsection called "elementary" – it means those whose possibilities are concentrated (symmetrically or asymmetrically, with wider or narrower dispersion) near some "modal" value – in our case near to 0.

It is also easily acceptable that even the shapes of the "anonymous" verbal variables, also mentioned in the previous subsection (like "many", "several", etc.) can fulfil conditions (1) – (4) with, perhaps, rather wide set

$$\{x \in R : \varphi_A(x) = 1\}.$$

Regarding some verbal variables, condition (4) can seem to be rather discutable. In fact, for the manipulations with fuzzy quantities expectable in the usual processing of verbal variables, condition (2) and related (4) does not seem to be quite unavoidable. It has some real sense for the products of fuzzy quantities, only. But on the other hand, it is not as nonrealistic as it could seem to be. Even verbal expressions like "many", "more than...", etc., have in the real applications their upper limits given by the common sense and generally limited extent of our practical quantitative experience. Nevertheless, in some models in which the very high numerical values of verbal variables are not significant, we may consider condition (4) in very liberal sense and to use, e. g., shapes modelling "much more than..." without formal mentioning the limit of possible values (we can see such approach, e. g., also in Example 4.3). Having introduced the concept of the "elementary" shape we can also easily expect that the more complex verbal quantitative structures (also mentioned in the previous subsection) will be characterized by some combinations of the elementary shapes. In one of the following subsections we will specify what all the expression "combinations" can mean.

In this moment, it is reasonable to remember the general note presented in the introductory paragraphs. We can see that the class of potentially possible shapes (i. e. functions fulfilling (1) – (4)) is infinitely large meanwhile the number of verbal expressions (words and combinations of words) which are to be modelled by them is finite and not very large. As we have already noted, this disproportion between real languages and theoretical tools for the abstract representation demands the application of subjective choice of actually used abstract models. This subjectivity looks unavoidable and its existence means that the practical applications of the fuzzy set theoretical model of verbal variables includes, besides mathematical exactness, also some kind of arbitrarity – let us say, some kind of art or creative view at the reality. This is, by the way, an inherent component of many applications of fuzzy set theoretical models, and also of some other mathematical models in general.

The formal superposition of the crisp core and shape will be defined in one of the following subsections of this chapter. Here, we only show the way in which it can be done. In the simplified model where we suppose the validity of shapes independent on the actual crisp value we can proceed as follows.

Let A be a verbal variable with crisp core x_A and shape φ_A. Then the fuzzy quantity a modelling A has the membership function μ_a defined by

$$\mu_a(x) = \varphi_A(x - x_A) \quad \text{for } x \in R. \tag{5}$$

It is easy to see the following statement.

Remark 2.1. Membership function μ_a defined by (5) fulfills properties (1), (2) as follows from (1) and (4) and from (5).

The simple procedure (5) how to get from a crisp core and shape of a verbal variable to the membership function of a fuzzy quantity representing it generally shows the effective way but it need not be fully satisfactory. Namely, there exists a quite strong intuitive feeling that the bound between shape φ_A and membership function μ_a should somehow respect the range of values which are covered by the verbal variable. That, e. g., "approximately 10" includes rather different extent of possible values than "approximately 10 000", or "a bit more than 1" is a much narrower interval than "a bit more than 1 000". Formula (5) does not respect this common expectation. In the next subsection we suggest a concept which can bridge this gap.

2.3 Scales

As we have already mentioned above, one verbal expression has rather different (narrower or wider) interpretation in dependence on the actual crisp value to which it is related. Usually, the higher the value of the crisp core is the wider is the range of possible values which are covered by the verbal expression. This proportionality cannot be accepted dogmatically. Some sources of data (e. g., estimations of some experts) can be especially qualified or more exact (i. e., they admit narrower range of possible values) in the neighborhood of some crisp cores, other sources (e. g., estimations of other experts, or statistical analyses, or experimental measurements) can be especially reliable for other crisp cores and their neighborhoods. This heuristical idea can be formalized by means of the concept of scale. By *scale* we mean every real-valued function $f : R \to R$ which fulfills

$$f(0) = 0, \tag{6}$$

$$f(x) \text{ is continuous,} \tag{7}$$

$$f(x) \text{ is nondecreasing,} \tag{8}$$

$$\lim_{x \to -\infty} f(x) = -\infty, \quad \lim_{x \to \infty} f(x) = +\infty. \tag{9}$$

The scale defined in this way is to reflect the local exactness of fuzzy data related to particular parts of the real line. For this purpose, the most significant property of scales appears their gradient, i. e., their first derivative if it exists. In the next subsection we will present exact formal relation between scales and membership functions generated by them.

Here, we only remember a few elementary properties of the class of scales, namely its closedness regarding some operations. All following properties are evident consequences of the elementary analytical properties following from (6) – (9).

Remark 2.2. Let f, g be scales. Then also their composition $f(g(x))$, sum $(f + g)(x) = f(x) + g(x)$ and the product $\operatorname{sgn}(x) \cdot f(x) \cdot g(x)$, $x \in R$, are scales.

Having introduced the concept of scale we have completed the system of components characterizing the verbal variable. The crisp core and shape (specifying its "main" numerical value and distribution of its uncertainty) were completed by scale, determining the qualities (accuracy, reliability or similar properties) of the source of the verbal data. Disposing with these three components we can submit complete formal relation between verbal variables and fuzzy quantities representing them.

Also in the case of scales we can observe certain disproportion between the infinite set of real-valued functions fulfilling (6) – (9) and much more limited demands for characterization of the source of the verbal data. Using its verbal specification, we will be probably limited to expressions like "very reliable", "quite reliable", "not very exact", "subjectively influenced", and several others. Even if we do not support our characterization of the sources by such verbal descriptions, our ability to distinguish between various qualities of the source will be much poorer than the offered set of functions. It means that the choice of a scale characterizing individual source of data will be subjected to certain degree of subjectivity unseparable from the bound of real situation and its abstract formal model.

2.4 Complete Representation of Verbal Variables

After introducing the concept of scale we have completed all three components fully characterizing an elementary verbal variable – crisp core, shape and scale. If they are formally represented by formal objects – real number, real-valued function fulfilling (1) – (4), and real-valued function fulfilling (6) – (9) – then it is not difficult to construct the fuzzy quantities which represent them and which can be exactly processed. It is quite natural to expect that this formal representation of verbal variable by a fuzzy quantity will be based on a formula somehow generalizing (5).

Let us consider a verbal variable A with crisp core x_A and uncertainty characterized by a shape φ_A. Let the source producing A be, regarding its qualities, characterized by a scale f_A. Then A is represented by fuzzy quantity a with membership function μ_a such that for any $x \in R$

$$\mu_a(x) = \varphi_A\left(f_A(x) - f_A(x_A)\right). \tag{10}$$

This fuzzy quantitative representation will be considered in the next subsections and chapters. To stabilize the terminology, we say that if the above case is considered then the triple (x_A, φ_A, f_A) *characterizes* or *describes* the verbal variable A and that A is *represented* by the fuzzy quantity a with membership function μ_a fulfilling (10).

Lemma 2.1. Let A be a verbal variable described by the triple (x_A, φ_A, f_A) and let μ_a be a mapping defined for every $x \in R$ by (10). Then μ_a is membership function of a fuzzy quantity.

Proof. We have to prove that μ_a defined by (10) fulfills conditions (1) and (2). First, the properties of φ_A imply that μ_a maps the real line in the closed unit interval $[0, 1]$. Further, (1) is obviously fulfilled with the modal value x_A of μ_a, i. e. $\mu_a(x_a) = 1$. Finally, condition (4) together with the continuity and unlimitedness of f_A mean that (2) is valid, as well. □

Remark 2.3. Conditions (2) and (3) together with the monotonicity of the scale f_A mean that the membership function μ_a defined by (10) is nondecreasing for $x \leq x_A$ and nonincreasing for $x \geq x_A$.

Now, we have constructed a direct way from the verbal quantitative expression to its fuzzy set theoretical representation.

2.5 Several Illustrative Examples

The relatively general models of shapes and scales admit quite wide arbitrarity in choosing their actual forms reflecting specific demands of particular practical situations. Let us show a few examples of the possible forms of those functions.

Examples of Shapes Typical examples of shapes describing the verbal expressions like "approximately" or "about", etc. can be, probably, symmetric functions like

$$\varphi_{\text{approximately}}(x) = \max(0, 1 - |x|), \tag{11}$$

or

$$\varphi_{\text{about}}(x) = \max\left(0, 1 - \frac{|x|}{3}\right), \tag{12}$$

or

$$\varphi_{\text{nearly}}(x) = \max\left(0, 2 \cdot \left(e^{-|x|} - \frac{1}{2}\right)\right), \tag{13}$$

or

$$\varphi_{\text{closely to}}(x) = 1 - \sqrt{1-(x+1)^2} \text{ for } x \in [-1,0],$$
$$= 1 - \sqrt{1-(x-1)^2} \text{ for } x \in [0,1], \quad (14)$$
$$= 0 \qquad \text{otherwise.}$$

The shapes, of course, need not be symmetric, if the modelled verbal expressions are not symmetric, e. g.

$$\varphi_{\text{approximately but rather more than}}(x) = \quad (15)$$
$$= \max(0,\ 1-|x|) \qquad \text{for } x < 0,$$
$$= \max\left(0,\ 1-\frac{|x|}{2}\right) \qquad \text{for } x \geq 0,$$

$$\varphi_{\text{almost}}(x) = \quad (16)$$
$$= \max\left(0,\ 2\cdot\left(e^x - \frac{1}{2}\right)\right) \qquad \text{for } x \leq 0,$$
$$= 0 \qquad \text{for } x > 0.$$

As we have already mentioned in the previous discussions, also the "anonymous" verbal variables can be connected with some shapes. Due to the actual modelled situation it can be, e. g.,

$$\varphi_{\text{few}}(x) = 0 \qquad \text{for } x < 0,\ \text{or } x > 7,$$
$$= 1 \qquad \text{for } x \in [0,5], \quad (17)$$
$$= -\frac{x}{2} + \frac{7}{2} \text{ for } x \in [5,7].$$

The shape characterizing "many" will obviously depend on the modelled real problem. For some of them it can be

$$\varphi_{\text{many}}(x) = 0 \qquad \text{for } x < -10^3 \text{ or } x > 10^{12},$$
$$= x\cdot 10^{-3}+1 \qquad \text{for } x \in [-10^3, 0],$$
$$= 1 \qquad \text{for } x \in [0, 10^7], \quad (18)$$
$$= -x/(10^{12}-10^7) + 10^{12}/(10^{12}-10^7) \qquad \text{for } x \in [10^7, 10^{12}],$$

in other cases it may be, e. g.,

$$\varphi_{\text{many}}(x) = 0 \qquad \text{for } x < -10 \text{ or } x > 110,$$
$$= x/10+1 \qquad \text{for } x \in [-10,0],$$
$$= 1 \qquad \text{for } x \in [0,100], \quad (19)$$
$$= -x/10+11 \text{ for } x \in [100,110]$$

or in some cases it can cover another (may be even partly negative) fuzzy interval fulfilling properties of the shape. Especially in the case of "anonymous" verbal variables the final formal representation of the quantitative expressions necessarily depends on the choice of the scale and also on the (rather arbitrarily) chosen fictive crisp core.

Examples of Scales We have already mentioned that the gradient of a scale function reflects the accuracy of the source of the verbal data. Generally, we can expect that the first derivative of a realistic scale function will decrease with the increasing distance from 0, even if some declinations of this rule, which were discussed in subsection 2.3, are also acceptable. It is also relatively obvious that the scale $f(x) = x$ can be considered for in certain sense neutral. It does not influence the exactness of the generated fuzzy quantity and turns (10) into (5). First derivative of $f(x)$ which is greater than 1 means that the verbal data produced by their source in the neighborhood of this value x are "sharper" than the "neutral" quality is, and vice versa.

Let us present some of possible examples of scales. The first one, the "neutral" $f(x) = x$, was already mentioned. Others are, e. g.,

$$\begin{aligned} f(x) &= x \qquad\qquad \text{for } x \in [-10, 10], \\ &= x/2 - 5 \text{ for } x < -10, \\ &= x/2 + 5 \text{ for } x > 10, \end{aligned} \tag{20}$$

or

$$\begin{aligned} f(x) &= \ln(x+1) \qquad \text{for } x \geq 0, \\ &= -\ln(-x+1) \text{ for } x \leq 0. \end{aligned} \tag{21}$$

The previous scale reflects the generally accepted development of subjective evaluation of the utility of quantitative amounts of goods, and many other types of dependence of subjectively felt "importance" of quantities in the dependence on their nominal values. The following type of the scale

$$f(x) = \text{sgn}(x) \cdot x^2 \tag{22}$$

looks rather irrational and the principle that the quantitative values are the more exact or certain the greater they are, is hardly expectable in the reality. On the other hand, scale

$$\begin{aligned} f(x) &= x \qquad\qquad \text{for } x < 10, \\ &= 3x - 20 \quad \text{for } x \in [10, 20], \\ &= x/2 + 30 \text{ for } x \in [20, 60], \\ &= x/3 + 40 \text{ for } x > 60 \end{aligned} \tag{23}$$

characterizes the source of data (maybe a measurement procedure) which is neutral (or not interesting as it is not used) for values less than 10, relatively very exact for values between 10 and 20, much less exact for values greater than 20, where it is even more uncertain (and, consequently, rarely used) for values greater than 60.

Probably most vague are the verbal data of "anonymous" character like "many", "several", etc. Their sources display high degree of uncertainty in the range of values which they characterize only by vague words. So, e. g., a source which distinguishes only several positive numbers and any greater value characterizes like "many" could be characterized by a scale

$$\begin{aligned} f(x) &= x && \text{for } x < 6, \\ &= x/2 + 3 && \text{for } x \in [6, 20], \\ &= x/10^2 + 9 && \text{for } x > 20. \end{aligned} \tag{24}$$

Similarly, a source which distinguishes only some range of numbers where smaller values are described as "few" and greater as "much" can be characterized by the scale (here, we suppose that negative values are irrelevant)

$$\begin{aligned} f(x) &= x && \text{for } x < 0, \\ &= x/10 && \text{for } x \in [0, 5] \\ &= x - 9/2 && \text{for } x \in [5, 25], \\ &= x/10 + 18 && \text{for } x > 25. \end{aligned} \tag{25}$$

Of course, the scales need not be part-wise linear. The previous example could be transformed into a more realistic one if it includes fluent transitions between the main trends of the scale, e. g.,

$$\begin{aligned} f(x) &= x && \text{for } x < 0, \\ &= 2x && \text{for } x \in [0, 1], \\ &= x^2 + 1 && \text{for } x \in [1, 2], \\ &= 5 + \ln(4x - 7) && \text{for } x > 2. \end{aligned} \tag{26}$$

Due to (8) the scale need not be strictly increasing. For example,

$$\begin{aligned} f(x) &= 2x \text{ for } x \leq 5, \\ &= 10 \text{ for } 5 \leq x \leq 10, \\ &= x \quad \text{for } x \geq 10. \end{aligned} \tag{27}$$

3 Elementary Processing of Verbal Variables

Having described the basic components of (elementary) verbal variables and their formal representations, and having discussed their interpretation, we can turn our attention to their processing. Its principles are expected to respect the duality between the quantitative and qualitative approach to verbal variables and to reflect the character of their combination.

The formally advanced methods of the manipulation with verbal variables and their fuzzy set theoretical representatives, based mostly on the concept of triangular norm and conorm, are presented in Chapters 5 and 6 of this contribution and in other parts of the publication. In this chapter, we summarize the simple elementary processing of shapes and scales which rather illustrates their behaviour and helps to show the motivation for their introduction and investigation.

Evidently, the definition of verbal variable concept and its components themselves would be of no sense if we do not suppose their processing aiming to the solution of some more or less practical problems or to the derivation of interesting properties of vagueness existing in our description of the real world.

3.1 Methodological Background

Wishing to formalize the processing of verbal variables we are to decide, first, for which purpose or purposes it is to serve. Generally, there are two of them.

First, we can process simple verbal variables to construct the composed ones. We have mentioned some examples of such procedures in the preceeding sections, namely in Section 2.1. So, verbal variable "a bit more than 8" can be constructed as a sum (in this moment we do not specify the definition of "sum") of "8" and "a bit", or verbal variable "something between 10 and 15" can be constructed as a logical conjunction of "something more than 10" and "something less than 15" (where each of them itself could be composed verbal variable). These two examples, however simple, illustrate one important fact – the processing of verbal variables is not a uniform nor strictly standardized procedure. It has to respect the semantic structure of relation between the input variables and due to it to be either of algebraical (algebraically operational) or logical (logically operational) character. This necessity to decide sometimes during the first steps of the analysis of the problem what is the prevailing character of the operations to be realized over the verbal quantities is unavoidable. It belongs to the opportunities of the subjectivity and personal intuition to enter the formal mathematical methods of handling vague verbal variables. Other opportunities of this kind were discussed also in the previous chapters.

Let us explicitely remember an intuitively percepted fact. The construction of more complex verbal variables from the simple ones by means of some (algebraical or logical) operations also means that each composed verbal variable can be decomposed into its simple components with crisp core, shape, and scale of the elementary structure fulfilling (1) – (4) and (6) – (9). It means that also any other processing of verbal variables, which will be discussed in the next paragraphs, can be reduced to procedures dealing with simple verbal variables and their components of the elementary type.

Second purpose of the processing of formal variables is their input into some computational algorithms. In the following paragraphs we are interested mainly in this topic.

Now, it is the time to present the main reason why the analytical concept of decomposition of verbal variable into crisp core, shape and scale was suggested. In the traditional approach to vague (i. e., verbal) quantities each verbal variable A is directly modelled by a fuzzy quantity a with membership function μ_a, and it is processed by means of one traditional method (using the extension principle) in spite of the fact that it includes both – quantitative and qualitative – features. The disadvantages of this approach were mentioned in the introductory sections of this contribution. Another one can be added – the computations due to the extension principle (especially if non-trivial membership functions are considered) are very far from simplicity. If they are to be repeated whenever an algorithm is repeatedly used for always new input fuzzy data then the whole procedure appears to be not very comfortable nor user-friendly. It is worth observing that in many cases the input fuzzy data mentioned above differ exclusively in what we call here "crisp core". Instead of "about 8" we compute with "about 12" or "something near to 10" is substituted by "something near to 25".

Taking these arguments into consideration, we can suggest the following approach to the computation of algorithms with fuzzy inputs.

Let us consider an algorithm dealing with variables which can be represented by vague data. We denote them by A, B, C, $D, \dots$ and suppose that they are, generally, verbal variables represented by fuzzy quantities a, b, c, $d, \dots$ with membership functions $\mu_a, \mu_b, \mu_c, \mu_d, \dots$ As shown in the introductory paragraphs of this section, we can suppose without lack of generality, that the verbal variables A, B, C, D already are the simple ones (with values surrounded in a neighborhood of some modal value), and that their eventual composition into more complex verbal variables was already included into the considered algorithm. Then we can proceed with respect to (10) and to suppose that there exist crisp cores $x_A, x_B, x_C, x_D, \dots$, shapes $\varphi_A, \varphi_B, \varphi_C, \varphi_D, \dots$ and scales $f_A, f_B, f_C, f_D, \dots$ such that for any $x \in R$

$$\mu_a(x) = \varphi_A(f_A(x) - f_A(x_A)), \qquad \mu_b(x) = \varphi_B(f_B(x) - f_B(x_B)),$$
$$\mu_c(x) = \varphi_C(f_C(x) - f_C(x_C)), \qquad \mu_d(x) = \varphi_D(f_D(x) - f_D(x_D)),$$

etc. It means that we have decomposed the sequence of vague input data A, B, C, $D, \dots$ into sequences of their structural components

$$x_A, x_B, x_C, x_D, \dots, \quad \varphi_A, \varphi_B, \varphi_C, \varphi_D, \dots, \quad f_A, f_B, f_C, f_D, \dots.$$

Each of these sequences of input components can be processed separately with respect to their formal structure and semantical meaning.

Evidently, the processing of crisp cores repeats the deterministic operations due to the considered algorithm, and it results into a crisp number $x_0 \in R$.

The processing of shapes would respect the either algebraic or logical operations given by the considered algorithm as well as their heuristic relations and it results to a shape function φ_0.

Finally, also the scales can be combined, usually with respect to the relation between the reliabilities of the sources, and an output scale f_0 can be derived.

Having done all these separate procedures, we can use (10) and derive the resulting verbal variable Ω defined by the triple of components (x_0, φ_0, f_0) and characterized by fuzzy quantity σ with membership function μ_σ such that (due to (10))

$$\mu_\sigma(x) = \varphi_0(f_0(x) - f_0(x_0)), \quad x \in R. \tag{1}$$

The possible (or, better said, interpretationally admissible) operations with shapes and scales and their elementary properties are summarized and discussed in the following three chapters.

As we have already mentioned above, the manipulation with crisp cores is a deterministic matter completely respecting the considered (originally deterministic) algorithm and not being significant from the fuzzy set theoretical point of view.

3.2 Processing of Shapes

We have defined the concept of shape as a real-valued function $\varphi : R \to [0, 1]$ fulfilling (1)–(4). We know that it is in some sense normalized pattern of membership functions with possibilities somehow concentrated in a neighborhood of the significant modal value equal to 0. We also know that membership functions of actual fuzzy quantities can be derived directly from the shapes and crisp cores by means of (10).

It is natural to demand that any rational operation with shapes results into a shape, as well. Even under this limitation there exist infinitely many such operations of composition of shapes where many of them have attractive or interesting properties. Let us limit our examples to several binary composition operations, even if the higher -arities are also easily imaginable.

Let φ_A, φ_B be shapes. Then we can define a mapping $\varphi_0 : R \to [0, 1]$, e. g., by means of the extension principle for addition

$$\varphi_0(x) = \sup_{y \in R} \left[\min(\varphi_A(y), \varphi_B(x - y))\right], \quad x \in R \tag{2}$$

or by means of truncated or normalized convolution (compare with [29,28,10], e. g.),

$$\varphi_0(x) = \min\left(1, \int \varphi_A(y) \cdot \varphi_B(x - y)\, \mathrm{d}y\right) \tag{3}$$

or

$$\varphi_0(x) = \int \varphi_A(y) \cdot \varphi_B(x - y)\, \mathrm{d}y \Big/ \sup_{x \in R} \left(\int \varphi_A(y) \cdot \varphi_B(x - y)\, \mathrm{d}y\right), \tag{4}$$

if the respective integrals exist and do not vanish, if the integral in (3) related to $x = 0$ is not less than 1, and if the supremum in (4) is achieved for $x = 0$. Other composition operators reflect rather fuzzy logical relations between the input shapes, e. g.,

$$\varphi_0(x) = \max(\varphi_A(x), \varphi_B(x)), \quad x \in R, \tag{5}$$
$$\varphi_0(x) = \min(\varphi_A(x), \varphi_B(x)), \quad x \in R, \tag{6}$$

or they can express another type of relation between the input shapes, usually reflecting some interpretations of the modelled reality. For example, for $x \in R$,

$$\varphi_0(x) = \varphi_A(x) \cdot \varphi_B(x), \tag{7}$$
$$\varphi_0(x) = \varphi_A(x) + \varphi_B(x) - \varphi_A(x) \cdot \varphi_B(x), \tag{8}$$
$$\varphi_0(x) = \alpha \cdot \varphi_A(x) + (1 - \alpha) \cdot \varphi_B(x), \quad \alpha \in [0, 1], \tag{9}$$

etc.

Each composition is to respect not only the semantic or logical structure of the modelled relations but also the interpretational demands of the given and described reality.

The following simple statement about the presented examples can be easily derived.

Remark 3.1. The compositions of shape functions presented in (2) – (9) preserve the definitoric properties of shapes, i. e., the resulting shapes φ_0 fulfil properties (1) – (4).

Generally, we denote by $\circ$ the binary composition operation of shapes which preserves (1) – (4), and if φ_A, φ_B are shapes then $(\varphi_A \circ \varphi_B)(x) = \varphi_A(x) \circ \varphi_B(x)$ denotes the composed shape. We say that the operation $\circ$ is *expansive* iff $(\varphi_A \circ \varphi_B)(x) \geq \max(\varphi_A(x), \varphi_B(x))$ for any $x \in R$ and any pair of shapes φ_A, φ_B. We say that $\circ$ is *restrictive* iff $(\varphi_A \circ \varphi_B)(x) \leq \min(\varphi_A(x), \varphi_B(x))$ for any $x \in R$ and any pair of shapes φ_A, φ_B.

Remark 3.2. It is easy to verify that the composition operations (6) and (7) are restrictive.

Remark 3.3. It can be easily verified that the composition operations (2), (5) and (8) are expansive.

Lemma 3.1. Let $\varphi_1, \varphi_2, \varphi_3, \dots$ be a sequence of shapes and let $\circ$ be a composition operation preserving (1) – (4). Let us denote by $\varphi_1', \varphi_2,', \dots$ the sequence of shapes generated as follows

$$\varphi_1'(x) = \varphi_1(x), \quad \varphi_k'(x) = (\varphi_{k-1}' \circ \varphi_k)(x), \quad x \in R, \ k = 2, 3, \dots.$$

If the composition operation $\circ$ is either expansive or restrictive then the sequence $\varphi_1', \varphi_2', \varphi_3', \dots$ is point-wise convergent.

Proof. The statement of this lemma evidently follows from the fact that for every $x \in R$ the sequence $\varphi'_1(x), \varphi'_2(x), \varphi'_3(x), \ldots$ is either decreasing (if $\circ$ is restrictive) or increasing (if $\circ$ is expansive) and that it is limited from below by 0 and from above by 1. □

Remark 3.4. If the assumptions of Lemma 3.1 are fulfilled, if, moreover, the composition operation $\circ$ is the product (7) and if for all $i = 1, 2, \ldots$. $\varphi_i = \varphi$ for some shape φ such that $\varphi(x) = 1$ iff $x = 0$ then the sequence $\varphi'_1, \varphi'_2, \varphi'_3, \ldots$ converges to a "deterministic" shape $\varphi_0 : R \to [0, 1]$ such that $\varphi_0(0) = 1$, $\varphi_0(x) = 0$ for $x \neq 0$.

The examples of composition operations over the class of shapes, given in this section, farely do not extend all rationally admissible or interpretable forms of such operation. It is possible to study their individual properties like commutativity (all compositions presented above are), associativity (e. g., for some shapes (3), (4) or (9) for $\alpha = (0, 1)$ are not), as well as the global properties of their class, i. e., their superposition, convergence of their repetitive applications, and others. It appears to be rational to study these problems in specific situations determined by actual application of this method.

3.3 More About Combined Verbal Variables and Their Shapes

We have already mentioned in several previous sections that the processing of "complex" verbal variables formalized as a combination of some simple ones can be considered for a specific type of realizing a computational algorithm or some of its steps. We have introduced examples of "something between... and..." as fuzzy logical conjunction of "something more than..." and "something less than...", "approximately natural multiple of 10" as a fuzzy logical disjunction of "approximately 10" or "approximately 20" or "approximately 30",..., etc., and "much more than x" as a fuzzy sum of crisp "x" and anonymous verbal variable "much". On the general level, this approach is formally correct and admissible. If we go into practical details of its actual realization we see that it deserves some comments and a bit deeper analysis.

The main specificity of these combined verbal variables is given by their position in the procedure of calculation due to an algorithm. Usual practical algorithms in which verbal variables are used are originally deterministic algorithms (e. g., optimization algorithms, computation of unknown quantities, etc.) which are used for vague input data given in the form of verbal variables. If these verbal variables are in the simple form respecting (10) then the computation of the algorithm is divided into its (deterministic) realization with crisp cores and some manipulation with (fuzzy-like) shapes. The existence of exactly one modal value – the crisp core – is consistent with the general paradigm of such approach. On the other hand, the combined verbal variables, like those ones in our examples, break this basic pattern of the input data. They can be decomposed into something like "crisp core" and "shapes", eventually also scales but those components need not fulfil the assumptions accepted in the previous chapters. It regards, especially, the shapes. Their natural "crisp core" explicitely named in the verbal expression need not be the modal value of the related fuzzy quantity ("much more than..."). It is something similar to the situation where the modal value of shape is not in 0. They need not have any modal value at all (in some instations of "something between... and..."). The membership function of the related fuzzy quantity need not be as monotonous as (2) and (3) demand ("approximately natural multiple of 10") or its support set need not be limited (the same example).

These combined verbal variables and with them corresponding fuzzy quantities can be transformed into transparent fuzzy algebraic or fuzzy logical combinations of simple verbal variables and standard fuzzy quantities, as shown in the previous chapters and sections. The problem to be discussed is how to include the combined variables and quantities into the considered algorithms. We can, generally, consider at least two methods. Both of them assume to decompose, in the first step, the combined verbal variable into several (maybe even infinitely many – see "approximately natural multiple of 10") simple verbal variables. Due to the former of both methods, we can use the procedure described in Section 3.1 for each of them separately. The obtained (simple) output verbal variables, constructed by means of (1) can be re-combined into a resulting combined verbal variable using the initial procedure of decomposition in the opposite sequence of steps. So, e. g., the input verbal datum "something between x and y" can be decomposed into "something more than x" and "something less than y", the computational algorithm can be repeatedly realized with both of them, two results – verbal variables U and V represented by fuzzy quantities u and v with membership functions μ_u, μ_v – are obtained, and the definitive result is a verbal variable W represented by fuzzy quantity w with μ_w constructed as a fuzzy logical conjunction

$$\mu_w(x) = \min(\mu_u(x), \mu_v(x)), \quad x \in R. \tag{10}$$

This method looks simple and lucid but it is based on the assumption that the initial decomposition procedure and the relations generated by it are not changed during the calculation procedure of the considered algorithm.

The latter method demands to extend some concepts of the model presented in the sections of Chapter 2, namely the concept of shape function. We can assume that, besides the elementary shapes fulfilling (1) – (4) there exist also *generated shapes* $\varphi^* : R \to [0, 1]$ which can be constructed from the elementary ones by means of (fuzzy) algebraical or (fuzzy) logical operations.

If $\varphi_1, \varphi_2, \varphi_3, \ldots$ is a (finite or infinite) sequence of elementary shapes, if $x_0^{(1)}, x_0^{(2)}, x_0^{(3)}, \ldots$ are real numbers and if $\omega_2, \omega_3, \ldots$ are binary composition operations with shapes (e. g., some of those mentioned in Section 3.2), then φ^* is a general shape if it can be constructed by means of the finite or infinite sequence of procedures

$$\begin{aligned}
\varphi_1^*(x) &= \varphi_1(x) \\
\varphi_k^*(x) &= \varphi_{k-1}^*(x)\,\omega_k \varphi_k(x * x_0^{(k)}), \quad k = 2,3,\ldots, \\
&\ldots\ldots\ldots \\
\varphi^*(x) &= \lim_{k=1,\ldots} \varphi_k^*(x)
\end{aligned} \tag{11}$$

where $x * x_0^{(k)}$ is an algebraic operation (summation, multiplication,...) and we assume that the limit exists. Each actual application of this construction procedure demands specific verification of the correctness of particular steps of this method (e. g., with respect to associativity and distributivity of applied operations) but as a general scheme it can be acceptable, namely if we consider the fact that in most of the real applications the referred sequence is finite and mostly not very long. This method extends the apparatus of used concepts in a significant way but it is methodologically purer and more consistent with our approach to the considered problem. The following steps are quite obvious. We can realize the given algorithm quite analogously to the procedure described in Section 3.1. We only accept the fact that the procedure with shapes can be used also for generalized shapes or mixtures of the elementary and generalized ones. The result will be (expectably) a generalized shape φ_0^* which can be substituted into (1) instead of φ_0, and a (rather more general) membership function μ_0 will be constructed. The fuzzy quantity characterized by it represent the resulting verbal variable.

The method discussed above deserves two comments. First, the main methodological difference of this method in comparison with the procedure described in the previous sections is in much greater freedom regarding the modal values of the generalized shapes. They need not be equal to 0, there can exist more than one modal value of φ^* or, vice versa, there need not exist any (if φ^* was constructed as minimum of φ_A^*, φ_B^* with different modal values.) This freedom in modal values opens also some new possibilities in the manipulation with "crisp cores". Now, they need not necessarily be the modal values of the membership functions of the fuzzy quantities characterizing the generalized verbal variables. Anyhow, even in this case they play an important role of in some sense significant "*reference values*" to which the values of the membership functions are somehow related. Especially in the case when the operations in (11) are fuzzy logical (minimum, maximum,...) it is rational to choose the "crisp reference value" attached to φ^* as, e. g., minimal of the crisp cores connected with $\varphi_1, \varphi_2, \varphi_3, \ldots$ in the corresponding verbal variables. Second, the infinity of the sequence of the elementary (and, consequently, generalized) shapes in (11) seems to be computationally ineffective. In fact, it is justifiable only by the existence of verbal expressions like "approximately natural multiple by..." which can rather rarely appear as input datum of an effective algorithm. Even in such case, it is quite admissible to accept the assumption about the limitedness of extent of possible values of real data (cf. Section 2.1, paragraphs about "anonymous" verbal variables). Then the infinite sequences needed for (11) turn into finite ones and some of the risks connected with its application vanish.

To illustrate the above explanations we can analyze two simple examples, already briefly mentioned above.

Example 3.1. The combined verbal variable "something between 10 and 20" was already interpreted as a logical conjunction of "something more than 10" and "something less than 20". Let us suppose for the sake of simplicity that in this actual situation "a bit more" and "a bit less" includes also equality. Then our verbal variable "something between... and..." is a generalized verbal variable constructed in the following way. Let us denote elementary verbal variables and their characteristic triples

$$A = \text{“a bit more than 10”} \ldots (x_A, \varphi_A, f_A)$$

$$\begin{aligned} x_A = 10, \quad \varphi_A(x) &= 0 && x < 0, \\ &= 1 && x \in [0,\, 1/3], \\ &= -\frac{3x}{2} + \frac{3}{2} && x \in [1/3,\, 1], \\ &= 0 && x > 1 \end{aligned}$$

$$B = \text{“a bit less than 20”} \ldots (x_B, \varphi_B, f_B)$$

$$\begin{aligned} x_B = 20, \quad \varphi_B(x) &= 0 && x < -1, \\ &= \frac{3x}{2} + \frac{3}{2} && x \in [-1,\, -1/3], \\ &= 1 && x \in [-1/3,\, 0], \\ &= 0 && x > 0. \end{aligned}$$

For the sake of simplicity we assume that in both cases

$$f_A(x) = f_B(x) = x/10, \quad x \in R.$$

We will see that this (rather arbitrary) choice would be in some sense harmonized with our interpretation of the expression "something between" – the subjectivity of the model formulation, so typical for fuzzy sets applications, appears here, as well.

Then we can construct the generalized verbal variable

$$C = \text{“something between 10 and 20”} \ldots (x_C, \varphi_C^*, f_C)$$

with generalized shape φ_C^* for "something between" as follows. We have a sequence (pair) of elementary shapes φ_A, φ_B, we have real numbers $x_0^{(A)} = 0$,

$x_0^{(B)} = 1$ and the fuzzy logical operation of minimum (6). (The choice of numbers $x_0^{(A)}$, $x_0^{(B)}$ is motivated by the type of the scales f_A, f_B and by the expected extent of the fuzzy quantities modelling A and B). For the algebraical operation $*$ over real numbers in (11) we choose $-$. Then

$$\varphi_C^*(x) = \min\left(\varphi_A(x), \varphi_B(x - x_0^{(B)})\right), \quad x \in R. \tag{12}$$

We can see that $\varphi_C^*(x) < 1$ for all $x \in R$, i. e. it has no modal value. Further, we put quite naturally $f_C(x) = f_A(x) = f_B(x)$, $x \in R$. In our case, we may choose the "crisp reference value of the generalized verbal variable C as"

$$x_C = \min(x_A, x_B) = 10$$

which choice is senseful with regard to the method of computation of φ_C^*. Then we get the membership function μ_c of the fuzzy quantity c representing the generalized verbal variable C as follows

$$\mu_c(x) = \varphi_C^*\left(f_C(x) - f_C(x_C)\right), \quad x \in R,$$

i. e.

$$\begin{aligned}
\mu_c(x) &= 0 && \text{for } x < 10 \text{ or } x > 20, \\
&= 3x/20 - 3/2 && \text{for } x \in [10, 15], \\
&= -3x/20 + 3 && \text{for } x \in [15, 20].
\end{aligned}$$

□

Example 3.2. Another, rather simpler, example shows the verbal variable "a bit more than 10" when the equality is not admitted. Then we have to do with generalized verbal variable whose "reference" value does not belong to its possible values. If we denote

$C =$ "a bit more than 10"

then we can see that it is a sum of elementary verbal variables "10" and "a bit more than" where the first one is deterministic and the second one has a vague meaning. Let us denote

$A =$ "10" $\ldots (x_A, \varphi_A, f_A)$

$$\begin{aligned}
x_A = 10,\ \varphi_A(0) &= 1, \\
\varphi_A(x) &= 0 \quad \text{for } x \neq 0,
\end{aligned}$$

and $\quad B =$ "a bit more than" $\ldots (x_B, \varphi_B, f_B)$

$$\begin{aligned}
x_B = \varepsilon > 0, \quad \varphi_B &= 0 && \text{for } x \le -\varepsilon \text{ or } x \ge 1,\ \varepsilon < 1, \\
&= x/\varepsilon + 1 && \text{for } x \in [-\varepsilon, 0], \\
&= 1 - x && \text{for } x \in [0, 1].
\end{aligned}$$

and for the sake of simplicity we put

$$f_A(x) = f_B(x) = x, \quad x \in R.$$

Then we can combine A and B as a sum due to, for example, the extension principle (2). We put $x_0^{(A)} = 0$, $x_0^{(B)} = \varepsilon$ and for the algebraic operation $*$ in (11) we use the contraction $-$. It means that the operation ω_2 used in (11) is the extension principle and

$$\begin{aligned} \varphi_C^*(x) &= \varphi_A(x)\,\omega_2\,\varphi_B(x-\varepsilon) = \\ &= \sup_{y \in R}\,[\min(\varphi_A(y), \varphi_B(x-y-\varepsilon))] = \\ &= 0 \qquad \text{for } x \leq 0 \text{ or } x \geq 1+\varepsilon, \\ &= x/\varepsilon \qquad \text{for } x \in [0,\varepsilon], \\ &= 1+\varepsilon-x \quad \text{for } x \in [\varepsilon, 1+\varepsilon]. \end{aligned} \tag{13}$$

We have obtained a generalized shape function with one modal value for $x = \varepsilon > 0$, which means that condition (1) is not fulfilled. Anyhow, $x = 0$ is an important "reference value" as φ_C^* represents a normalized shape for the generalized verbal variable "a bit more than". The fuzzy quantity c representing the verbal variable C = "a bit more than 10" has a membership function μ_c which can be derived from the triple (x_C, φ_C^*, f_C). Here x_C is the reference value, $x_C = 10$, φ_C^* was calculated by (13) and f_C can be quite naturally putted $f_C(x) = f_A(x) = f_B(x) = x$, $x \in R$. Then

$$\begin{aligned} \mu_C(x) &= \varphi^*\left(f_C(x) - f_C(x_C)\right) \;=\; \varphi^*(x - x_C) \;=\; \varphi^*(x-10) \\ &= 0 \qquad \text{for } x \leq 10,\ x \geq 11+\varepsilon, \\ &= x/\varepsilon - 10/\varepsilon \text{ for } x \in [10,\ 10+\varepsilon] \\ &= 11+\varepsilon-x \text{ for } x \in [10+\varepsilon, 11+\varepsilon]. \end{aligned}$$

It is worth mentioning that the value $\varepsilon > 0$ which in certain sense represent the most accepted numerical expression of the anonymous specification "a bit" is to be chosen for every actual application of the procedure to real modelled situation. □

3.4 Processing of Scales

The last component of the triple defining the standard verbal variables is the scale. We already know that the scales reflect something like reliability or specific accuracy of verbal data whose normalized uncertainty is described by the shape and which are located at a specific place of the real line. This place is given by the crisp core of the verbal variable.

The main property of scale – their representation of accuracy of data sources – develops during the computation process when various verbal variables are processed and their accuracy is combined. In general, two main approaches to this combination of accuracies are imaginable. The former one respects only the relation between the particular accuracies without any regard to the actual computational procedures which are done with the processed verbal variables, especially with their crisp cores. The latter approach is based on the pressumption that the combination of scales is to reflect the algebraical procedures done with the verbal variables (their crisp cores) and that the character of these procedures influences the accuracy of their otputs. Here, we briefly mention both of them.

The first approach includes only (maybe subjective) evaluation of mutual rates between reliabilities of the input verbal data sources, pesimism or optimism regarding their superpositions. If A, B are verbal variables with scales f_A, f_B, respectively, and C is a result of some binary operation with A and B (i. e., some arithmetic operation with crisp cores and composition of shapes) then the scale f_C can be derived from f_A and f_B, e. g., by some of the following operations

$$f_C(x) = \alpha \cdot f_A(x) + (1-\alpha) \cdot f_B(x), \quad x \in R,\ \alpha \in [0,1], \tag{14}$$

$$f_C(x) = \max\left(f_A(x), f_B(x)\right), \tag{15}$$

$$f_C(x) = \min\left(f_A(x), f_B(x)\right), \tag{16}$$

or by an operation of similar style, eventually by any general aggregation operator $f_C = \mathrm{Ag}(f_A, f_B)$. Especially (14) offers a rational method how to combine both scales. Another possibility is to suppose that the resulting scale will in every part of the real line copy the qualities of the either more or less reliable input source. In formulae this idea looks as follows. Let there exist a sequence (in practical applications finite) of real numbers

$$y_1, y_2, \ldots, y_{n-1}, y_n \tag{17}$$

such that $y_i < y_{i+1}$ for $i = 1, \ldots, n-1$, the first derivatives of f_A and f_B exist on the intervals (y_i, y_{i+1}), $i = 1, \ldots, n-1$, and let there exist a subsequence of (17)

$$z_1, z_2, \ldots, z_{m-1}, z_m \tag{18}$$

such that the first derivatives fulfill

$$f'_A(x) \geq f'_B(x) \text{ for } x \in (z_1, z_2),$$

$$f'_A(x) < f'_B(x) \text{ for } x \in (z_2, z_3),$$

and generally,

$$f'_A(x) \geq f'_B(x) \text{ for } x \in (z_{2i+1}, \ldots, z_{2(i+1)}),$$

$$f'_A(x) < f'_B(x) \text{ for } x \in (z_{2j}, z_{2j+1}),$$

$$i = 0, 1, \ldots,\quad j = 1, 2, \ldots .$$

Then we can put

$$\begin{aligned} f_C(x) &= f_A(x) + k_i \text{ for } x \in (z_{2i+1}, \ldots, z_{2(i+1)}), \\ f_C(x) &= f_B(x) + \ell_j \text{ for } x \in (z_{2j}, z_{2j+1}), \end{aligned} \tag{19}$$

where k_i, ℓ_j are real constants guaranteeing the continuity of f_C. Construction (19) reflects the optimistic expectation that for every segment of R the higher accuracy of verbal data prevails. We can also choose the pesimistic variant of this construction and put

$$\begin{aligned} f_C(x) &= f_A(x) + k_j^* \text{ for } x \in (z_{2j}, \ldots, z_{2j+1}), \\ f_C(x) &= f_B(x) + \ell_i^* \text{ for } x \in (z_{2i+1}, z_{2(i+1)}), \end{aligned} \tag{20}$$

where the constants k_j^* and ℓ_j^* are chosen to guarantee the continuity, and $i = 0, 1, \ldots, j = 1, 2, \ldots$. In this case we suppose that the accuracy of the less reliable source of verbal data prevail in the accuracy of the result.

The properties of the resulting scales derived by some of the above procedures are quite obvious and differ due to the actual procedure of their construction. The constructions by (16) and (15) are commutative and associative, and the commutativity and associativity are fulfilled also by constructions due to (19) and (20).

The second approach to the derivation of the scale of the resulting verbal variable is based on the paradigm that the accuracy of a result of computation depends not only on the accuracy of inputs but also on the arithmetic operations which were realized. Let us consider some simple arithmetic and algebraic operations realized with crisp cores: addition, multiplication by a scalar coefficient, and product, and illustrate by them the approach discussed here.

Let f_A, f_B be scales of verbal variables A and B, and let us denote by f_A^{-1}, f_B^{-1} the pseudoinverse in the sense of [20], i.e. $f^{-1} : [-\infty, \infty] \to [-\infty, \infty]$,

$$f^{-1}(t) = \sup(x \in [-\infty, \infty] : f(x) < t),$$

functions to f_A, f_B, i.e., $f_A(f_A^{-1}(x)) = x$ for any $x \in R$. Then we can suggest the scale of the sum of verbal variables by means of (in shortened symbolics)

$$f_{A+B}(x) = \left[f_A^{-1} + f_B^{-1}\right]^{-1}(x), \tag{21}$$

which means that for every $x \in R$ we find all $x_\alpha \in R$ such that $f_A(x_\alpha) = y$ (due to (7), (8) and (9) there exists at least one such x_α and if there exist two of them, x_α, x'_α, $x_\alpha < x'_\alpha$, then for all $x \in [x_\alpha, x'_\alpha]$, $f_A(x) = y$). Analogously, we find (at least one) $x_\beta \in R$ such that $f_B(x_\beta) = y$. Summing up all such pairs of x_α, x_β we get $x = x_\alpha + x_\beta$ (even here, x is a least one and if there are more of them, they form a closed interval) and for all such x, $f_{A+B}(x) = y$. Note that formula (21) is correct even if the scale functions f_A of f_B are not strictly increasing, compare [18–20].

Example 3.3. Let us consider simple scales

$$f_A(x) = \alpha \cdot x, \quad f_B(x) = \beta \cdot x, \quad \alpha, \beta > 0.$$

Then the method described above leads to the scale of the sum $A + B$ is

$$f_{A+B}(x) = \frac{1}{\frac{1}{\alpha} + \frac{1}{\beta}} = \frac{\alpha \cdot \beta}{\alpha + \beta}. \qquad \square$$

Remark 3.5. Formula (21) is equivalent with

$$f_{A+B}(x) = \sup_{y \in R} [\min(f_A(y), f_B(x - y))]$$

(see [19]).

Lemma 3.2. Let f_A, f_B be scales, and let f_{A+B} be a real-valued function constructed by means of the procedure described above. Then f_{A+B} is also a scale fulfilling (6) – (9).

Proof. Though the statement follows from results of [18–20], we give here an alternative proof. The above procedure, in accordance with the continuity of f_A, f_B and with their universality means that $f_{A+B}(x)$ is defined for any $x \in R$. Moreover, $f_{A+B}(0) = 0$ as follows from (6) and (21) (or, better, from its interpretation). As f_A and f_B are continuous then also f_{A+B} is continuous. This conclusion follows from the fact that for every $y \in R$ and $\varepsilon > 0$ there exists $\delta > 0$ such that for all $x^*_\alpha \in R$ for which $f_A(x^*_\alpha) = y$, for all x^*_β for which $f_B(x^*_\beta) = y$ and for all $x \in (x^*_\alpha - \delta, x^*_\alpha + \delta) \cup (x^*_\beta - \delta, x^*_\beta + \delta)$ the values $f_A(x) \in (y - \varepsilon, y + \varepsilon)$, $f_A(x) \in (y - \varepsilon, y + \varepsilon)$. It means that also $f_{A+B}(x) \in (y - \varepsilon, y + \varepsilon)$ for $x \in (x^*_\alpha + x^*_\beta - \delta, x^*_\alpha + x^*_\beta + \delta)$ and f_{A+B} is continuous. The procedure of construction of f_{A+B} immediately means that it is nondecreasing. Also the validity of (9) for f_{A+B} follows from its validity for f_A and f_B. $\square$

Analogously, we may proceed for the construction of the scale of a product of verbal variable A and a crisp real number $r \in R$, $r \neq 0$. We denote it $f_{r \cdot A}$ and its construction is abbreviately expressed by

$$f_{r \cdot A}(x) = \operatorname{sign}(r) \cdot [r \cdot f^{-1}]^{-1}(x). \tag{22}$$

This scheme means that for every $y \in R$ we find all $x_A \in R$ such that $f_A(x_A) = y$ (we already know that there exists at least one such x_A and if there more than one then they form a closed interval). Then we calculate the product

$$|r| \cdot x_A = x'_A$$

and put $f_{r \cdot A}(x'_A) = y$. Even in this case the following statement holds.

Lemma 3.3. If f_A is the scale of a verbal variable A and $r \neq 0$, $r \in R$, is a real number then the function $f_{r \cdot A}$ constructed by means of the above procedure is a scale function.

Proof. The validity of properties (6) – (9) immediately follows from its construction procedure. □

Finally, let us use the approach considered here for the construction of the scale of product of two verbal variables A and B. Their scales are denoted by f_A, f_B, respectively, and the scale of the product is denoted by $f_{A \cdot B}$. The procedure can be in a simplified form (the simplification consists, analogously to the previous cases, in the correctness of the next formula for strictly increasing scales, only)

$$f_{A \cdot B}(x) = \left[\operatorname{sign}(x) \cdot f_A^{-1} \cdot f_B^{-1}\right]^{-1}(x). \tag{23}$$

This formula represents the following procedure. For every $y \in R$ we find the values x_α and x_β for which $f_A(x_\alpha) = y = f_B(x_\beta)$. We know that they may form a closed intervals, eventually (and usually) degenerated into single values. We compute the product

$$\operatorname{sign}(y) \cdot x_\alpha \cdot x_\beta = x_{\alpha\beta}$$

and put

$$f_{A \cdot B}(x_{\alpha\beta}) = y. \tag{24}$$

This procedure, as well as the previous ones, defines the function $f_{A \cdot B}$ uniquely, and the following statement can be formulated.

Lemma 3.4. Let f_A, f_B be scales of verbal variables A and B, respectively, and let the function $f_{A \cdot B}$ be constructed by means of the previous procedure. Then $f_{A \cdot B}$ is a scale.

Proof. Also in this case the construction procedure produces a function $f_{A \cdot B}$ fulfilling (6) – (9). □

It is usually possible to proceed analogously also in the case of other operations. The just described approach to the construction of scale functions belonging to the result of some operation with verbal variables has at least one significant property. It respects the fact, so accented by the extension principle (3) (see also (4), (5) or (2)), that some operations with vague data in some sense cummulate their vagueness. To model this cummulation as a property of the sources of data appears to be, at least in some situations, more adequate to the essential structure of the modelled situation than to include it in the complex combination of possible values as it is done in the extension principle. At least, everyone who wants to apply the presented model of verbal variable has the possibility to choose not only if but also in which of the offered ways he wants to include the relation between vagueness and operations with the verbal variables into his procedure. He can use either the above operations with scales or he can choose some adequate operation with shapes.

The processing of scales related to (23) is more precisely treated in Section 6.2.

3.5 Orderings

The ordering relations are especially important if some optimization problems are to be solved, and in this sense they are significant also for the processing of verbal variables. We already know that verbal variables are formally represented by fuzzy quantities. It means that all existing approaches to the ordering of fuzzy quantities (let us remember [31,16] and also a simple specimen of the ordering shown in Section 1.1) are directly applicable to the specific case of the verbal variables. Anyhow, the presentation of verbal variables as compositions of crisp cores, shapes and scales offers an opportunity to study also further approaches to their ordering.

If we look into the existing works about ordering of fuzzy numbers or fuzzy quantities (their very representative summary is given in [16]) then we can be surprised by a great amount of different approaches to this problem. Their variability indicates certain perplexity regarding the problem what does the ordering of fuzzy quantities mean. It means, what is to be compared? Quantitative values of the quantities, distributions of their vagueness, some combination of both aspects? The complex inner structure of fuzzy quantities admits different answers on this question, and, consequently, numerous different approaches to the concept of ordering relation between fuzzy quantities and their diversity.

The model of verbal variables presented in this work offers the possibility to separate the concepts of ordering related to particular components of the verbal variable structure – its crisp core, shape and scale. Each of them can respect the specific property serving as the criterion of the respective ordering. This decomposition of the ordering problem cannot fully substitute the global orderings, e. g., the one mentioned in Section 1.1 and defining the ordering relation as a fuzzy relation between fuzzy quantities. Anyhow, the orderings of particular components of verbal variable can complete the relation defined between corresponding fuzzy quantities and add qualitatively new views on the structure of relations between verbal variables (see, e. g., ([15,14])).

The ordering of crisp cores obviously represents the quantitative ordering, e. g., the ordering of numerical values of verbal variables. It is the usual ordering of real numbers and it compares the numerical values to which the verbal variables are related. This ordering is quite natural and brings no problems if it regards simple verbal variables (and related fuzzy quantities) with exactly one modal value which is identical with the crisp core. If the combined verbal variables with more modal values (e. g., "approximately natural multiple of 10") or with may be none modal value (e. g., "something between 5 and 10") then the ordering becomes less obvious, and it is necessary to decide which crisp cores of the simple verbal values composing the considered combined ones (see Section 3.3) are to be compared, or which of their numerical combinations represents their "typical" quantitative value.

If we are to compare the shapes then we have, first, to make clear which property of verbal variable we compare, and what are to be the criteria for this comparison. We know that the shapes represent the normalized distribution of uncertainty connected with the verbal expression and its vagueness. It means that their characteristics which can be compared are, e. g., the extent of possible uncertainty (i. e., the extent of the support of the shape function), the degree of uncertainty (i. e. the functional inequality between shapes), or some of their modifications like the extents of some α-cuts of the shapes, etc. More formally, we can compare the sets

$$\{x \in R : \varphi(x) > 0\}, \tag{25}$$

which means to compare the extents of uncertainty, the sets

$$\{x \in R : \varphi(x) > \alpha\}, \quad \alpha \in (0,1) \tag{26}$$

which means to compare the extents of "important" uncertainty where α determines the degree of the "importance". Or we can also point-wise compare the shapes. If φ_A, φ_B are two shapes then we can test if

$$\varphi_A(x) \geq \varphi_B(x) \quad \text{for all } x \in R. \tag{27}$$

All these comparisons (more exactly – the ordering relations following from them) are partial only, but they can offer valuable information about the character of uncertainty connected with particular verbal variables.

In fact, some of the other ordering relations suggested for the membership functions of fuzzy quantities can be used for comparing shapes. Only some of them as the fact that $\varphi(0) = 1$ for any simple shape makes some others of the ordering methods useless. It concerns also the fuzzy ordering scheme mentioned in Section 1.1.

All approaches mentioned above and based on formulae (25), (26) and (27) lead to relations having the usual properties. If φ_A, φ_B are shapes of verbal variables A, B, respectively, then we say that *the extent* of uncertainty of A is larger or equal to the one of B and write $\varphi_A \geq_{(0)} \varphi_B$ iff

$$\{x \in R : \varphi_A(x) > 0\} \supset \{x \in R : \varphi_B(x) > 0\}. \tag{28}$$

Then it is easy to verify the following statement.

Remark 3.6. For relation (28) and any of verbal variables A, B, C with triple of shapes $\varphi_A, \varphi_B, \varphi_C$

$$\varphi_A \geq_{(0)} \varphi_A, \quad \varphi_A \geq_{(0)} \varphi_B \quad \text{and} \quad \varphi_B \geq_{(0)} \varphi_C \Rightarrow \varphi_A \geq_{(0)} \varphi_C$$

if $\varphi_A \geq_{(0)} \varphi_B$ and $\varphi_B \geq_{(0)} \varphi_A$ then

$$\{x \in R : \varphi_A(x) > 0\} = \{x \in R : \varphi_B(x) > 0\}.$$

Similarly, if $\alpha \in (0,1]$ then we say that the *α-extent* of A is larger or equal to the one of B and write $\varphi_A \geq_{(\alpha)} \varphi_B$ iff

$$\{x \in R : \varphi_A(x) \geq \alpha\} \supset \{x \in R : \varphi_B(x) \geq \alpha\}. \tag{29}$$

Then it is again easy to verify the following statements.

Remark 3.7. If φ_A is a shape then

$$\{x \in R : \varphi_A(x) > 0\} = \lim_{\alpha \to 0} \{x \in R : \varphi_A(x) \geq \alpha\}$$

in the usual fuzzy set theoretical sense.

Remark 3.8. If $\beta > 0$ and $\varphi_A \geq_{(\alpha)} \varphi_B$ for all $\alpha \in (0, \beta)$ then $\varphi_A \geq_{(0)} \varphi_B$.

Remark 3.9. If $\alpha \in (0, 1]$ and $\varphi_A, \varphi_B, \varphi_C$ are shapes of verbal variables A, B, C, respectively, then

$$\varphi_A \geq_{(\alpha)} \varphi_A, \quad \varphi_A \geq_{(\alpha)} \varphi_B \quad \text{and} \quad \varphi_B \geq_{(\alpha)} \varphi_C \quad \text{then} \quad \varphi_A \geq_{(\alpha)} \varphi_C,$$

if $\varphi_A \geq_{(\alpha)} \varphi_B$ and $\varphi_B \geq_{(\alpha)} \varphi_A$ then

$$\{x \in R : \varphi_A(x) \geq \alpha\} = \{x \in R : \varphi_B(x) \geq \alpha\}.$$

Finally, we say that A is *uniformly larger* than B and write $\varphi_A \geq \varphi_B$ iff

$$\varphi_A(x) \geq \varphi_B(x) \quad \text{for all } x \in R.$$

The next statements are obvious.

Remark 3.10. If φ_A, φ_B are shapes of the verbal variables A, B, respectively, and $\varphi_A \geq \varphi_B$ then $\varphi_A \geq_{(\alpha)} \varphi_B$ for all $\alpha \in (0, 1]$ and, consequently, $\varphi_A \geq_{(0)} \varphi_B$.

Remark 3.11. If $\varphi_A, \varphi_B, \varphi_C$ are shapes of verbal variables A, B, C, respectively, then evidently $\varphi_A \geq \varphi_A$, $\varphi_A \geq \varphi_B$ and $\varphi_B \geq \varphi_C$ implies $\varphi_A \geq \varphi_C$, and if $\varphi_A \geq \varphi_B$ and $\varphi_B \geq \varphi_A$ then $\varphi_A(x) = \varphi_B(x)$ for all $x \in R$.

Let us, finally, consider the possibility of comparison of scales. They, as we already know, characterize the quality of the source of verbal data. This quality is connected with the gradient of the shape in particular segment of the real line or, better, in neighborhood of particular elements of the domain of the scale. It means that for the scale f_A of a verbal variable A, not the numerical values $f_A(x)$ but the values of its first derivative $f'_A(x)$ for particular x are significant. Consequently, these first derivatives are to be compared. If f_A, f_B are scales of verbal variables A, B, respectively, and $x \in R$ then we say that f_A is at least as reliable as f_B in x iff $f'_A(x)$ and $f'_B(x)$ exist and $f'_A(x) \geq f'_B(x)$.

The following statements are evident.

Remark 3.12. Let $x \in R$ and f_A, f_B, f_C be scales of verbal variables A, B, C, respectively, and let the first derivatives $f'_A(x)$, $f'_B(x)$, $f'_C(x)$ exist. Then

$f'_A(x) \geq f'_A(x)$, $f'_A(x) \geq f'_B(x)$ and $f'_B(x) \geq f'_C(x)$ implies $f'_A(x) \geq f'_C(x)$,

and if $f'_A(x) \geq f'_B(x)$ and $f'_B(x) \geq f'_A(x)$ then $f'_A(x) = f'_B(x)$.

Remark 3.13. If f_A, f_B are scales of verbal variables A, B, respectively, and if f_C is constructed by the procedure described by (19) then, evidently, $f'_C(x) \geq f'_A(x)$ and $f'_C(x) \geq f'_B(x)$ for all $x \in R$ for which the derivatives exist. If f_C is the scale constructed by the procedure described by (20) then, for all $x \in R$ for which the derivatives exist, $f'_A(x) \geq f'_C(x)$ and $f'_B(x) \geq f'_C(x)$.

Remark 3.14. If f_A, f_B are scales of verbal variables A, B, respectively, if $C = A + B$ and f_C is constructed by means (21) then for all $x \in R$ for which the derivatives exist

$$f'_A(x) \geq f'_C(x), \quad f'_B(x) \geq f'_C(x).$$

Remark 3.15. If $r \in R$, $r \neq 0$, is real number, f_A is a scale of verbal variable A, and if $C = r \cdot A$ is verbal variable and f_C is constructed by means of (22) then for all $x \in R$ for which the relevant derivatives exist

$$f'_C(x) \geq f'_A(x) \quad \text{if } r \leq 1, \qquad f'_A(x) \geq f'_C(x) \quad \text{if } r \geq 1.$$

4 Three Illustrative Examples

The model described in the previous chapters can be, in its completeness, illustrated by several simple examples. They use the general ideas of the above approach to verbal vague data to describe the possible solutions of three elementary problems regarding the management of uncertainty.

Example 4.1. A traditional cooking instruction demands to add three handfuls of flour. The cook has added two handfuls, each about 50 grams. To be sure that he does not add too much of flour, he makes an effort to estimate the third handful also as usually (it means about 50 grams) but as much as possible not more than usual mean amount (it means not significantly more than 50 grams). The cook is able to estimate the amounts not greater than 50 grams quite well, and his estimation is less reliable for larger amounts. Are we able to describe the (evidently vague) total amount of flour added to the prepared food, using the above vague descriptions?

We are to characterize the "sum" of verbal quantities

$$A = \text{"about 50 grams"},$$

$$B = \text{"about 50 grams"},$$

$$C = \text{"about 50 grams and if possible not more"}.$$

Each of them is defined by the respective triple (x_A, φ_A, f_A), (x_B, φ_B, f_B), (x_C, φ_C, f_C) where

$$x_A = x_B = x_C = 50,$$

$\varphi_A = \varphi_B$ is the shape for "about", let us define it for any $x \in R$ by

$$\begin{aligned}\varphi_A(x) &= x+1 \text{ for } x \in [-1,0],\\ &= 1-x \text{ for } x \in [0,1],\\ &= 0 \qquad \text{otherwise.}\end{aligned}$$

The shape φ_C describes the logical conjunction of "about" and "if possible not more than", which can be constructed, for any $x \in R$, as

$$\varphi_C(x) = \min(\varphi_A(x), \varphi_D(x))$$

where $\varphi_D(x)$ represents the second statement, and

$$\begin{aligned}\varphi_D(x) &= (x+11)/10 \text{ for } x \in [-11,-1],\\ &= 1 \qquad \text{for } x \in [-1,0],\\ &= 1-10x \qquad \text{for } x \in [0,\ 1/10],\\ &= 0 \qquad \text{otherwise.}\end{aligned}$$

Then for $x \in R$

$$\begin{aligned}\varphi_C(x) &= x+1 \quad \text{for } x \in [-1,0],\\ &= 1-10x \text{ for } x \in [0,\ 1/10],\\ &= 0 \qquad \text{otherwise.}\end{aligned}$$

The scales reflect in all three cases the estimation abilities of one person. It means

$$f_A(x) = f_B(x) = f_C(x) \quad \text{for all } x \in R.$$

The described properties of the cook can be, e. g., represented by

$$\begin{aligned}f_A(x) &= 2x \qquad \text{for } x \le 50,\\ &= x+50 \text{ for } x \ge 50,\end{aligned}$$

(the values for $x \le 0$ are not significant). The fuzzy quantities defined by these verbal quantities will be denoted a, b, c with membership functions μ_a, μ_b, μ_c, where for any $x \in R$

$$\mu_a(x) = \varphi_A(f_A(x) - f_a(x_A)), \quad \mu_b(x) = \varphi_B(f_B(x) - f_B(x_B)),$$
$$\mu_c(x) = \varphi_C(f_C(x) - f_C(x_C)),$$

respectively. It means that

$$\begin{aligned}\mu_a(x) = \mu_b(x) &= 2x - 99 \quad &&\text{for } 49.5 \le x \le 50,\\ &= 51 - x &&\text{for } 50 \le x \le 51,\\ &= 0 &&\text{otherwise,}\end{aligned}$$

$$\begin{aligned}\mu_c(x) &= 2x - 99 \quad &&\text{for } 49.5 \le x \le 50,\\ &= 501 - 10x &&\text{for } 50 \le x \le 50.1,\\ &= 0 &&\text{otherwise.}\end{aligned}$$

If we aim to compute the fuzzy quantity $a \oplus b \oplus c$ corresponding to the verbal quantity "sum of A, B and C", then we can proceed in two ways.

Either, we can compute the sum $a \oplus b \oplus c$ by means of the extension principle (4) and then the membership function of this new fuzzy quantity is

$$\begin{aligned}\mu_{a\oplus b\oplus c}(x) &= 2x/3 - 99 \quad &&\text{for } 148.5 \le x \le 150,\\ &= (1521 - 10x)/21 &&\text{for } 150 \le x \le 152.1,\\ &= 0 &&\text{otherwise.}\end{aligned}$$

Or, we can use the generation procedure described in the previous sections. Then we construct three components x_0, φ_0, f_0 defining the verbal quantity "sum of A, B and C". It is evident that

$$x_0 = x_A + x_B + x_C = 150.$$

The structure of the analyzed situation allows us to conclude that the shape φ_0 is to be adequate to the verbal expression "about but if possible not more" which reflects the logical conjunction of the verbal expressions represented by φ_A, φ_A and φ_C and, in formulae,

$$\varphi_0(x) = \min\left(\varphi_A(x), \varphi_B(x), \varphi_C(x)\right) = \varphi_C(x), \quad x \in R.$$

Regarding the combination of scales $f_A = f_B = f_C$, resulting into f_0, we can choose among two procedures described in Section 3.4. If we follow the method based on the combination of derivatives then the situation is very simple. The equality between f_A, f_B, f_C means that also $f_0 = f_A$ (see (18), (19), (20)).[1]

Or, we can use the combination of scales respecting the realized algebraical operations (and the increase of the uncertainty connected with them) reflected by (21). Then we construct the scale function which we denote by f_0^*, where for any $x \in R$

$$f_0^*(x) = \left[f_A^{-1} + f_B^{-1} + f_C^{-1}\right]^{-1}(x),$$

[1] Let us mention that also the procedures formalized by (14), (15) and (16) and marginally mentioned in Section 3.4 lead to an identical result.

which means

$$f_0^*(x) = 2x/3 \qquad \text{for } x \le 150,$$
$$= x/3 + 50 \text{ for } x > 150.$$

In the first case, the triple (x_0, φ_0, f_0) generates the fuzzy quantity e with membership function μ_e such that for any $x \in R$

$$\mu_e(x) = \varphi_0\left(f_0(x) - f_0(x_0)\right),$$

which means

$$\mu_e(x) = \varphi_0(x - 150) = x - 149 \qquad \text{for } 149 \le x \le 150,$$
$$= 1501 - 10x \text{ for } 150 \le x \le 150.1,$$
$$= 0 \qquad \text{otherwise.}$$

In the second case, the triple (x_0, φ_0, f_0^*) generates the fuzzy quantity e^* with membership function μ_e^* such that

$$\mu_e^* = \varphi_0\left(f_0^*(x) - f_0^*(x_0)\right)$$

which means

$$\mu_e^*(x) = (2x/3 - 100) + 1 = 2x/3 - 99 \qquad \text{for } 148.5 \le x \le 150,$$
$$= 1 - 10(50 + x/3 - 100) = 501 - 10x/3 \text{ for } 150 \le x \le 150.3,$$
$$= 0 \qquad \text{otherwise.}$$

If we measure the extent of uncertainty of the resulting fuzzy quantities by the length of their support interval, and this approach is evidently correct at least for the triangular fuzzy quantities considered here, then we get

— for $\mu_{a \oplus b \oplus c}$ the interval between 148.5 and 152.1 which means the extent of 3.6,
— for μ_e the interval between 149 and 150.1 which means the extent of 1.1
— for μ_e^* the interval between 148.5 and 150.3 which means the extent of 1.8.

We can see that the alternative approaches based on the generation procedure, offer much narrower extent of uncertainty than the classical extension principle. This conclusion is even more significant if we consider the fact that all steps of the generation procedure described in this example are well interpretable and quite realistic with respect to the modelled situation. □

We can see that the advantages of the method of generated verbal variables need not be as effective as shown in the preceding example, if the operation of product of crisp and fuzzy quantity is considered. This is illustrated by the following example.

Example 4.2. Let us try to solve the problem to describe approximately the capacity of a liquid being stored in a cylindric tank where the level of the liquid is only vaguely described. Let the ground of the tank be a circle with diameter well exactly measured as "2 meters" and the level of the liquid in the tank is estimated by a very rough method as "about 3 meters or rather a bit more". Let us formally describe the verbal variables entering this calculation. There exists only one of them, namely

$$B = \text{"about 3 meters or rather a bit more"},$$

and it is characterized by triple (x_B, φ_B, f_B).

Evidently, $x_B = 3$. Let us choose

$$\begin{aligned} &\varphi_B \text{ for "about... or rather a bit more"} \\ &\varphi_B = 1 + x/2 \quad \text{for } x \in [-2, 0], \\ &\quad = 1 \quad \text{for } x \in [0,\ 1/2] \\ &\quad = (5 - 2x)/4 \text{ for } x \in [1/2,\ 5/2] \\ &\quad = 0 \quad \text{for } x < -2 \text{ or } x > 5/2. \end{aligned}$$

Finally, the scale f_B is chosen as follows. The estimation of height of the liquid in the tank was characterized as very rough. Let us specify the situation a bit better. The estimator is able to estimate the level if it is not very high (let us suppose not more than 1 meter) and not very far from the top of the tank (let us say at least 5 meters). Inside the interval between 1 and 5 meters the estimation is not very good. In formulae

$$\begin{aligned} f_B(x) &= 3x \quad \text{for } x \in [0, 1] \\ f_B(x) &= (x + 5)/2 \text{ for } x \in [1, 5], \\ &= 3x - 10 \quad \text{for } x \geq 5, \end{aligned}$$

and it has no sense to consider $f_B(x)$ for $x < 0$. The capacity of the liquid in the tank is a variable C, where

$$C = \pi \cdot \left(\frac{d}{2}\right)^2 \cdot B = \pi \cdot B.$$

It can be computed in two ways. Either, we can use the triple (x_B, φ_B, f_B) to compute the fuzzy quantity b with membership function μ_b, by means of

$$\mu_b(x) = \varphi_B\left(f_B(x) - f_B(x_B)\right).$$

It is easy to verify that $\mu_b(x)$ is composed by segments connecting the points $(x, \mu_b(x)) : (2/3,\ 0), (1,\ 1/2), (3, 1),\ (4, 1),\ (5,\ 3/4),\ (11/2,\ 0)$, i. e.

$$
\begin{aligned}
\mu_b(x) &= 3x/2 - 1 && \text{for } x \in [2/3,\ 1],\\
&= x/4 + 1/4 && \text{for } x \in [1, 3],\\
&= 1 && \text{for } x \in [3, 4]\\
&= 2 - x/4 && \text{for } x \in [4, 5],\\
&= 33/4 - 3x/2 && \text{for } x \in [5,\ 11/2],\\
&= 0 && \text{otherwise.}
\end{aligned}
$$

If we proceed due to the classical extension principle then we compute the fuzzy quantity $c = \pi \cdot b$ by means of

$$\mu_c(x) = \mu_b(x/\pi).$$

It will be composed by segments connecting the points $(x, \mu_c(x)) : (2\pi(3, 0),$ $(\pi, 1/2),\ (3\pi, 1),\ (4\pi, 1),\ (5\pi, 3/4),\ (11\pi/2, 0))$, i. e.

$$
\begin{aligned}
\mu_c(x) &= 3x/(2\pi) - 1 && \text{for } x \in [2\pi/3, \pi],\\
&= x/(4\pi) + 1/4 && \text{for } x \in [\pi, 3\pi],\\
&= 1 && \text{for } x \in [3\pi, 4\pi],\\
&= 2 - x/(4\pi) && \text{for } x \in [4\pi, 5\pi],\\
&= 3/4 - 3x/(2\pi) && \text{for } x \in [5\pi, 11\pi/2],\\
&= 0 && \text{otherwise.}
\end{aligned}
$$

But we can also construct new verbal variable $C =$ "*about* $3 \cdot \pi\ m^3$ *or a bit more*" by means of using the verbal variable $B = (x_B, \varphi_B, f_B)$. Then $C = (x_C, \varphi_C, f_C)$ and its particular components can be specified in the following quite natural way

$$
\begin{aligned}
&x_C = \pi \cdot x_B = 3\pi,\\
&\varphi_C(x) = \varphi_B(x), \quad x \in R.
\end{aligned}
$$

This choice is really natural. As the verbal quantity B is the single source of vagueness in the modelled situation, it can be preserved and transmitted from the input to the output. The choice of the scale f_C of the result deserves a brief comment. We could respect the one of the methods presented in Subsection 3.4 choosing the input scale the derivative of which does respect our idea about the exactness of the resulting scale. In such case (as there exists only one input scale), $f_C(x) = f_B(x)$, $x \in R$. This procedure does not look very satisfactorily in our case. The operation $\pi \cdot B$ has shifted the possible values of B (namely the values $x \in R$ for which $f_C(x)$ is near enough to $f_C(x_C)$) from the segment of "bad reliability of the information" given by $f_B = f_C$ (i. e. the segment inside [1, 5]) to the segment of "good reliability" (i. e., above 5) without any corresponding qualitative change of the circumstances of the estimation.

It means that it is more natural to use formula (22) due to which the "units" of f_C change, in relation to f_B, proportionally to the deterministic change of the "measure" of the referred quantities. Then

$$f_C(x) = \text{sign}(\pi) \cdot \left[\pi \cdot f_b^{-1}\right]^{-1}(x) = f_B(x/\pi).$$

It means

$$\begin{aligned} f_C(x) &= 3x/\pi \qquad \text{for } x \in [0, \pi], \\ &= x/2\pi + 5/2 \text{ for } x \in [\pi, 5\pi], \\ &= 3x/\pi - 10 \quad \text{for } x \geq 5\pi. \end{aligned}$$

Then the membership function μ_c^* computed by means of

$$\mu_c^*(x) = \varphi_C\left(f_C(x) - f_C(x_C)\right)$$

is equal to the membership function μ_c computed above, i. e., it is composed from the segments connecting the points $(x, \mu_c^*(x)) : (2\pi/3, 0)$, $(\pi,\ 1/2)$, $(3\pi, 1)$, $(4\pi, 1), (5\pi,\ 3/4)$, $(11\pi/2,\ 0)$. In this example (using only single input) both approaches appear to be equivalent – if the scale f_C is constructed in the way used above.

Let us show, just for illustration, the resulting membership function for the case when the scale f_C was chosen to be equal to f_B. If we denote by μ_c^{**} the membership function derived for $x_C = 3\pi$, $\varphi_C = \varphi_B$, $f_C = f_B$ then

$$\begin{aligned} \mu_c^{**}(x) &= \frac{3x - 9\pi + 2}{18\pi - 28} \qquad \text{for } x \in [3\pi - 2/3,\ 9\pi - 10], \\ &= 1 \qquad \text{for } x \in [9\pi - 10,\ 3\pi + 1/6], \\ &= -3x/2 + (18\pi + 5)/4 \text{ for } x \in [3\pi + 1/6,\ 3\pi + 5/6]. \end{aligned}$$

We can see that, meanwhile the support set of $\mu_c = \mu_c^*$ was the interval $(2\pi/3,\ 11\pi/3)$, and its length is 3π, the support of μ_c^{**} is the interval $(3\pi - 2/3,\ 3\pi + 5/6)$ and its length is 3/2. □

Example 4.3. Let us imagine a hunter who is going to hunt for 6 days or, maybe, one day more. He expects to get about 5 animals and hopefully not less in every day of hunting. What are the expected total amounts of bagged animals?

It is evident that we deal with integer verbal (or fuzzy) quantities and that the result is the product of "5 or, maybe, one more" and "about 5 and hopefully not less". Let us formalize these verbal fuzzy quantities.

Regarding $A =$ "5 or, maybe, one more", its crisp core $x_A = 6$, evidently, and the shape φ_A may be putted as function of integer variable

$$\varphi_A(0) = 1, \quad \varphi_A(1) = 0.7, \quad \varphi_A(x) = 0 \quad \text{otherwise.}$$

Note, please, that this shape can be considered as a composition of two (deterministic) shapes "exactly" (denote it $\varphi_A^{(1)}$) and "may be exactly" (denote it $\varphi_A^{(2)}$) where

$$\varphi_A^{(1)}(0) = 1, \quad \varphi_A^{(1)}(x) = 0 \quad \text{for } x \neq 0,$$
$$\varphi_A^{(2)}(x) = 0.7 \cdot \varphi_A^{(1)}(x),$$
$$\varphi_A(x) = \max\left(\varphi_A^{(1)}(x), \varphi_A^{(2)}(x-1)\right),$$

which relation reflects the logical disjunction between components.

Regarding $B =$ "about 5 and hopefully not less" has the crisp core $x_B = 5$ and its shape φ_B can be putted as

$$\varphi_B(-1) = 0.1, \quad \varphi_B(0) = 1, \quad \varphi_B(1) = 0.6, \quad \varphi_B(2) = 0.2.$$

Even in this case the shape φ_A can be considered as a composition, in this case logical conjunction, of shapes representing "about" (denote it $\varphi_B^{(1)}$) and "hopefully not less" (denote it $\varphi_B^{(2)}$). If we respect the integer values of inputs then these components can be putted as

$$\varphi_B^{(1)}(-2) = 0.2, \quad \varphi_B^{(1)}(-1) = 0.6, \quad \varphi_B^{(1)}(0) = 1, \quad \varphi_B^{(1)}(1) = 0.6,$$
$$\varphi_B^{(1)}(2) = 0.2, \quad \varphi_B^{(2)}(-1) = 0.1, \quad \varphi_B^{(2)}(0) = 1, \quad \varphi_B^{(2)}(n) = 1,$$
$$\text{for } n = 1, 2, 3, \ldots.$$

To respect condition (4) we can limit the values of the parameter n from above, let us say by $n \leq 10$. The following procedure obviously means that this limitation cannot influence the result of the computation. The value $\varphi_B^{(2)}(-1) = 0.1$ reflects the uncertainty hidden in the formulation "hopefully...". Then

$$\varphi_B(x) = \min\left(\varphi_B^{(1)}(x), \varphi_B^{(2)}(x)\right).$$

Let us suppose that both scales of A and B express neutral objective knowledge of situation, i.e., $f_A(x) = f_B(x) = x$.

As usual, the verbal quantities $A = (x_A, \varphi_A, f_A)$ and $B = (x_B, \varphi_B, f_B)$ generate corresponding fuzzy quantities a and b, respectively, where

$$\mu_a(6) = 1, \quad \mu_a(7) = 0.7,$$
$$\mu_b(4) = 0.1, \quad \mu_b(5) = 1, \quad \mu_b(6) = 0.6, \quad \mu_b(7) = 0.2,$$
$$\text{and } \mu_a(x) = 0, \quad \mu_b(x) = 0 \text{ for other values of } x.$$

Then, we can use the classical extension principle, and to compute $a \odot b$ with membership function $\mu_{a\odot b}(x)$ such that

$$\mu_{a\odot b}(24) = 0.1, \quad \mu_{a\odot b}(28) = 0.1, \quad \mu_{a\odot b}(30) = 1, \quad \mu_{a\odot b}(35) = 0.7,$$
$$\mu_{a\odot b}(36) = 0.6, \quad \mu_{a\odot b}(42) = 0.6, \quad \mu_{a\odot b}(49) = 0.2.$$

This result is not very satisfactory from the interpretational point of view. Not only the wide extent of possible values of the result but also the alarming and hardly explicable gaps among these values do not correspond with common idea about the possible outcomes of such situation.

Let us apply the alternative approach suggested and discussed in the above sections. It means that we attempt to construct a new verbal quantity $C = A \cdot B$ as a product of A and B, The components of C will be denoted by (x_C, φ_C, f_C) where, quite naturally

$$x_c = x_A \cdot x_B = 30,$$
$$f_C(x) = f_A(x) = f_B(x) = x,$$

and φ_C is to be a composition of φ_A and φ_B. In the model of this type of problem the composition based on the extension-principle-like approach (2) appears natural. Hence, we put

$$\varphi_C(x) = \max_{\substack{y\in R \\ y\neq 0}} [\min(\varphi_A(y), \varphi_B(x/y))]$$

for $x \neq 0$ and

$$\varphi_C(0) = \max(\varphi_A(0), \varphi_B(0)).$$

Then

$$\varphi_C(-1) = 0.1, \quad \varphi_C(0) = 1, \quad \varphi_C(1) = 0.6, \quad \varphi_C(2) = 0.2$$

and $\varphi_C(x) = 0$ for other x.

If we construct fuzzy quantity c with membership function μ_c generated by (x_C, φ_C, f_C) then

$$\mu_c(x) = \varphi_C(f_C(x) - f_C(x_C)),$$

i. e.

$$\mu_c(29) = 0, \quad \mu_c(30) = 1, \quad \mu_c(31) = 0.6, \quad \mu_c(32) = 0.2$$

and $\mu_c(x) = 0$ for other x. The realism of such narrow extent of possible results might be, perhaps, discussed. Nevertheless, this results looks much more natural than the one derived by means of the classical processing of fuzzy quantities a and b. □

5 Triangular Norms

Fuzzy quantities discussed in previous chapters are fuzzy subsets of the real line. However, fuzzy set theory is a rich variety of approaches to the vagueness modelling which differs in sets of fuzzy connectives applied throughout the processing of involved fuzzy subsets. The original Zadeh's [52] proposal of the fuzzy set theory is related to the modelling of the intersection of fuzzy sets by min operator. Similarly, max operator models the union of fuzzy sets, while the complementation is (point-wisely) derived from the negation operator $N : [0,1] \to [0,1]$, $N(x) = 1 - x$. All ideas, results, formulae in Chapters 1–4 are related to this original Zadeh approach, especially the extension principle introduced in [52], see (1.3). However, since eighties, alternative models for fuzzy connectives were proposed and introduced [11,5,17,2]. The use of these connectives allow to model different attitude of vague notions in different areas, though modelled (possibly) by the same fuzzy sets. For example, the standard approach to the fuzzy sets results to the linear increase of the imprecizness during the processing. However, in several domains (especially in physics), the outcome imprecizness is just the maximal input imprecizness. Alternative approaches discussed in this and the next chapters allow to model also such phenomena.

5.1 Basic Connectives in Fuzzy Set Theory

Classical Cantorian set theory is closely related to the standard two-valued logic. The operations in the set theory are built point-wisely and then the intersection (of sets) is related to the conjunction (of logical terms), the union is related to the disjunction, the complementation to the negation. Obviously, some notions in the set theory have no direct counterpart in the logics (e. g., the symmetric difference of sets), and vice-versa (e. g., the implication). Nevertheless, when modelling vagueness by means of fuzzy sets, mostly we do not distinguish the fuzzy set operations and the fuzzy logic operations. Observe that the subject entitled "fuzzy logic" is commonly used for entire theory (and even applications) of the vagueness modelling and processing by means of fuzzy sets. Each crisp set $A \subset X$, X being our universe of discour, can be interpreted as a fuzzy subset of X with two-valued membership function $\mu_A = 1_A$, which is just the characteristic function of A,

$$1_A(x) = \begin{cases} 1 \text{ if } x \in A, \\ 0 \text{ if } x \notin A. \end{cases} \tag{1}$$

Observe that then, for each $x \in X$, A, $B \in \mathcal{P}(X)$ (the power set of X), $1_A(x)$ and $1_B(x)$ can be understood as a logical statement, and

$$1_{A\cap B}(x) = 1_A(x) \& 1_B(x), \tag{2}$$

$$1_{A\cup B}(x) = 1_A(x) \vee 1_B(x), \tag{3}$$

$$1_{A^c}(x) = \neg 1_A(x). \tag{4}$$

The logical connectives &, $\vee$ are binary operations on the set $\{0,1\}$ while the negation $\neg$ is a unary operation on $\{0,1\}$. Zadeh's proposal of fuzzy sets [52] was based on the generalization of (2), (3), (4), in which the operations on $\{0,1\}$ (i.e., two-valued logical connectives) were replaced by the next operations on $[0,1]$ (i.e., many-valued logical connectives, truth-values from $[0,1]$):

$$\& \cdots \min\ (\inf,\ \wedge), \tag{5}$$

$$\vee \cdots \max\ (\sup,\ \vee), \tag{6}$$

$$\neg \cdots N\ (N(x) = 1 - x). \tag{7}$$

Obviously, these generalizations cannot possess all properties of the original logical connectives. Especially, the law of non-contradiction and the excluded middle law are violated. Therefore, some alternative approaches to the fuzzy logic connectives (or, equivalently, to the fuzzy sets connectives) were introduced and applied. In any case, some classical properties are violated, and then it is a matter of the real field of application which of these connectives should be applied, following the preference of the properties we expect to be fulfilled. For interested reader we can recommend [7,8] to be more informed about the developments and streamings of the fuzzy logic. Taking as a basic logical connective the conjunction &, the relevant extensions to a binary operation on $[0,1]$ always require to be a monotone extension. Denote this extension as $C : [0,1]^2 \to [0,1]$ (C stands here for conjunction). Then necessarily C is non-decreasing, $C(1,1) = 1$ and $C(0,x) = C(y,0) = 0$ for all x, $y \in [0,1]$. Other properties of a conjunction C acceptable for fuzzy logic (fuzzy sets) theory (as discussed and advocated, e.g., in [7]), are the commutativity (symmetry) of C, its associativity, and 1 is supposed to be a neutral (unit) element of C, $C(1,x) = C(x,1) = x$ for all $x \in [0,1]$. Observe that these properties correspond to the monotonicity, commutativity and associativity of the (classical) intersection of sets and the neutrality of X with respect to the intersection. Summarizing, an appropriate operation $C : [0,1]^2 \to [0,1]$ to model a fuzzy conjunction is a triangular norm as introduced by Schweizer and Sklar [46] in the framework of statistical (probabilistic) metric spaces. More about t-norms we will discuss in the next section. Further, a (general) t-norm will be denoted by T. Then the intersection (related to T) of two fuzzy subsets a, $b \in \mathcal{F}(X)$ is a fuzzy subset $a \cap b \in \mathcal{F}(X)$ such that for the corresponding membership function we have

$$\mu_{a\cap b}(x) = T\left(\mu_a(x),\ \mu_b(x)\right), \quad x \in X. \tag{8}$$

Concerning the complementation in the fuzzy set theory, we have first to discuss the negation operators in the fuzzy logic. They can be deduced from the conjunction operator, i.e., from the relevant t-norm T (e.g., by residuation [8]), however, then some important (and required) properties may be lost. Therefore, as a negation in fuzzy logic modelling the complementation via (4), always a decreasing involutive operator $n(x) : [0,1] \to [0,1]$, is taking into account. However, due to [48], see also [43,21], any such operator is a simple transformation of the original Zadeh operator $N : [0,1] \to [0,1]$, $N(x) = 1-x$. Therefore, the negation N is accepted to model the complementation in the fuzzy set theory

$$\mu_{a^c}(x) = N(\mu_a(x)) = 1 - \mu_a(x), \quad a \in \mathcal{F}(X),\ x \in X. \tag{9}$$

Finally, the union in the fuzzy set framework is supposed to generalize (together with the intersection and negation) the classical de Morgan rule,

$$a \cup b = (a^c \cap b^c)\,, \quad a,\, b \in \mathcal{F}(X). \tag{10}$$

Consequently, this union can be built via (3) from a fuzzy disjunction operator $S : [0,1]^2 \to [0,1]$,

$$S(u,v) = N(T(N(u),\, N(v))) = 1 - T(1-u,\, 1-v), \quad u,\, v \in [0,1]. \tag{11}$$

Operators characterized by (11) (where T is a t-norm) are called triangular conorms, t-conorms in short, see [46,47,21]. Triangular conorms can be introduced independently of t-norms as non-decreasing, commutative and associative $[0,1]^2 \to [0,1]$ operators with neutral element 0. The union of fuzzy sets from $\mathcal{F}(X)$ can be then defined by

$$\mu_{a\cup b}(x) = S\left(\mu_a(x), \mu_b(x)\right), \quad a,\, b \in \mathcal{F}(X),\ x \in X. \tag{12}$$

Recall that the triplets (T, S, N) of fuzzy connectives related by (11) are called de Morgan triplets [7]. As an example, the original Zadeh connectives $(\min, \max, N)$ form a de Morgan triplet. For more details about fuzzy connectives, including those not mentioned in this short overview, we recommend [7,8,21].

5.2 Triangular Norms

As already mentioned, triangular norms were introduced by Schweizer and Sklar [46] in the framework of statistical (probabilistic) metric spaces when generalizing the classical triangle inequality of metrics. For an exhaustive overview of several classes, properties and applications of t-norms we recommend a recent monograph [21]. For the readers convenience, recall what t-norms are.

Definition 5.1. A mapping $T : [0,1]^2 \to [0,1]$ is called a triangular norm (t-norm in short) if it is

i) non-decreasing, $T(u,v) \le T(u',v')$ whenever $0 \le u \le u' \le 1$, $0 \le v \le v' \le 1$;
ii) commutative, $T(u,v) = T(v,u)$, $(u,v) \in [0,1]^2$;
iii) associative, $T(T(u,v),\, w) = T(u, T(v,w))$, $(u,v,w) \in [0,1]^3$;
iv) 1 is the neutral element, $T(1,u) = T(u,1) = u$, $u \in [0,1]$.

Triangular norms can be non-continuous, even (Borel) non-measurable. However, up to the weakest t-norm $T_D : [0,1]^2 \to [0,1]$ (drastic product),

$$T_D(u,v) = \begin{cases} 0 & \text{if } (x,y) \in [0,1[^2, \\ \min(x,y) & \text{otherwise} \end{cases} \tag{13}$$

which is non-continuous on the boundary $]0,1] \times \{1\} \cup \{1\} \times]0,1]$ of the unit square, the majority of applications deal with the continuous t-norms. This is, e.g., the case of the strongest, and unique idempotent t-norm T_M = min. Each t-norm T is bounded (as a function of two variables) by T_D and min, $T_D \le T \le \min$. Up to T_M, two basic examples of continuous t-norms appearing in several applications are the product t-norm T_P and the Lukasiewicz t-norm T_L,

$$T_P(u,v) = uv, \quad (u,v) \in [0,1]^2; \tag{14}$$

$$T_L(u,v) = \max(0, u+v-1), \quad (u,v) \in [0,1]^2. \tag{15}$$

These two t-norms T_P and T_L are examples of Archimedean continuous t-norms, while T_D is an Archimedean non-continuous t-norm. Recall that a t-norm $T : [0,1]^2 \to [0,1]$ is Archimedean if and only if for any $u, v \in]0,1[$ there is some $n \in N$ such that $u_T^{(n)} < v$, where $u_T^{(1)} = u$ and $u_T^{(n)} = T(u, u_T^{(n-1)})$ for $n \ge 2$, $u \in [0,1]$. For a continuous t-norm T, its Archimedean property is equivalent with the diagonal inequality $T(u,u) < u$ for all $u \in]0,1[$. In several applications of specific triangular norms, the knowledge of their structure plays an important role. Therefore we recall some characterization and representation results from the theory of t-norms. For the relevant proofs we recommend [21]. The next result is due to Ling [27].

Theorem 5.1. A mapping $T : [0,1]^2 \to [0,1]$ is a continuous Archimedean t-norm if and only if there is a continuous strictly decreasing function $t : [0,1] \to [0,\infty]$, $t(1) = 0$, so that

$$T(u,v) = t^{(-1)}(t(u) + t(v)), \quad (u,v) \in [0,1]^2, \tag{16}$$

where the pseudo-inverse $t^{(-1)} : [0,\infty] \to [0,1]$ is given by

$$t^{(-1)}(z) = \sup(u \in [0,1] : t(u) > z). \tag{17}$$

Observe that the function t related to t-norm T via (16) is called an additive generator (of T) and that, for a continuous Archimedean t-norm T, the additive generator t is unique up to a positive multiplicative constant. Already mentioned continuous Archimedean t-norms T_P and T_L are related to the additive generators $t_P : [0,1] \to [0,\infty]$, $t_P(u) = -\log u$, and $t_L : [0,1] \to [0,1]$, $t_L(u) = 1-u$. Among several other continuous Archimedean t-norms recall the parametric family $(T_\lambda^Y)_{\lambda \in]0,\infty[}$ of Yager's t-norms [51] generated by additive generators $t_\lambda^Y : [0,1] \to [0,\infty]$, $t_\lambda^Y(u) = (1-u)^\lambda$,

$$\begin{aligned} &T_\lambda^Y(u,v) \\ &= \max\left(0, 1 - ((1-u)^\lambda + (1-v)^\lambda)^{1/\lambda}\right), \quad (u,v) \in [0,1]^2. \end{aligned} \tag{18}$$

Another important parametric family of t-norms is the Aczél–Alsina family [1] $(T_\lambda^{AA})_{\lambda\in]0,\infty[}$ related to the additive generators $t_\lambda^{AA} : [0,1] \to [0,\infty]$, $t_\lambda^{AA}(u) = (-\log u)^\lambda$,

$$\begin{aligned} &T_\lambda^{AA}(u,v) \\ &= \exp\left(-((-\log u)^\lambda + (-\log v)^\lambda)^{1/\lambda}\right). \end{aligned} \tag{19}$$

Observe that $T_1^Y = T_L$ while $T_1^{AA} = T_P$. Further, the limit members of both families are $T_0^Y = T_0^{AA} = T_D$ and $T_\infty^Y = T_\infty^{AA} = T_M$. From the algebraic point of view, continuous Archimedean t-norms with unbounded additive generators are called strict t-norms, and they all are isomorphic to T_P. The remaining continuous Archimedean t-norms are characterized by bounded additive generators. They are called nilpotent t-norms, and they all are isomorphic to T_L. Note that even non-continuous t-norms may be generated. So, for example, the drastic product T_D is generated by an additive generator $t_D : [0,1] \to [0,\infty]$,

$$t_D(u) = \begin{cases} 2-u \text{ if } u < 1, \\ 0 \qquad \text{if } u = 1. \end{cases}$$

Further, for each t-norm T generated via (16) by an additive generator $t : [0,1] \to [0,\infty]$ such that for all u, $v \in [0,1]$, $t(u)+t(v) \in \operatorname{Ran} t$ or $t(u)+t(v) > t(0)$, we have that the genuine n-ary extension of T is given by

$$T(u_1,\ldots,u_n) = t^{(-1)}\left(\sum_{i=1}^n t(u_i)\right), \quad (u_1,\ldots,u_n) \in [0,1]^n. \tag{20}$$

An exhaustive overview of generated t-norms, including several interesting results and applications, can be found in [22]. Several other generalizations of additive generators can be found in [49,50]. The class of all continuous t-norms is fully characterized by the next theorem, shown for the first time by Ling [27], see also [21].

Theorem 5.2. A continuous mapping $T : [0,1]^2 \to [0,1]$ is a triangular norm if and only if there is a system $(]a_k,b_k[)_{k\in\mathcal{K}}$ of pairwisely disjoint open subintervals of $[0,1]$ and the relevant system of continuous Archimedean t-norms $(T_k)_{k\in\mathcal{K}}$ so that T is the ordinal sum $(\langle a_k,b_k,T_k\rangle)_{k\in\mathcal{K}}$,

$$T(u,v) = \begin{cases} a_k + (b_k - c_k)\,T_k\left(\dfrac{u-a_k}{b_k-a_k}, \dfrac{v-a_k}{b_k-a_k}\right) \\ \qquad\qquad\qquad\qquad \text{if } (u,v) \in]a_k,b_k[^2, \\ \min(u,v) \qquad\qquad\qquad\qquad\qquad \text{otherwise.} \end{cases} \tag{21}$$

Observe that if $\mathcal{K}$ in Theorem 5.2 is empty, then the relevant ordinal sum is just $T_M = \min$. Further, an ordinal sum $T = (\langle a_k,b_k,T_k\rangle)_{k\in K}$ acts on

$[a_k, b_k]^2$ just as a linear transformation of T_k. If $t_k : [0,1] \to [0,\infty]$, $k \in \mathcal{K}$, are the relevant additive generators of continuous Archimedean t-norms T_k, $k \in \mathcal{K}$, define $g_k : [a_k, b_k] \to [0,\infty]$ by $g_k(u) = t_k\left(\frac{u-a_k}{b_k-a_k}\right)$. Then (21) can be rewritten into

$$T(u,v) = \begin{cases} g_k^{(-1)}(g_k(u) + g_k(v)) & \text{if } (u,v) \in]a_k, b_k[^2, \\ \min(u,v) & \text{otherwise.} \end{cases} \tag{22}$$

As an example of a nonidempotent (i. e., different from min) continuous t-norm T which is not Archimedean, consider the ordinal min $T = (\langle 0,\ 0.5,\ T_L \rangle, \langle 0.5,\ 1,\ T_L \rangle)$. Then

$$T(u,v) = \begin{cases} \max(0,\ u+v-0.5) & \text{if } (u,v) \in]0,\ 0.5[^2, \\ \max(0.5,\ u+v-1) & \text{if } (u,v) \in]0.5,\ 1[^2, \\ \min(u,v) & \text{otherwise.} \end{cases} \tag{23}$$

Observe also that up to an isomorphism, each continuous t-norm is built from $\min = T_M$, T_P (as a representative of strict t-norms), and T_L (as a representative of nilpotent t-norms). Therefore, when applying a general continuous t-norm in a processing of fuzzy sets, the basic and most important are the cases related to T_M, T_P and T_L, respectively. We briefly mention also triangular conorms. Because of their duality with t-norms given in (11), all notions and results for t-norms can be directly rewritten for t-conorms (with some duality modification when necessary). So, e. g., the weakest t-conorm is $S_M = \max$ (dual t-conorm to min), while the strongest t-conorm is S_D, drastic sum dual to the drastic product T_D, i. e., for any t-conorm S we have $\max \leq S \leq S_D$. Observe that

$$S_D(u,v) = \begin{cases} 1 & \text{if } (u,v) \in]0,1]^2 \\ \max(u,v) & \text{otherwise.} \end{cases} \tag{24}$$

Continuous Archimedean t-conorm S is characterized by the diagonal inequality $S(u,u) > u$, $u \in]0,1[$. Moreover, each continuous Archimedean t-conorm S is generated by an increasing (continuous) additive generator $s : [0,1] \to [0,\infty]$, $s(0) = 0$, so that

$$S(u,v) = s^{(-1)}(s(u) + s(v)), \quad (u,v) \in [0,1]^2, \tag{25}$$

where the pseudo-inverse $s^{(-1)} : [0,\infty] \to [0,1]$ is given by, see also [20],

$$s^{(-1)}(z) = \inf(u \in [0,1] : s(u) > z). \tag{26}$$

Observe that the difference between (17) and (26) when defining a pseudo-inverse of t and s, respectively, is caused by the different type of the monotonicity of t and s, respectively. Dual t-conorms to strict t-norms are called strict

t-conorms. They are characterized by unbounded additive generators and they all are isomorphic to the probabilistic sum S_P (dual to T_P)

$$S_P(u,v) = u + v - uv, \quad (u,v) \in [0,1]^2. \tag{27}$$

Similarly, nilpotent t-conorms are characterized by bounded additive generators and they all are isomorphic to the bounded sum S_L (Lukasiewicz t-conorm dual to T_L),

$$S_L(u,v) = \min(1,\, u+v), \quad (u,v) \in [0,1]^2. \tag{28}$$

Further, a t-norm T is generated by an additive generator t if and only if its dual t-conorm S is generated by an additive generator s, $s(u) = t(1-u)$, $u \in [0,1]$. Also each continuous t-conorm S can be represented as an ordinal sum $(\langle c_k, d_k, S_k)_{k\in\mathcal{K}}$, where S_k are continuous Archimedean t-conorms, and

$$S(u,v) = \begin{cases} c_k + (d_k - c_k)\, S_k\left(\dfrac{u-c_k}{d_k-c_k}, \dfrac{v-c_k}{d_k-c_k}\right) & \\ \qquad\qquad\qquad\qquad\qquad\text{if } (u,v) \in]c_k, d_k[^2, & \\ \max(u,v) \qquad\qquad\qquad\quad \text{otherwise.} & \end{cases} \tag{29}$$

Observe that if $T = (\langle a_k, b_k, T_k\rangle)$, the corresponding dual t-conorm S is also an ordinal sum, $S = (\langle c_k, d_k, S_k\rangle)$, where $c_k = 1 - b_k$, $d_k = 1 - a_k$, S_k is dual t-conorm to T_k, $k \in \mathcal{K}$. Finally, we present some other examples of continuous t-conorms. The parametric family $(S_\lambda^Y)_{\lambda\in]0,\infty[}$ of Yager's continuous Archimedean t-conorms is given by

$$S_\lambda^Y(u,v) = \min\left(1,\, (u^\lambda + v^\lambda)^{1/\lambda}\right), \quad \lambda \in]0,\infty[,\; (u,v) \in [0,1]^2. \tag{30}$$

The dual t-conorm S to t-norm T given by (23) is $S = (\langle 0,\, 0.5,\, S_L\rangle,$ $\langle 0.5,\, 1,\, S_L\rangle)$,

$$S(u,v) = \begin{cases} \min(0.5,\, u+v) & \text{if } (u,v) \in]0,\, 0.5[^2, \\ \min(1,\, u+v) & \text{if } (u,v) \in]0.5,\, 1[^2, \\ \max(u,v) & \text{otherwise.} \end{cases} \tag{31}$$

5.3 Triangular Norm-based Extension Principle and the Processing of Fuzzy Quantities

Zadeh [53] introduced the extension principle as a method of extending the classical crisp mappings $f : X \to Y$ to the mappings $\tilde{f} : \mathcal{F}(X) \to \mathcal{F}(Y)$ with fuzzy inputs and fuzzy outputs. As far as the aim of this contribution is restricted to the fuzzy quantities and their processing, we restrict our consideration on real functions and their fuzzifying only. Let $f : \operatorname{Dom} f \to R$ be a real function with $\operatorname{Dom} f \subset R^n$, i. e., f is a function of n-variables. Following the Zadeh extension principle, an extension $\tilde{f} : \mathcal{F}(R^n) \to \mathcal{F}(R)$ of the original crisp real function f is defined as follows:

for any n-dimensional fuzzy quantity $a \in \mathcal{F}(R^n)$, $\tilde{f}(a) \in \mathcal{F}(R)$ is a fuzzy quantity with membership function

$$\mu_{\tilde{f}(a)}(x) = \\ = \sup\left(\mu_a((x_1, \ldots, x_n)) : (x_1, \ldots, x_n) \in \operatorname{Dom} f,\ f(x_1, \ldots, x_n) = x\right) \quad (32)$$

with the convention $\sup \emptyset = 0$. Evidently, for any point $(x_1, \ldots, x_n) \in \operatorname{Dom} f$ represented as a fuzzy quantity $\langle x_1, \ldots, x_n \rangle$, see (6), $\tilde{f}(\langle x_1, \ldots, x_n \rangle) = \langle f(x_1, \ldots, x_n) \rangle$. In practical applications, rarely we deal with a priori given fuzzy subsets of R^n. Mostly, the vague information in each component results to the knowledge of fuzzy quantities $a_1, \ldots, a_n$ representing the (unknown) crisp quantities $x_1, \ldots, x_n$, respectively. Then the global n-dimensional fuzzy quantity $a \in \mathcal{F}(\tilde{R})$ should be derived. From logical point of view, a represents the information that

$$\text{“}x_1 \text{ is } a_1\text{”}\&\text{“}x_2 \text{ is } a_2\text{”}\& \cdots \&\text{“}x_n \text{ is } a_n\text{”} = \text{“}x \text{ is } a\text{”}, \quad (33)$$

where “$x_i = a_i$” should be read that the i'th coordinate of our global (n-dimensional) fuzzy quantity a is the (one-dimensional) fuzzy quantity a_i. Zadeh [53] continued by identifying of $a_i \in \mathcal{F}(R)$ with its n-dimensional cylindric representation $\tilde{a}_i \in \mathcal{F}(R^n)$,

$$\mu_{\tilde{a}_i}(x_1, \ldots, x_i, \ldots, x_n) = \mu_{a_i}(x_i). \quad (34)$$

Now, the fuzzy set $a \in \mathcal{F}(R^n)$ is simply the fuzzy intersection of cylindric fuzzy sets $\tilde{a}_i \in \mathcal{F}(R^n)$, $a = \tilde{a}_1 \cap \cdots \cap \tilde{a}_n$. Depending on the t-norm T chosen to model the fuzzy intersection, we finally obtain the membership function of the fuzzy quantity a in the form

$$\begin{aligned}\mu_a(x_1, \ldots, x_n) &= T\left(\mu_{\tilde{a}_1}(x_1, \ldots, x_n), \ldots, \mu_{\tilde{a}_n}(x_1, \ldots, x_n)\right) \quad (35)\\ &= T\left(\mu_{a_1}(x_1), \ldots, \mu_{a_n}(x_n)\right).\end{aligned}$$

Observe that Zadeh in his original approach [53] has dealt with his standard fuzzy intersection as introduced in [52], i. e., with $T = \min$. Combining (32) and (35), we finally come to T-based extension principle for fuzzy quantities.

Definition 5.2. Let T be a given t-norm and let $f : \operatorname{Dom} f \to R$, $\operatorname{Dom} f \subset R^n$, be a real function of n-variables. Then the mapping $f_T : (\mathcal{F}(R))^n \to \mathcal{F}(R)$ given by

$$f_T(a_1, \ldots, a_n) = b, \quad a_1, \ldots, a_n, b \in \mathcal{F}(R),$$

where

$$\mu_b(x) = \sup\Big(T(\mu_{a_1}(x_1), \ldots, \mu_{a_n}(x_n)) : (x_1, \ldots, x_n) \in \\ \in \operatorname{Dom} f,\ f(x_1, \ldots, x_n) = x\Big) \quad (36)$$

is called T-based fuzzy extension of f.

Obviously, $f_T(\langle x_1\rangle,\ldots,\langle x_n\rangle) = \langle f(x_1,\ldots,x_n)\rangle$ for any $(x_1,\ldots,x_n) \in \operatorname{Dom} f$, i.e., f_T is really an extension of f. Moreover, observe that the extension principle as given in (3) and further applied in (4), (5), etc., is a particular case of min-based extension principle (this was the original Zadeh's proposal in [53]). Applying (36) to simple algebraic operations, the T-based addition of fuzzy quantities a, b (denoted by $a \oplus_T b$) is a fuzzy quantity and its membership function is

$$\mu_{a\oplus_T b}(x) = \sup_{y\in R}\left[T(\mu_a(y),\mu_b(x-y))\right], \tag{37}$$

and their product $a \odot_T b$ is a fuzzy quantity with membership function

$$\begin{aligned}\mu_{a\odot_T b}(x) &= \sup_{\substack{y\in R\\ y\neq 0}}\left[T(\mu_a(y),\mu_b(x/y))\right] \quad \text{if } x \neq 0,\\ &= \max(\mu_a(0),\mu_b(0)) \qquad\qquad \text{if } x = 0.\end{aligned} \tag{38}$$

Observe that each t-norm T acts on the boundaries of the unit square $[0,1]^2\setminus\,]0,1[^2$ equally as min. Consequently, any T-based processing involving either unary operations or processing with only one fuzzy quantity has to act equally as the original min-based approach as given in Chapter 1. So, e. g., $\langle r\rangle \odot_T a = \langle r\rangle \odot a$ for any t-norm T, $r \in R$ and fuzzy quantity a.

Though the formulae for T-based addition (37) and multiplication (38) of fuzzy quantities are rather transparent and simple, similarly as the relevant specific min-based formulae (4) and (5), their real application can lead to computational complexity not allowing the real time use. This was already observed in the first chapters of this contribution, and as an alternative solution, the separate processing of single components of complex verbally generated fuzzy quantities was proposed. However, for some specific types of fuzzy quantities and specific triangular norms, the computation of output fuzzy quantities in (37) or (38) can be efficiently reduced. An exhaustive and updated overview on T-based fuzzy arithmetics and related results can be found in [5]. We recall here some of them. Majority of industrial applications working with the fuzzy quantities (with bounded support) work with triangular (trapezoidal) fuzzy numbers. The advocating of these fuzzy quantities can be found, among others, in [44,45]. A triangular fuzzy number a can be represented by a triplet (x_a,α,β), $x_a \in R$, $\alpha > 0$, $\beta > 0$, with corresponding membership function

$$\mu_a(x) = \begin{cases} \dfrac{x - x_a + \alpha}{\alpha} & \text{if } x \in\,]x_a - \alpha, x_a],\\[2mm] \dfrac{x_a - x + \beta}{\beta} & \text{if } x \in\,]x_a, x_a + \beta[,\\[2mm] 0 & \text{otherwise.}\end{cases} \tag{39}$$

Observe that x_a is the unique modal value of a, and that the support of a is $]x_a - \alpha, x_a + \beta[$. A trapezoidal fuzzy number a can be represented by a quadruplet $(x_a, y_a, \alpha, \beta)$, $x_a < y_a$, $\alpha > 0, \beta > 0$, with corresponding membership function

$$\mu_a(x) = \begin{cases} 1 & \text{if } x \in [x_a, y_a], \\ \dfrac{y_a - x + \beta}{\beta} & \text{if } x \in]y_a, y_a + \beta[, \\ \dfrac{x - x_a + \alpha}{\alpha} & \text{if } x \in]x_a - \alpha, x_a[, \\ 0 & \text{otherwise.} \end{cases} \tag{40}$$

The interval of modal values is now $[x_a, y_b]$ and the support is $]x_a - \alpha, y_b + \beta[$. Both triangular and trapezoidal fuzzy numbers are briefly called TFN. The constants α and β by a TFN a are called the left and the right spread, respectively. As a limit case, we can admit also $\alpha = 0$ and/or $\beta = 0$, as well as $(x_a, x_a, \alpha, \beta) = (x_a, \alpha, \beta)$ (i. e., crisp numbers are limit TFN as well as crisp closed intervals, triangular fuzzy numbers are limit trapezoidal fuzzy numbers, etc.). Observe that among already treated fuzzy quantities, TFN are, e. g., given in (11), (12), (15), (17), (18), (19), etc. The T-sum of TFN $a_1 \oplus_T \cdots \oplus_T a_n = a$ is again a TFN in the next cases – see in [24–26,38,12,39–42].

i) $T = \min$;
ii) $T = T_\lambda^Y$, $\lambda \in]0, \infty[$;
iii) $T \leq T_L$.

Suppose that $a_i = (x_{a_i}, y_{a_i}, \alpha_i, \beta_i)$ (including the possibilities $x_{a_i} = y_{a_i}$, $\alpha_i = 0$, $\beta_i = 0$), $i = 1, \ldots, n$, and $a = (x_a, y_a, \alpha, \beta) = a_1 \oplus_T \cdots \oplus_T a_n$. Then $x_a = \sum_{i=1}^n x_{a_i}$, $y_a = \sum_{i=1}^n y_{a_i}$, and:

$$\text{if } T = \min,\ \alpha = \sum_{i=1}^n \alpha_i,\ \beta = \sum_{i=1}^n \beta_i, \tag{41}$$

$$\text{if } T = T_\lambda^Y,\ \lambda > 1, \alpha = \left(\sum_{i=1}^n \alpha_i^q\right)^{1/q}, \tag{42}$$

$$\beta = \left(\sum_{i=1}^n \beta_i^q\right)^{1/q}, \text{ where } \frac{1}{\lambda} + \frac{1}{q} = 1;$$

$$\text{if } T \leq T_L \text{ including } T_\lambda^Y,\ \lambda \leq 1, \tag{43}$$
$$\alpha = \max(\alpha_1, \ldots, \alpha_n),\ \beta = \max(\beta_1, \ldots, \beta_n).$$

As we have seen, the modal values of the outcome a do not depend on t-norm T we use to model the fuzzy intersection (observe that this is a general rule). However, the spreads of the outcoming TFN a are strongly related to the T applied, varying from the maximal spread when $T = \min$ (outcoming spread is the sum of incoming spreads) to the minimal spreads when $T \leq T_L$ (outcoming spread equals the maximal incoming spread). Observe that the choice of the parameter λ when applying Yager t-norm T_λ^Y allows to control the outcoming spread. A typical example of fuzzy quantities with unbounded support are Gaussian fuzzy quantities $a = G(x_a, \sigma^2)$ with membership function

$$\mu_a(x) = \exp\left(-\frac{(x - x_a)^2}{\sigma^2}\right). \tag{44}$$

The T-sum of Gaussian fuzzy numbers $a_i = G(x_{a_i}, \sigma_i^2)$ is again a Gaussian fuzzy number $a = G(x_a, \sigma^2)$, $a_1 \oplus_T \cdots \oplus_T a_n = a$, if:

i) $T = \min$;
ii) $T = T_\lambda^{AA}$, $\lambda \in]0, \infty[$;
iii) $T \leq T_{0.5}^{AA}$.

In any case, $x_a = \sum_{i=1}^n x_{a_i}$, and for specific t-norms,

$$\text{if } T = \min, \quad \sigma^2 = \left(\sum_{i=1}^n \sigma_i\right)^2, \tag{45}$$

$$\text{if } T = T_\lambda^{AA},\ \lambda > 0.5, \quad \sigma^2 = \left(\sum_{i=1}^n \sigma_i^q\right)^{2/q}, \text{ where } \frac{2}{\lambda} + \frac{1}{q} = 1, \tag{46}$$

$$\text{if } T \leq T_{0.5}^{AA}, \quad \sigma^2 = \max(\sigma_1^2, \ldots, \sigma_n^2). \tag{47}$$

Especially, if $T = T_P$, then $\sigma^2 = \sum_{i=1}^n \sigma_i^2$, i. e.,

$$G(x_{a_1}, \sigma_1^2) \oplus_{T_P} \cdots \oplus_{T_P} G(x_{a_n}, \sigma_n^2) = G\left(\sum_{i=1}^n x_{a_i}, \sum_{i=1}^n \sigma_i^2\right). \tag{48}$$

Observe also that for a general continuous t-norm T, the relevant processing of fuzzy quantities can be splitted into min-based parts and into continuous Archimedean t-norms-based parts, see [3]. We give only one example related to the t-norm $T = (\langle 0, 0.5, T_L\rangle, \langle 0.5, 1, T_L\rangle)$, see (23). For

TFN $a_i = (x_{a_i}, y_{a_i}, \alpha_i, \beta_i)$, $i = 1, \ldots, n$, the outcoming fuzzy quantity $a = a_1 \oplus_T \cdots \oplus_T a_n$ has the membership function

$$\mu_a(x) = \begin{cases} 1 & \text{if } x \in \left[\sum_{i=1}^n x_{a_i}, \sum_{i=1}^n y_{a_i}\right] = [x_a, y_a], \\ 1 - \dfrac{x - y_a}{\max \beta_i} & \text{if } x \in \left]y_a, y_a + \frac{1}{2}\max \beta_i\right[, \\ 1 - \dfrac{(x_x - x)}{\max \alpha_i} & \text{if } x \in \left]x_a - \frac{1}{2}\max \alpha_i, x_a\right[\\ \dfrac{1}{2} & \text{if } x \in \left[x_a - \frac{1}{2}\sum_{i=1}^n \alpha_i,\ x_a - \frac{1}{2}\max \alpha_i\right] \cup \\ & \cup \left[y_a + \frac{1}{2}\max \beta_i, y_a + \frac{1}{2}\sum_{i=1}^n \beta_i\right], \\ \dfrac{1}{2} - \dfrac{x - y_a}{\max \beta_i} & \text{if } x \in \left]y_a + \frac{1}{2}\sum_{i=1}^n \beta_i, y_a \right. \\ & \qquad \left. + \frac{1}{2}\sum \beta_i + \frac{1}{2}\max \beta_i\right[, \\ \dfrac{1}{2} - \dfrac{x_a - x}{\max \alpha_i} & \text{if } x \in \left]x_a - \frac{1}{2}\sum \alpha_i - \frac{1}{2}\max \alpha_i, \right. \\ & \qquad \left. x_a - \left(\frac{1}{2}\right)\sum \alpha_i\right[, \\ 0 & \text{otherwise.} \end{cases} \quad (49)$$

6 Triangular Norm-based Processing of Verbal Fuzzy Quantities

Elementary processing of verbal variables as discussed in Chapter 3 was done in the framework of the standard Zadeh fuzzy set theory. Namely, the fuzzy conjunction whenever applied has been modeled by the strongest t-norm min. As already mentioned, this approach leads to the maximal uncertainty increase during the processing. Though this undesirable fact was partially eliminated by a separate processing of single components of more complex verbal quantities, see Examples 4.1 – 4.3, the use of some specific t-norm T (depending on the field of application) to model the fuzzy conjunction may further improve the likelihood of applied fuzzy quantities processing.

6.1 T-based Processing of Shapes

Recall that the concept of shape was given in Section 2.2 as a mapping $\varphi : R \to [0, 1]$ fulfilling (1) – (4). Obviously, each shape φ can be understood as a membership function of some fuzzy quantity p, $\varphi = \mu_p$. Because of (1), the point 0 is a modal value of any fuzzy quantity p related to a shape φ. The monotonicity (2) and (3) of shapes as well as the modality of 0 should be preserved during the processing of the shapes. Observe that these properties are preserved when applying T-based addition or multiplication of relevant fuzzy quantities (and thus reflecting the processing with verbal variables we intend to provide). Hence T can be arbitrary (preferably continuous) t-norm. Because of (37) and (38) this means, that the next formulae are good candidates for an appropriate combination of given shapes φ_A, φ_B during the verbal quantities processing (φ_0 represent the output shape):

$$\varphi_0(x) = \sup\left(T(\varphi_A(y), \varphi_B(x-y))\right), \quad x \in R; \tag{1}$$

$$\begin{aligned}\varphi_0(x) &= \sup_{\substack{y\in R\\ y\neq 0}}\left(T(\varphi_A(y), \varphi_B(x/y))\right), \quad x \in R,\ x \neq 0 \\ &= 1 \qquad \text{if } x = 0.\end{aligned} \tag{2}$$

For illustration, we give now some examples related to the shapes introduced in (11) – (12). Observe that $\varphi_{\text{approximately}}$ given in (11) can be viewed as a membership function of TFN $(0,1,1)$. Similarly, φ_{about} given in (12) corresponds to TFN $(0,3,3)$. Consequently, applying (41) – (43), we obtain, taking as a shape composition rule (1), that:

i) for $T = \min$

$$\begin{aligned}\varphi_{\text{approximately}} \circ \varphi_{\text{approximately}} &\sim \text{TFN } (0,2,2)\\ \varphi_{\text{approximately}} \circ \varphi_{\text{about}} &\sim \text{TFN } (0,4,4)\\ \varphi_{\text{about}} \circ \varphi_{\text{about}} &\sim \text{TFN } (0,6,6)\end{aligned}$$

ii) for $T = T_\lambda^Y$, $\lambda > 1$, $q = \frac{\lambda}{\lambda-1}$,

$$\begin{aligned}\varphi_{\text{approximately}} \circ \varphi_{\text{approximately}} &\sim \text{TFN } \left(0, 2^{\frac{1}{q}}, 2^{\frac{1}{q}}\right)\\ \varphi_{\text{approximately}} \circ \varphi_{\text{about}} &\sim \text{TFN } \left(0, (1+3^q)^{\frac{1}{q}}, (1+3^q)^{\frac{1}{q}}\right)\\ \varphi_{\text{about}} \circ \varphi_{\text{about}} &\sim \text{TFN } \left(0, 3\cdot 2^{\frac{1}{q}}, 3\cdot 2^{\frac{1}{q}}\right)\end{aligned}$$

iii) $T \leq T_L$,

$$\begin{aligned}\varphi_{\text{approximately}} \circ \varphi_{\text{approximately}} = \varphi_{\text{approximately}} &\sim \text{TFN } (0,1,1)\\ \varphi_{\text{approximately}} \circ \varphi_{\text{about}} = \varphi_{\text{about}} \circ \varphi_{\text{about}} = \varphi_{\text{about}} &\sim \text{TFN } (0,3,3).\end{aligned}$$

Similarly, taking as a shape composition rule (2), we have

i) for $T = \min$

$$\varphi_{\text{approximately}} \circ \varphi_{\text{approximately}}(x) = \max\left(0,\, 1 - \sqrt{|x|}\right), \quad x \in R;$$

$$\varphi_{\text{aproximately}} \circ \varphi_{\text{about}}(x) = \max\left(0,\, 1 - \sqrt{\frac{|x|}{3}}\right), \quad x \in R;$$

$$\varphi_{\text{about}} \circ \varphi_{\text{about}}(x) = \max\left(0,\, 1 - \sqrt{\frac{|x|}{3}}\right), \quad x \in R.$$

ii) $T = T_L$

$$\varphi_{\text{approximately}} \circ \varphi_{\text{approximately}} = \varphi_{\text{approximately}}$$

$$\varphi_{\text{approximately}} \circ \varphi_{\text{about}} = \varphi_{\text{about}} \circ \varphi_{\text{about}} = \varphi_{\text{about}}.$$

iii) $T = T_P$

$$\varphi_{\text{approximately}} \circ \varphi_{\text{approximately}}(x) = \left(\max\left(0,\, 1 - \sqrt{|x|}\right)\right)^2, \quad x \in R;$$

$$\varphi_{\text{approximately}} \circ \varphi_{\text{about}}(x) = \left(\max\left(0,\, 1 - \sqrt{\frac{|x|}{3}}\right)\right)^2, \quad x \in R;$$

$$\varphi_{\text{about}} \circ \varphi_{\text{about}}(x) = \left(\max\left(0,\, 1 - \frac{\sqrt{|x|}}{3}\right)\right)^2, \quad x \in R.$$

Point-wise composition of shapes as suggested in (5) – (9) can be effectively done by means of t-norms and t-conorms. Indeed, (5) corresponds to max and (8) to S_P t-conorms, while (6) is related to min and (7) to T_P triangular norms. In general, an expansive composition is related to t-conorms,

$$\varphi_A \circ \varphi_B = S(\varphi_A, \varphi_B), \tag{3}$$

while a restrictive composition is related to t-norms,

$$\varphi_A \circ \varphi_B = T(\varphi_A, \varphi_B). \tag{4}$$

Taking into account the fuzzy quantities representation of φ_A and φ_B by p and $q \in \mathcal{F}(R)$, respectively, the composition (3) is related to the S-based fuzzy union $p \cup q$, while (4) corresponds to the T-based fuzzy intersection $p \cap q$.

Observe also that if we consecutively apply (3) (or (4)) to shape $\varphi = \varphi^{(1)}$, $\varphi^{(n)} = \varphi^{(n-1)} \circ \varphi$, $n \geq 2$, then the limit shape $\varphi^{(\infty)}$ is:

— $\varphi^{(\infty)} = 1_{\text{supp}\varphi}$, i. e., $\varphi^{(\infty)}(x) = \begin{cases} 1 \text{ if } \varphi(x) > 0, \\ 0 \text{ otherwise,} \end{cases}$
whenever (3) is applied with an Archimedean t-conorm S;

— $\varphi^{(\infty)} = 1_{\mathrm{mod}\varphi}$, i. e. $\varphi^{(\infty)}(x) = \begin{cases} 1 \text{ if } \varphi(x) = 1, \\ 0 \text{ otherwise,} \end{cases}$
whenever (4) is applied with an Archimedean t-norm T;

— apply (4) with $T = (\langle 0,\, 0.5,\, T_L\rangle,\, \langle 0.5,\, 1,\, T_L\rangle)$ given in (23); then

$$\varphi^{(\infty)}(x) = \begin{cases} 1 & \text{if } \varphi(x) = 1 \\ 0.5 & \text{if } \varphi(x) \in [0.5,\, 1[, \\ 0 & \text{otherwise;} \end{cases}$$

— apply (3) with $S = (\langle 0,\, 0.5,\, S_L\rangle,\, \langle 0.5,\, 1,\, S_L\rangle)$ given in (31); then

$$\varphi^{(\infty)}(x) = \begin{cases} 1 & \text{if } \varphi(x) > 0.5, \\ 0.5 & \text{if } \varphi(x) \in]0,\, 0.5], \\ 0 & \text{otherwise.} \end{cases}$$

6.2 T-based Processing of Scales

The processing of scales discussed in Section 3.4 can be derived from the processing of specific shapes (with unbounded supports). Indeed, for a scale $f : R \to R$, define a real function $f^* : R \to [0,1]$ by

$$f^* = \frac{1}{1 + |f|}. \tag{5}$$

It is immediate that f^* fulfills the standard shape properties (1) – (3). Now, we can apply shape combination methods as given in Section 3.2 to f^* and g^* related via (5) to scales f and g, respectively, $h^* = f^* \circ g^*$. Then the scale combination h of f and g can be derived from

$$h(x) = \begin{cases} \dfrac{1}{h^*(x)} - 1 \text{ if } x \geq 0, \\ 1 - \dfrac{1}{h^*(x)} \text{ if } x < 0. \end{cases} \tag{6}$$

For example, applying (5) to f^* and g^*, $f^* \circ g^* = \max(f^*, g^*)$, we obtain just (16) scale combination, $h = \min(f, g)$. Similarly, (6) shape combination is related to the (15) scale combination, while (14) scale combination is related to a shape combination by means of a weighted harmonic mean. Not only point-wise combination methods of shapes and of scales are related. After a careful examination, we will find out that the sum based method for scale combinations as given in (21) is related to the min-based addition of fuzzy quantities given in (4), i. e., to the shape combination method (2). For more details in this context we recommend [18,19], where the min-based arithmetics on the fuzzy real line is discussed and related with so called pseudo-inverses (introduced in Section 5.2).

Remark 6.1. In some specific case, the shape processing may mix the values of incoming shapes in positive reals with values in negative reals (for example, by a T-based multiplication of shapes). It seems to be intuitively sound not to mix the scale values in positive reals with the scale values in negative reals. To avoid possible violating of this principle, we propose to treat shapes f^*, g^* corresponding to scales f, g, respectively, separately on positive and on negative parts of the real line, whenever necessary.

Scale processing based on the shape processing methods as given in (5) and (6) can be successfully applied to the methods introduced in the previous Section 6.1. First of all, we can exploite the point-wise expansive composition (3). As in the Section 3.4, for given scales f_A and f_B, suppose that f_C is the resulting scale combining f_A and f_B. Then

$$\begin{aligned} f_C(x) &= \frac{1}{S\left(\frac{1}{1+f_A(x)}, \frac{1}{1+f_B(x)}\right)} - 1 \quad \text{if } x \geq 0, \\ &= 1 - \frac{1}{S\left(\frac{1}{1-f_A(x)}, \frac{1}{1-f_B(x)}\right)} \quad \text{otherwise.} \end{aligned} \tag{7}$$

For example, if we take the probabilistic sum S_P, see (27), and we choose the linear scales $f_A(x) = ax$, $f_B(x) = bx$, $a > 0$, $b > 0$, then

$$f_C(x) = abx \, \frac{|x|}{1 + (a+b)\,|x|}.$$

The same scales processed by means of the bounded sum S_L, see (28), lead to the output scale f_C,

$$f_C(x) = \begin{cases} \frac{abx^2-1}{2+ax+bx} & \text{if } x > \frac{1}{\sqrt{ab}}, \\ 0 & \text{if } x \in \left[-\frac{1}{\sqrt{ab}}, \frac{1}{\sqrt{ab}}\right], \\ \frac{1-abx^2}{2-ax-bx} & \text{if } x < -\frac{1}{\sqrt{ab}}. \end{cases}$$

Similarly, we can start from a restrictive composition of shapes (4) based on some t-norm T. Continuing in our example with linear scales f_A and f_B, for the product t-norm T_P we get (in (7) we replace S by T_P)

$$f_C(x) = x(a + b + ab|x|).$$

Remark 6.2. When starting from the Lukasiewicz t-norm T_L, see (15), we get $|f_C(x)| = \infty$ whenever $|x| \geq \sqrt{ab}$. As all scales should be real functions, T_L cannot be applied for the scale processing. In general, we can process the scales (when applying the point-wise combination) only by means of a t-norm T with no zero divisors, i. e., $T(u, v) > 0$ whenever $u > 0$, $v > 0$.

An interesting generalization of the sum-related processing of scales introduced in (21) is related to the T-based sum of fuzzy quantities (37), i. e., to the shape composition described in (1). For a given t-norm T and the scales f_A, f_B, the resulting composed scale f_C is given by

$$f_C(x) = \begin{cases} \dfrac{1}{\sup_{y\in[0,x]}\left[T\left(\frac{1}{(1+f_A(y))}, \frac{1}{1+f_B(x-y)}\right)\right]} - 1 \text{ if } x \geq 0, \\ 1 - \dfrac{1}{\sup_{y\in[x,0]}\left[T\left(\frac{1}{(1-f_A(y))}, \frac{1}{1-f_B(x-y)}\right)\right]} \text{ if } x < 0. \end{cases} \tag{8}$$

Now, any t-norm T can be applied. For the linear scales f_A and f_B related to $a > 0$, $b > 0$, respectively, and the Lukasiewicz t-norm T_L, (8) results to $f_C(x) = \min(a, b)\, x$. For the product t-norm T_P, we get the same output scale f_C. This is a consequence of the fuzzy arithmetics results given in [42,40], from which we will see that f_A and f_B are processed to f_C, $f_C(x) = \min(a, b)\, x$, for all t-norms between the drastic product and t-norm generated by an additive generator $t : [0, 1] \to [0, \infty]$, $t = f_A^{-1}$. Note that this is always the case if we process f_A and f_B related to some fixed strictly monotone scale f, $f_A(x) = f(ax)$, $f_B(x) = f(bx)$. In our case, $t(x) = \frac{1}{a}\left(\frac{1}{x} - 1\right)$ is an additive generator of the Hamacher product $T_H : [0, 1]^2 \to [0, 1]$, see [9],

$$T_H(u, v) = \frac{uv}{u + v - uv} \quad \text{with convention } \frac{0}{0} = 0.$$

Recall the Dombi parametric family of t-norms $(T_\lambda^D)_{\lambda\in]0,\infty[}$ introduced in [4], which is generated by additive generators $t_\lambda^D : [0, 1] \to [0, \infty]$, $t_\lambda^D(x) = \left(\frac{1}{x} - 1\right)^\lambda$. For more details about these t-norms we recommend [21]. Taking a Dombi t-norm T_λ^D with $\lambda > 1$, and applying (8) to the linear scales f_A and f_B, the output scale f_C is given by

$$f_C(x) = \frac{ab}{(a^q + b^q)^{1/q}}\, x, \quad \text{where } \frac{1}{\lambda} + \frac{1}{q} = 1.$$

Observe that the limit member of the Dombi family is $T_\infty^D = \min$, which case corresponds to $q = 1$, i. e., $f_C(x) = \left(\frac{ab}{a+b}\right) x$. However, this is just the result obtained when applying (21).

Another combination of scales can be based on the T-based multiplication of fuzzy quantities (38), see (2) for the shape processing. Observe that when $T = \min$, because of Klement's results in [18,19], we obtain already introduced formula (23). In general, for given t-norm T and scales f_A, f_B, the resulting output scale f_C is then given by (taking into account also Remark 6.1)

$$f_C(x) = \begin{cases} \dfrac{1}{\sup_{y\in]0,\infty[}\left[T\left(\frac{1}{1+f_A(y)}, \frac{1}{1+f_B(x/y)}\right)\right]} - 1 & \text{if } x > 0, \\ 0 & \text{if } x = 0, \\ 1 - \dfrac{1}{\sup_{y\in]-\infty,0[}\left[T\left(\frac{1}{1-f_A(y)}, \frac{1}{1-f_B\left(\frac{-x}{y}\right)}\right)\right]} & \text{if } x < 0. \end{cases} \tag{9}$$

Take, e. g., the scales f_A and f_B given by $f_A(x) = \text{sign}(x)\left(\exp\left|\frac{x}{a}\right| - 1\right)$, $f_B(x) = \text{sign}(x)\left(\exp\left|\frac{x}{b}\right| - 1\right)$, $a > 0$, $b > 0$. Then for the product t-norm T_P, the output scale f_C obtained by means of (9) is given by

$$f_C(x) = \text{sign}(x)\left(\exp\left(2\sqrt{\frac{|x|}{ab}}\right) - 1\right).$$

7 Conclusions

The above paragraphs were written to suggest an alternative view on the modelling of the vague quantitative data by means of the fuzzy set theoretical tools. The main goal of their submission is not to substitute the already classical processing of fuzzy quantities but to offer an analysis of the essence of the structure of fuzzy data which can influence – and, maybe, even simplify – their practical handling in complex computational algorithms.

Particular properties and consequences of the proposed approach were already discussed in the previous chapters. Here, in the conclusive remarks, we want to remember two significant facts connected with the role of human subjectivity included into the practical application of this model.

In rather narrower sense, there exists certain kind of complementarity between the scale functions and shapes. Uncertainty of verbal data can be represented by both the quality of their resource (scale) and the vagueness of the linguistic description (shape). Due to the composition rule (10), the mentioned two representations of uncertainty influence the membership function of the resulting fuzzy quantity. To keep its structure in rational limits, it is necessary to keep this complementarity in mind when the actual model of vague variable is constructed.

This fact is a special case of much wider role of subjectivity in the submitted model. It was already discussed in the heuristic paragraphs of the previous chapters but it is useful to mention it briefly and in a summarized form also in these conclusive comments. The model of verbal variable presented above includes several steps opening a wide space for the assertion of the subjectivity of the constructor of the actual model. These steps are, namely,

— the choice of the scale function,
— the choice of the method of composition of the scales,
— the choice of shapes,
— the choice of the particular operations composing the shapes,
— the choice of detailed structure of the decomposition of combined shapes into the elementary ones.

The situation is even more complicated by the fact that each of these choices influences, by means of the composition rule (10), the effects of the other ones from our list (as it was shown in the previous paragraphs in the case of complementarity of scales and shapes). In fact, this subjectivity does not fully result from our model of verbal data. It is only a formalized and analytically presented subjectivity which is in a more hidden form encoded in any representation of quantitative vagueness by means of fuzzy quantities. In this sense, the model of derivation of fuzzy quantities from vague verbal variables, presented in the previous chapters, makes also the subjectivity of the model construction more distinct and better understandable.

The analysis of the structure of the vagueness and subjectivity, forming an unseparable component of mathematical processing of uncertain quantitative data and presented in this contribution is (and cannot be) complete. Anyhow, its aim is to open new possibilities of the modelling of the phenomenon of uncertainty connected with numerical objects. In this sense it, perhaps, deserves eventual further study of the concepts suggested here, including eventual bridges between subjective evaluation of the given problems and its formal mathematical representation.

References

1. J. Aczél and C. Alsina: Characterization of some classes of quasilinear functions with applications to triangular norms and to synthesizing judgments. Methods Oper. Res. 48 (1984), 3–22.
2. C. Alsina, E. Trillas and L. Valverde: On some logical connectives for fuzzy sets theory. BUSEFAL 3 (1980), 18–29.
3. D. DeBaets and A. Marková–Stupňanová: Analytical expressions for the addition of fuzzy intervals. Fuzzy Sets and Systems 91 (1997), 203–213.
4. J. Dombi: A general class of fuzzy operators, De Morgan class of fuzzy operators and fuzziness measures induced by fuzzy operators. Fuzzy Sets and Systems 8 (1982), 149–163.
5. D. Dubois, E. Kerre, R. Mesiar and H. Prade: Fuzzy interval analysis. In: D. Dubois, H. Prade (Eds.): Fundamentals of Fuzzy Sets. Handcontributions on Fuzzy Sets, Vol. 1, Kluwer Acad. Publ., Dordrecht 2000, 483–581.
6. D. Dubois and H. Prade: New results about properties and semantics of fuzzy set – theoretic operators. In: P. Wang, S. Hcnag, editors, Fuzzy Sets: Theory and Applications to Policy Analysis and Information Sytems. Plenum Press, new York 1980, 59–75.

7. S. Gottwald: Fuzzy Sets and Fuzzy Logic. Foundations of Application – from a Mathematical Point of View. Mathematical Foundations and Applications. Vieweg, Braunschweig 1993.
8. P. Hájek: Metamathematics of Fuzzy Logic. Kluwer Acad. Publ, Dordrecht 1998.
9. H. Hamacher: Über logische Aggregationen nicht-binär explicierter Entscheidungskriterien. Rita G. Fischer Verlag, Frankfurt 1978.
10. B. Harman: Sum and product of the modified real fuzzy numbers. Kybernetika, supplement 1992.
11. U. Höhle: Probabilistisch kompakte L-unscharfe Mengen. Manuscripta Math. 26 (1978), 345–356.
12. D. H. Hong: On shape preserving additions of fuzzy intervals. J. Math. Anal. Appl., in press.
13. J. Jacas, J. Recasens: Fuzzy numbers and equality relations. In: Transactions of 2nd IEEE Internat. Conf. on Fuzzy Systems 1993, IEEE Neural Network Council 1993.
14. E. E. Kerre, M. Mareš, R. Mesiar: Orderings of generated fuzzy quantities. In: Proceedings of IPMU'98, La Sorbone, Paris 1998, Vol. I, 250–254.
15. E. E. Kerre, M. Mareš, R. Mesiar: Generated fuzzy quantities and their orderings. International Journal of Uncertainty, Fuzziness and Knowledge–Based Systems. To appear.
16. E. E. Kerre, X. Wang: Reasonable properties for the ordering of fuzzy quantities. Part I, Part II. Fuzzy Sets and Systems. To appear.
17. E. P. Klement: Fuzzy σ-algebras and fuzzy measurable functions. Fuzzy Sets and Systems 4 (1980), 83–93.
18. E. P. Klement: Integration of fuzzy valued functions. Revue Roumaine Math. Pures Appl. 30 (1985), 375–384.
19. E. P. Klement: Strong law or large numbers for random variables with values in the fuzzy real line. IFSA Commun. Math. Chapt. (1987), 7–11.
20. E. P. Klement, R. Mesiar and E. Pap: Quasi- and pseudo-inverses of monotone functions, and the construction of t-norms. Fuzzy Sets and Systems 104 (1999), 3–13.
21. E. P. Klement, R. Mesiar, E. Pap: Triangular Norms. Kluwer Acad. Publ., Dordrecht 2000.
22. E. P. Klement, R. Mesiar, E. Pap: Generated triangular norms. Kybernetika 36 (2000), 363–377.
23. G. J. Klir, Bo Yuan (Eds.): Fuzzy Sets, Fuzzy Logic, and Fuzzy Systems. Selected papers by Lotfi A. Zadeh. World Scientific, Singapore, London 1996.
24. A. Kolesárová: Additions preserving the linearity of t-norms. Tatra Mt. Math. Publ. 6 (1995), 75–81.
25. A. Kolesárová: Similarity preserving t-norm based addition of fuzzy numbers. Fuzzy Sets and Systems 91 (1997), 215–229.
26. A. Kolesárová: Triangular norm-based addition preserving linearity of T-sums of linear fuzzy intervals. Mathware and Soft Computing 5 (1998), 91–98.
27. C. M. Ling: Representation of associative functions. Publ. Math. Debrecen 12 (1965), 189-212.
28. M. Mareš: How to handle fuzzy quantities. Kybernetika 13 (1977), 1, 23–40.
29. M. Mareš: Addition of rational fuzzy quantities: Convolutive approach. Kybernetika 25 (1989), 1, 1–12.

30. M. Mareš: Computation Over Fuzzy Quantities. CRC–Press, Boca Raton 1994.
31. M. Mareš: Weak arithmetics of fuzzy numbers. Fuzzy Sets and Systems 91 (1997), 143–154.
32. M. Mareš: Verbal quantitative information in economic reality. Acta Oeconomica Pragensia 5 (1997), 1, 121–134.
33. M. Mareš: Critical path method with verbal inputs. In: Proceedings ICOTA'98, Curtin University of Technology, Perth 1998, Vol. 1, 634–640.
34. M. Mareš: Fuzzy Coalitional Games. Physica–Verlag, Heidelberg. To appear.
35. M. Mareš, R. Mesiar: Composition of shape generators. Acta Mathematica et Informatica Universitatis Ostraviensis 4 (1996), 1, 37–45.
36. M. Mareš, R. Mesiar: Calculation over verbal variables. In: J. Kacprzyk, L. A. Zadeh (Eds.): Computing With Words in Information/Intelligent Systems, Physica–Verlag, Heidelberg 1999, 409–427.
37. M. Mareš, R. Mesiar: Vagueness of verbal variable. In: R. Ribeiro, J. Kacprzyk, R. R. Yager, L. A. Zadeh (Eds.): Soft Computing in Financial Engineering. Physica–Verlag, Heidelberg 1999, 3–20.
38. A. Marková: Idempotents of the T-addition of fuzzy numbers. Tatra Mt. Math. Publ. 12 (1997), 67–72.
39. A. Marková: T-sum of $L-R$ fuzzy numbers. Fuzzy Sets and Systems 85 (1996), 379–384.
40. A. Marková–Stupňanová: A note to the addition of fuzzy intervals based on the continuous Archimedean t-norm. Fuzzy Sets and Systems 91 (1997), 253–258.
41. R. Mesiar: Shape preserving additions of fuzzy intervals. Fuzzy Sets and Systems 86 (1997), 73–78.
42. R. Mesiar: Triangular norm-based additions of fuzzy intervals. Fuzzy Sets and Systems 91 (1997), 231–237.
43. H. T. Nguyen and E. Walker: A First Course in Fuzzy Logic. CRC Press, Boca Raton 1997.
44. W. Pedrycz: Why triangular membership functions. Fuzzy Sets and Systems 64 (1994), 21–30.
45. J. Pančochář, M. Drahoš, J. Vrba: Fuzzy number as a product of geometrical construction. Fuzzy Sets and Systems 83 (1996), 43–50.
46. B. Schweizer, A. Sklar: Statistical metric spaces. Pacific. J. Math. 10 (1960), 313–334.
47. B. Schweizer, A. Sklar: Probabilistic Metric Spaces. North–Holland, New York 1983.
48. E. Trillas: Sobre functiones de negación en la teoría de conjuntas difusos. Stochastica 3 (1979), 47–60.
49. P. Viceník: A note on generators of t-norms. BUSEFAL 75 (1998), 33–38.
50. P. Viceník: Generated t-norms and the Archimedean property. Proc. EUFIT'99, Aachen 1999, CD-rom.
51. R. R. Yager: On a general class of fuzzy connectives. Fuzzy Sets and Systems 4 (1980), 235–242.
52. L. A. Zadeh: Fuzzy sets. Information and Control 8 (1965), 3, 338–353.
53. L. A. Zadeh: The concept of a linguistic variable and its applications to approximate reasoning. Information Sciences, Part I: 8 (1975), 199–249; Part II: 8 (1975), 301–357; Part III: 9 (1975), 43–80.

FSC
www.fsc.org
MIX
Papier | Fördert gute Waldnutzung
FSC® C083411

Kolibri 360 GmbH
Ferdinand-Jühlke-Straße 7
99095 Erfurt, Deutschland
produktsicherheit@kolibri360.de